T0296507

LONDON MATHEMATICAL SOCIETY LECTURE NOTE SERIES

Managing Editor: Professor N.J. Hitchin, Mathematical Institute,
University of Oxford, 24–29 St Giles, Oxford OX1 3LB, United Kingdom

The titles below are available from booksellers, or, in case of difficulty, from Cambridge University Press.

London Mathematical Society Lecture Note Series. 276

Singularities of Plane Curves

Eduardo Casas-Alvero
University of Barcelona

CAMBRIDGE
UNIVERSITY PRESS

CAMBRIDGE UNIVERSITY PRESS
Cambridge, New York, Melbourne, Madrid, Cape Town, Singapore, São Paulo

Cambridge University Press
The Edinburgh Building, Cambridge CB2 2RU, UK

Published in the United States of America by Cambridge University Press, New York

www.cambridge.org
Information on this title: www.cambridge.org/9780521789592

First published 2000

A catalogue record for this publication is available from the British Library

Library of Congress Cataloguing in Publication data

Casas-Alvero, E. (Eduardo), 1948-
Singularities of plane curves / Eduardo Casas-Alvero.
  p. cm.
Included bibliographical references and index.
ISBN 0 521 78959 1 (pb)
1. Curves, Plane. 2. Singularities (Mathematics) I. Title.

QA565 .C37 2000
516.3'5–dc21   00-027671

ISBN-13 978-0-521-78959-2 paperback
ISBN-10 0-521-78959-1 paperback

Transferred to digital printing 2005

To Mercè, Carles and Eduard

Contents

Preface

The singularities of algebraic and analytic varieties constitute an old and today very active field of research which combines techniques and viewpoints from different mathematical fields such as Geometry, Algebra, Topology and Function Theory. Doubtless, the oldest and best understood singularities are those of plane curves. Some curves with singular points appear in the work of the ancient Greek geometers, and the first contribution to a systematic study of plane curve singularities is due to Isaac Newton. Even if some relevant questions still remain open, nowadays, after the work of geometers such as Puiseux, Smith, Noether, Halphen, Enriques and Zariski, there is a well established theory for the analysis and classification of the singularities of plane curves. I have intended in this book to give a precise and detailed account of most of the main facts of this theory from a decidedly geometrical viewpoint which, I hope, will not completely hide its more algebraic or topological aspects. Infinitely near points have been taken as the cornerstone of the presentation: main description and classification of singularities are stated using infinitely near points, and other properties and invariants, such as characteristic exponents, Milnor number, discriminantal index, semigroups, polar quotients, etc., are related to them.

Infinitely near points are a nice and old idea for describing singularities. They appear in the work of M. Noether and their geometry was extensively developed by Enriques ([35], Book IV). Since then the geometry of infinitely near points has seldom appeared in the mathematical literature, maybe due to the rather obscure and non-intrinsic way the Italians used to introduce infinitely near points, or also to the fact that they do not have a straightforward translation in terms of rings and ideals. Besides the book by Enriques and Chisini, other references are the survey in the first chapter of Zariski's book on surfaces [92], chapter XI of the classical book on curves by Semple and Kneebone [73], a nice paper by Van der Waerden [83] and section 5 of Zariski's paper on saturation [93]. Fortunately infinitely near points may be introduced today in a very precise and clear way. I believe that they give a very appealing picture of how singularities of plane curves behave, and hope that this book will contribute to show it. Recent updating and development of Enriques' theory of infinitely near imposed singularities (virtual multiplicities) has lead to a better understanding of polar curves, linear systems and ideals of $\mathbb{C}\{x, y\}$ most of whose fruits, I think, are still to come. Furthermore, the use of infinitely near points and their properties, such as proximity, satellitism, etc., in the study of singularities in a wider context (varieties other than plane curves, foliations...) is just beginning and seems to be a very promising approach. This adds a further interest to the study of infinitely near points developed here.

Most of the results presented in this book have already appeared in recent or classical literature in a more or less explicit way, and so they may hardly be considered as essentially new. I have tried to precise and update the oldest ones and to organize all of them in a self-contained and comprehensive exposition centred on infinitely near points, which perhaps is new. Nevertheless, most of the proofs given here are new, either because preceding proofs do not fit in the context, or, as for many of the classical results, because no proof according to the modern standards was available. Among other results whose proofs I believe are new, let me quote in particular the Enriques theorem relating characteristic exponents and infinitely near points on an irreducible germ of curve (5.5.1), the theorem about the generators of the semigroup of a branch (5.8.2), whose proof uses the way they are defined from the branch rather than their particular values, a restricted version of the Lê-Ramanujam's μ-constant theorem (7.3.7), the determination of the E-sufficiency degree (7.5.1 and 7.6.1) and Zariski's theorem on factorization of complete ideals, which is proved using clusters of base points (8.4.13). If some of the results in the book may be qualified as new, they come from further development of Enriques' theory of virtual multiplicities and its application to linear systems. Among them, I would like to mention the determination of the (infinitely near) singular points of a curve from the base points of its jacobian system (8.6.4).

Singularities of plane curves is a rather elementary subject in the sense that it may be developed without using very complicated techniques from Algebra, Algebraic Geometry or Topology. I have tried to keep within this elementary frame making the book accessible to graduate students. Algebraic prerequisites are the most common facts on polynomials, series, commutative rings and ideals. Analytic prerequisites are reduced to some basic notions on complex analytic functions including implicit and inverse map theorems. Neither Weierstrass preparation and division theorems for two variables, nor the algebraic properties of the rings of convergent series are assumed as prerequisites, as they will be easily derived (for two variables) from Puiseux's theorem in chapter 1. Some knowledge of topics from algebraic geometry, as those usually contained in a first course on algebraic curves ([37], for instance), is maybe advisable even if not strictly needed. In this way I hope this book will be useful to both the graduate students, as a first step into the field of singularities, and the non-specialists that need to know the basic facts about singularities of plane curves.

Before giving a short description of the contents of the different chapters, maybe it is fair to say something on what is not to be found here. Throughout the book the base field is the complex one and the frame will be more analytic than algebraic: this is perhaps the best for an introductory book as it makes things easier and allows objects to stay close to the intuition (curves have points, topology is the classical one, series may be convergent, etc.), but obviously this does not cover the case of non-zero characteristic. Many results in the book are still true in the abstract case of an arbitrary algebraically closed base field, but not all of them: for instance Newton–Puiseux's algorithm and Puiseux's theorem in chapter 2 do not hold in positive characteristic. The book by Campillo [16] is advisable for a further reading covering the particular phenomena due to the positive characteristic. On the other hand, the reader will not find here a topological study of plane curve singularities. Fortunately, the topological side is maybe the best covered in the literature: the nice book by Milnor [60] is an obligate reading for anyone interested in singularities, and furthermore, among others, there are the books by Brieskorn and Knörrer [13], Eisenbud and Neumann [33] and Dimca [30], the latter being not specifically devoted to plane curves. Even within the more geometrical frame, the present book is far from being complete. In particular Zariski saturation theory [91] is missing. This is an interesting way of understanding the classification of plane curve singularities but unfortunately it is beyond the scope of this book, as it is an essentially non-planar approach.

Switching to what may be found in the book, chapter 0 is of introductory nature: basic facts that are needed in subsequent chapters are quickly presented there, mainly to set definitions and notations, as most of them should be familiar to the reader. Often proofs are not given, but the reader is referred to other sources.

Chapter 1 is mainly devoted to give a constructive proof of the Puiseux theorem on the roots of a power series in two variables. From it we obtain the main algebraic properties of the ring of convergent power series in two variables, including Weierstrass' theorems.

Chapter 2 presents the properties of germs of curve that follow from Puiseux's theorem, namely the decomposition of a germ into irreducible ones and the parameterization of an irreducible germ by means of its Puiseux series. We use such parameterizations to introduce the intersection multiplicity of germs of curve and to prove its main properties.

Chapter 3 contains the basic facts of the geometry of infinitely near points, the definition of equisingularity (the main notion of equivalence of singularities) and the description of infinitely near points and equisingularity classes by means of Enriques diagrams. Two sections are devoted to Northcott's neighbouring rings, which are an algebraic counterpart of the infinitely near points on a curve and allow us to introduce and compute the order δ of a singularity. A section about the Artin theorem for plane curves closes the chapter.

Chapter 4 deals with the conditions of asking curves to go through certain infinitely near points with given multiplicities (virtual multiplicities) and the families they give rise to. Certain inequalities (proximity inequalities) determine whether these conditions may be fulfilled. In case of the proximity inequalities being not satisfied, an effective algorithm, named unloading, gives the generic behaviour of the curves to which the conditions are imposed. Dual graphs and their relationship to Enriques diagrams are presented in section 4.4. The last two sections are devoted to the adjoint curves and their characterization by means of the conductor ideal, and to the $Af + B\varphi$ theorem of M. Noether, both for arbitrary singularities.

Chapter 5 relates the two main classical ways of describing and classifying singularities, namely the infinitely near points (M. Noether, Enriques) and the Puiseux series and their characteristic exponents (Smith, Halphen). The link is given by the Enriques theorem, which is proved in section 5.5. Further sections are devoted to semigroups associated with irreducible germs and to Abhyankar's approximate roots of polynomials.

Chapter 6 is devoted to polar germs, which are a very sharp tool for analyzing singularities. Their first properties lead to the Plücker formula and to the definition and computation of the Milnor number. Behaviour and equisingularity classes of polar germs are then considered and used for uncovering some properties of germs of curve that depend on their (local) isomorphism classes and not only on their equisingularity classes. Next, decompositions of the polar germs and polar invariants for both irreducible an non-irreducible germs of curve are presented.

Chapter 7 contains a local study of linear families of germs of curve and their base points including a sort of local Bertini theorem about non-existence of variable infinitely near multiple points and a restricted version (for linear families) of the μ-constant theorem. Properties of linear systems are then applied to sufficiency problems, that is, to the determination of the equisingularity class or the isomorphism class of a germ of curve by finitely many monomials of its local equation: in particular we present results of Samuel about algebraicity of analytic germs, as well as results of Teissier and Kuo-Lu determining equisingularity-sufficiency in terms of polar invariants.

Chapter 8 presents a geometric theory of valuations of the ring of convergent series $\mathbb{C}\{x, y\}$. Next, the relationship between complete ideals (those defined by valuative conditions) and the base points of their corresponding linear systems is shown: this relationship allows us to give an explicit proof of Zariski's theorem about unique factorization of complete ideals. Back to polar germs, the last two sections deal with aspects closely related to its base points and the linear structure of the jacobian system.

An appendix includes a couple of affine global results that are presented as applications of the local theory. They are the theorem of Abhyankar-Moh on immersions of the affine line in the affine plane and Jung's theorem about generators of the group of algebraic automorphisms of the affine plane.

Exercises are proposed at the end of each chapter. They include extensions of the theory already developed, as well as applications of local results to global algebraic geometry. Results proved in exercises are used in other exercises, but not in the text.

I am not able to quote all sources which in one way or another have influenced this book, so let me just quote two of them which I believe are the most important. The first one is the already quoted book by Enriques and Chisini *Teoria geometrica delle equazioni e delle funzioni algebriche* [35], my first reading on singularities and still a frequent one: in spite of many obscure proofs and some mistakes, this is a book full of ideas where I managed for the first time to understand infinitely near points not just as intermediate steps in a desingularization procedure, but as the points of an infinitesimal space where the local geometry of singularities may be displayed. The second one is the entire work of Zariski on singularities as it is the source of all modern work, at least on the more algebraic side. In particular, without his very precise critical and foundational work I would have not been able to understand and handle most of the Italian ideas. Other specific original sources of the most important results will be quoted within the text as far as I am aware of them, but since I am afraid that these quotations may be incomplete, I apologize in advance for the missing ones. I have tried to make a consistent and proper use of classical nomenclature, keeping classical conventions when possible and using the old names only for notions which really agree with the classical ones.

During the preparation of this book I have circulated preliminary versions to many colleagues and students, who made very valuable comments. I want to thank all of them, and in particular M. Alberich, A. Campillo, F. Delgado, J. Ma. Giral, O. Lavila, V. Navarro, A. Nobile, R. Peraire, J. Roé and G. Welters: their encouragement, help and suggestions largely improved the final version.

I wish also to thank the staff of Cambridge University Press for their kind attention and careful work, and the D.G.I.C.Y.T. of the Spanish Governement and the D.G.U. of the Catalan Governement for financial support.

Barcelona, May 2000.

Chapter 0

Preliminaries

This chapter is devoted to setting our general assumptions and conventions, to fixing notations and to recalling some basic notions and results in the form to be used throughout this book. The reader is assumed to be familiar with some very basic notions relating to analytic functions of several variables, such as germs of functions, varieties, manifolds, and analytic maps, including the inverse and implicit mapping theorems. For them he is referred to the first chapters of any book on analytic functions of several variables, such as for instance [41] or [40]. We will also make use of some rather elementary facts from ring and ideal theory: they can be found in general books such as [48] and, of course, also in those specifically devoted to commutative Algebra, [58], [8] or [32], for instance.

Throughout the book we will denote by $\mathbb{Z}$ the ring of integers, by $\mathbb{N}$ the set of the natural or positive integers and by $\mathbb{R}$ and $\mathbb{C}$ the fields of the real and complex numbers. We will use the symbol ∞ with the usual algebraic rules and the total order of $\mathbb{Z}$ extended so that $\infty \geq n$ for any $n \in \mathbb{Z}$. Domain means connected non-empty open set and, unless otherwise stated, all neighbourhoods of points will be assumed to be open and connected.

0.1 Projective spaces

A *projective space* of dimension d over a field K is a set $\mathbb{P}$ together with a exhaustive map $\pi : F - \{0\} \longrightarrow \mathbb{P}$, where F is a $(d+1)$-dimensional K-vector space and $\pi(v) = \pi(w)$ if and only if $v = aw$ for some $a \in K$. We often say that π defines a structure of projective space on $\mathbb{P}$. Once a basis in F is fixed the components of a non-zero vector v will be taken as homogeneous coordinates of $\pi(v)$, homogeneous coordinates of a point being thus determined up to a multiplicative constant. We will usually write $[x_0, \ldots, x_n]$ for the point of homogeneous coordinates $x_0, \ldots, x_n$. *Projectivities* (sometimes also called linear projectivities) are the maps between projective spaces induced by linear isomorphisms between the corresponding vector spaces, or, equivalently, invertible maps that are linear in homogeneous coordinates.

If $\mathbb{P}_1$ is a one-dimensional projective space and z_0, z_1 are projective coordinates on $\mathbb{P}_1$, the one to one map

$$\mathbb{P}_1 \longrightarrow \mathbb{C} \cup \{\infty\}$$
$$[z_0, z_1] \longmapsto z_0/z_1$$

which is assumed to send the point $[1, 0]$ to ∞, will be called an *absolute coordinate* on $\mathbb{P}_1$.

0.2   Power series

We will denote, as customary, by $K[x_1, \ldots, x_n]$ and $K[[x_1, \ldots, x_n]]$, respectively, the rings of polynomials and formal power series in the variables $x_1, \ldots, x_n$ with coefficients in the ring K. In the case $K = \mathbb{C}$, we denote by $\mathbb{C}\{x_1, \ldots, x_n\}$ the ring of the convergent power series. If s is a power series we use the notation $o(s)$ for the *order* of s: if $s \neq 0$, $s = \sum_{i > 0} s_i$ with each s_i a homogeneous polynomial of degree i, $o(s) = \min\{i | s_i \neq 0\}$ and $o(0) = \infty$. Recall that o satisfies the characteristic properties of valuations, namely

$$o(s + s') \geq \min\{o(s), o(s')\} \quad \text{and} \quad o(ss') = o(s) + o(s').$$

If $s \in K[[x_1, \ldots, x_n]]$, $n > 1$, we will write $o_{x_i}(s)$ for the order of s considered as a series in the single variable x_i whose coefficients are series in the remaining variables.

Assume that K is a field. Then a series s either in $K[[x_1, \ldots, x_n]]$ or in $\mathbb{C}\{x_1, \ldots, x_n\}$ is invertible if and only if $o(s) = 0$, that is, s has non-zero independent term. The non-invertible elements are thus those in the ideal $(x_1, \ldots, x_n)$ in either $K[[x_1, \ldots x_n]]$ or $\mathbb{C}\{x_1, \ldots, x_n\}$. Therefore, this ideal is the only maximal one and both rings are local rings.

If s is a series in a single variable, $s \in K[[x]]$, we write also $o_x(s)$ for $o(s)$. If K is a field it is clear that one may write $s = x^{o(s)}u$, where u is an invertible series. Furthermore, if $K = \mathbb{C}$ and s is convergent, then u is convergent too. It follows that both the rings $K[[x]]$ and $\mathbb{C}\{x\}$ are principal, their ideals being those generated by the powers of x.

We will use *substitution* of series into series in both the formal and convergent cases. If $f = f(x_1, \ldots, x_n)$ is a series in n variables $x_1, \ldots, x_n$ and $g_i = g_i(y_1, \ldots, y_m)$ are non-invertible series in m variables, $i = 1, \ldots, n$, there is a well determined series in the variables $y_1, \ldots, y_m$, currently denoted by $f(g_1, \ldots, g_n)$, which is obtained by substituting g_i for x_i in f. Substitution induces a morphism of K-algebras

$$K[[x_1, \ldots, x_n]] \longrightarrow K[[y_1, \ldots, y_m]]$$
$$f \longmapsto f(g_1, \ldots, g_n)$$

which is an isomorphism if K is a field, $m = n$ and the degree-one parts of the g_i are linearly independent linear forms. (see [95], Vol. II, Ch. VII, for instance).

Notice that the last condition may be equivalently stated by saying that the jacobian determinant of the g_i is invertible.

If $K = \mathbb{C}$ and the series f and g_i, $i = 1, \ldots, n$ are convergent, then they define holomorphic maps $\bar{f} : V \longrightarrow \mathbb{C}$, $p \mapsto f(p)$ and $\psi : U \longrightarrow \mathbb{C}^n$, $q \mapsto (g_1(q), \ldots, g_n(p))$, where V and U are suitable neighbourhoods of the origin in $\mathbb{C}^n$ and $\mathbb{C}^m$, respectively. After a suitable shrinking of U one may consider the composite map $\bar{f} \circ \psi$ which is a holomorphic function. Its series expansion at the origin is just $f(g_1, \ldots, g_n)$ which, therefore, is convergent too. Thus, if the g_i are convergent, the above substitution morphism restricts to a morphism between the corresponding rings of convergent series: if convergent series are identified with germs of holomorphic functions, then this morphism is just the *pull-back morphism* ψ^* associated with ψ, namely $\psi^*(\bar{f}) = \bar{f} \circ \psi$. Using the inverse mapping theorem one may easily prove that the substitution morphism is an isomorphism if and only if $n = m$ and the jacobian determinant $\partial(g_1, \ldots, g_n)/\partial(y_1, \ldots, y_n)$ is invertible, just as in the formal case.

0.3   Surfaces, local coordinates

We are specially interested in irreducible and smooth analytic surfaces, since on they are lying the curves we will study in this book. Unless otherwise stated, the word *surface* will mean connected two-dimensional complex analytic manifold throughout the book. Note that a domain in a surface is itself a surface. Because of its own definition, surfaces are covered by open sets, each of which is actually analytically isomorphic to an open subset of $\mathbb{C}^2$. Then, for any fixed point O on a surface there is an open neighbourhood U and functions $x, y : U \longrightarrow \mathbb{C}$ that give an analytic isomorphism $p \longmapsto (x(p), y(p))$ onto an open subset U' of $\mathbb{C}^2$ and furthermore $x(O) = y(O) = 0$: a such pair of functions will be called a system of *local coordinates* (on S) at O. After choosing a system of local coordinates, x, y, at O, we will often identify each point $p \in U$ and its image $(x(p), y(p)) \in U'$ and so we will write just (x, y) for the point whose local coordinates are x, y. The functions holomorphic (or analytic) in a neighbourhood of O in S are obtained as the compositions $f(x, y)$ of the local coordinates and the functions f holomorphic in a neighbourhood of $(0, 0)$ in $\mathbb{C}^2$. Thus, locally at O one may thing of these functions just as functions of two variables x, y, holomorphic in a neighbourhood of the origin, and each of them is represented, in a neighbourhood of O, as the sum of an uniquely determined convergent power series in x, y. A second pair of holomorphic functions x', y', defined in an open neighbourhood of O and such that $x'(O) = y'(O) = 0$, is another system of local coordinates at O if and only if the jacobian determinant $\partial(x', y')/\partial(x, y)$ is not zero at O.

0.4   Morphisms

Holomorphic maps of surfaces will be also called *analytic morphisms* or just *morphisms* for short, and biholomorphic maps will be also called *analytic isomorphisms* or just *isomorphisms*. Saying that $\varphi : S' \longrightarrow S$ is a morphism means thus that for each $O' \in S$ there are local coordinates x', y' in a neighbourhood U' of O', local coordinates x, y in a neighbourhood U' of $O = \varphi(O')$, and functions f_1, f_2 analytic in U' such that $\varphi(U') \subset U$ and the restriction of φ to U' is given by the rule $(x', y') \mapsto (f_1(x', y'), f_2(x', y'))$. In these conditions we will say that φ is represented in U, or locally at O, by the equations

$$x = f_1(x', y'), \quad y = f_2(x', y').$$

The inverse map theorem asserts that the above morphism φ restricts to an isomorphism between suitable neighbourhoods of O' and O if and only if the jacobian determinant $\partial(f_1, f_2)/\partial(x', y')$ is non-zero at O. In such a case it is said that O' is a *non-critical point* of φ. The reader may notice that the set of non-critical points of φ is, by ts own definition, an open subset of S'. Its complement, the set of *critical points* of φ, is thus closed. We will see next that it is either nowhere dense or the whole of S'. Indeed, call X the interior of the set of critical points: it is open and so, S' being assumed to be connected, it will be enough to see that it is also closed. Take any $O' \in S'$ and assume that φ is represented in a neighbourhod U of O' by the equations $x = f_1(x', y')$, $y = f_2(x', y')$. If there is a point $q \in U \cap X$, then $\partial(f_1, f_2)/\partial(x', y')$, which is analytic in U, identically vanishes in a neighbourhood of q in U, and hence in the whole of U. Thus $O' \in X$ and X is closed. All morphisms we will consider thuroughout this book will have some non-critical point and thus a nowhere dense set of critical points.

If S is a surface, an *S-surface* or a *surface over S* will mean a pair (S', φ') where S' is a surface and φ' a morphism $\varphi' : S' \longrightarrow S$, its *structural morphism*. An *S-morphism* (or *morphism over S*) between S-surfaces $(S_1, \varphi_1), (S_2, \varphi_2)$ is just an ordinary analytic morphism between the surfaces $\varphi : S_1 \longrightarrow S_2$ that commutes with their structural morphisms: $\varphi_1 = \varphi_2 \circ \varphi$.

Assume that S and S' are surfaces and that $\varphi : U \longrightarrow U'$ is an analytic isomorphism between domains $U \subset S$, $U' \subset S'$ whose graph is closed in $S \times S'$. Then, by identifying in the disjoint union $S \amalg S'$ the points p and $\varphi(p)$, for $p \in U$, we get a connected Hausdorff topological space $S \amalg_\varphi S'$ which has a well determined structure of analytic surface for which the natural maps $S \longrightarrow S \amalg_\varphi S'$ and $S' \longrightarrow S \amalg_\varphi S'$ induce isomorphisms between S and S' and their respective images ([41], V.5). We will refer to this surface as the surface obtained by *patching together* S and S' along φ.

0.5   Local rings

Fix a point O in a surface S. The germs at O of the functions holomorphic in a neighbourhood of O describe a local ring which will be called the *local ring of S at O* or else the *local ring of O on S*. It will be denoted by $\mathcal{O}_{S,O}$, and also by $\mathcal{O}_O$, or just by $\mathcal{O}$ if S and O are clear from the context. Similarly, we denote by $\mathcal{M}_{S,O}$ (or $\mathcal{M}_O$, or just $\mathcal{M}$) the maximal ideal of $\mathcal{O}_{S,O}$. The elements of $\mathcal{M}_{S,O}$ are the germs of functions f for which $f(O) = 0$. The residual field $\mathcal{O}_{S,O}/\mathcal{M}_{S,O}$ is just the field of complex numbers $\mathbb{C}$, the germ of any function f being congruent modulo $\mathcal{M}_{S,O}$ with that of the constant function $f(O)$. If x, y are local coordinates at O, the representation of each holomorphic function near O by a convergent series leads to an isomorphism of local rings between the local ring of S at O and that of convergent power series in x, y, $\mathcal{O}_{S,O} \simeq \mathbb{C}\{x, y\}$, and in particular the germs of any pair of local coordinates at O are a pair of generators of $\mathcal{M}_{S,O}$. In the sequel, since no confusion can be made, we will use the same notations for the local coordinates and for their germs at O. Notice that it is clear from the identification $\mathcal{O}_{S,O} \simeq \mathbb{C}\{x, y\}$ that all powers of $\mathcal{M}_{S,O}$ have finite codimension as linear subspaces of $\mathcal{O}_{S,O}$. Main algebraic properties of the local rings $\mathcal{O}_{S,O}$ will be obtained in chapter 1, after proving Puiseux's theorem.

Assume that S and S' are surfaces, $\varphi : U' \longrightarrow U$ is a morphism from a neighbourhood U' of a point O' in S' into a neighbourhood U of O in S and $\varphi(O') = O$. By composing the representatives of the germs of functions with φ, one gets a *pull-back morphism* of local rings $\varphi^* = \varphi^*_{O,O'} : \mathcal{O}_{S,O} \longrightarrow \mathcal{O}_{S',O'}$. Notice that φ^* is a local morphism, that is, $\varphi^*(\mathcal{M}_{S,O}) \subset \mathcal{M}_{S',O'}$. Obviously φ^* depends only on the germ of φ at O': restrictions of φ to smaller neighbourhoods of O' induce the same pull-back morphism.

It is easy to see that all pull-back morphisms $\varphi^*_{O',O}$ are monomorphisms, but for the case of φ having no non-critical point. Indeed, assume that g is analytic in a neighbourhood V of O and has non-zero germ at O. This implies that $T = \{p \in V | g(p) = 0\}$ is a nowhere dense subset of V. If $g \circ \varphi$ identically vanishes in a neighbourhod V' of O', then $\varphi(V') \subset T$. This forces all points in V' to be critical and so (see section 0.4) φ has no non-critical points.

The identity map clearly induces the identity map between local rings, $Id^*_S = Id_{\mathcal{O}_{S,O}}$. If ψ is another morphism from a neighbourhood of a third point $O'' \in S''$ into U', $\psi(O'') = O'$, then $(\varphi \circ \psi)^* = \psi^* \circ \varphi^*$. In particular, if O' is a non-critical point of φ, then φ^* is an isomorphism of local rings. The converse is also true, see 0.6 below.

0.6 Tangent and cotangent spaces

The complex vector space $\Omega_{S,O} = \mathcal{M}_{S,O}/\mathcal{M}^2_{S,O}$ is (or may be identified with) the cotangent space of S at O, differentiation maps the germ f to the class df of $f - f(0) \mod \mathcal{M}^2_{S,O}$. If x, y are local coordinates at O, then $\mathcal{M}_{S,O} = (x, y)$, and so the classes dx, dy of x, y in $\Omega_{S,O}$ (the differentials of x, y at O, in fact) are a basis of $\Omega_{S,O}$ over $\mathbb{C}$. For any $f \in \mathcal{O}_{S,O}$, $df = (\partial f/\partial x)_O dx + (\partial f/\partial y)_O dy$.

We denote by $T_{S,O}$ the tangent space to S at O. No matter how the tangent space is defined (one way is to take it as the dual of the cotangent space), $\Omega_{S,O}$ is its dual space, and hence the elements of $\Omega_{S,O}$ are linear forms on $T_{S,O}$.

The one-dimensional linear subspaces of $T_{S,O}$ will be called the *tangent lines* or *tangent directions* to S at O, each non-zero tangent vector determines a such tangent line and the whole of tangent lines to S at O is a one dimensional complex projective space, usually called the *pencil of tangent lines at O*. Notice that, since the spaces are of dimension two, also each non-zero form in the cotangent space determines a tangent line at O, namely its own kernel. Once a system of local coordinates x, y at O has been chosen, one usually takes basis dx, dy for $\Omega_{S,O}$ and its dual for $T_{S,O}$.

Let $\varphi : U' \longrightarrow U \subset S$ be, as above, a morphism between neighbourhoods of $O' \in S'$ and $O \in S$, $\varphi(O') = O$: we denote by $d\varphi$ and $\partial\varphi$ the associated morphisms $d\varphi : \Omega_{S,O} \longrightarrow \Omega_{S',O'}$ and $\partial\varphi : T_{S',O'} \longrightarrow T_{S,O}$. The first one is just that induced by φ^*, while the second one is its dual. If φ is represented locally at O' by the equations $x = f_1(x', y')$, $y = f_2(x', y')$, then the matrix of $\partial\varphi$ is the jacobian matrix of f_1, f_2 valued at O'. Thus O' is a non-critical point of φ if and only if $\partial\varphi$ (or $d\varphi$) is an isomorphism. This is the case if φ^* is an isomorphism of local rings, as claimed at the end of section 0.5 above.

We will also consider the graduate ring $\mathcal{G}_{S,O}$ of $\mathcal{O}_{S,O}$, namely

$$\mathcal{G}_{S,O} = \bigoplus_{i \geq 0} \mathcal{M}_{S,O}^i / \mathcal{M}_{S,O}^{i+1}.$$

It is a graduate ring and its pieces of degrees zero and one are $\mathbb{C}$ and $\Omega_{S,O}$ respectively. The graduate ring $\mathcal{G}_{S,O}$ is a polynomial ring in two variables as it may be easily seen by identifying $\mathcal{O}_{S,O}$ with the ring of convergent power series in local coordinates x, y: then it is clear that each homogeneous piece $\mathcal{M}_{S,O}^i / \mathcal{M}_{S,O}^{i+1}$ is freely generated over $\mathbb{C}$ by the classes of the monomials of degree i in x, y which in turn are the monomials of degree i in the classes dx, dy of x, y. It turns out, in particular, that the elements of $\mathcal{G}_{S,O}$ are the polynomial functions on the tangent space $T_{S,O}$. If f is a non-zero germ of holomorphic function at O, then one has $f \in \mathcal{M}^e - \mathcal{M}^{e+1}$ for some $e \geq 0$. One says then that e is the *order* of f and that the class $[f]$ of f in $\mathcal{M}^e / \mathcal{M}^{e+1}$ is the *initial form* or the *leading form* of f. Of course $[f][g] = [fg]$.

Using coordinates gives an easier (but not intrinsic) description of the graduate ring. Fix local coordinates x, y at O, defined in an open neighbourhood U of O, and use them to identify U with an open neighbourhood U' of the origin in $\mathbb{C}^2$. Then we take $\mathbb{C}^2$ as its own tangent space at the origin and identify the local coordinates x, y with their differentials dx, dy. We have thus $\mathcal{O}_{S,O} = \mathbb{C}\{x, y\}$ and $\mathcal{G}_{S,O} = \mathbb{C}[x, y]$. If s is a convergent series in x, y, say $s = \sum_{i \geq e} s_i$, each s_i being a homogeneous polynomial in x, y of degree i and $s_e \neq 0$, then the order and the initial forms of s are just its order and initial form as a series, namely e and s_e. Nevertheless, it should be noticed that the linear structure of $\mathbb{C}^2$ has an intrinsic meaning when $\mathbb{C}^2$ is identified with the tangent space, but the same linear structure has no intrinsic meaning if it is translated to U: for instance the linear character of, say, $x + y$ is intrinsic if $x + y$ is taken as an initial form (i.e., x, y are understood as forms on the tangent space) but it depends on the choice of the coordinates if $x + y$ is considered as a germ of function.

0.7 Curves

Our main interest in this book is the local study of analytic curves lying on a (smooth) surface: these curves are locally flat in the sense that they are lying on surfaces which are locally isomorphic to planes. Just for this, from the local viewpoint there is no difference between these curves and those lying on a true plane $\mathbb{C}^2$ and so our study is certainly not more general than that of the germs of curves at the origin of $\mathbb{C}^2$. Nevertheless, since blowing up will give rise to surfaces other than the plane, and to (global) curves on them, it will be easier to place ourselves in the just apparently more general frame of studying local properties of curves on smooth surfaces.

Even if they are not our main field of interest, we need to consider curves some of whose parts are counted with multiplicities. Because of this we take the curves (and later their germs) as defined by their systems of equations, rather than just as sets of points: an *analytic curve* (or just *curve*, for short) ξ is defined in a non-empty open subset U of a surface S by a system $(U_i, f_i)_{i \in I}$, where each U_i is a domain in U, $U = \bigcup_i U_i$, each f_i is a non-zero holomorphic function in U_i, and for each $i, j \in I$ for which $U_i \cap U_j \neq \emptyset$, there is a function $u_{i,j}$, holomorphic and with no zeros in $U_i \cap U_j$ such that $f_i = u_{i,j} f_j$ in $U_i \cap U_j$. We will call f_i an *equation* of ξ in U_i, or just a *local equation* of ξ. Two systems $(U_i, f_i)_{i \in I}$, $(V_j, g_j)_{j \in J}$ give rise to the same curve if and only if f_i/g_j is holomorphic and has no zeros in $U_i \cap V_j$ if $U_i \cap V_j \neq \emptyset$, for $(i,j) \in I \times J$. Notice that the definition includes all the (non-necessarily reduced) algebraic curves on S if S itself is algebraic. In particular we have the affine and projective plane algebraic curves if S is the affine or the projective plane over $\mathbb{C}$. Our interests being local, our curves will be mostly defined by a single equation in a single domain. Then if f is holomorphic and not identically zero in a domain U, we just say the curve $\xi : f = 0$ or the curve $f = 0$ to mean the curve ξ which has equation f in U. In particular, if x, y are local coordinates at O, the curves $y = 0$ and $x = 0$ will be called the *x-axis* and the *y-axis*, or the *first* and *second axis*, respectively.

A point p *lies on* (or belongs to) the curve ξ defined by $(U_i, f_i)_{i \in I}$ if and only if p is in one of the U_i and then $f_i(p) = 0$, the same condition being then obviously satisfied for any other index j for which $p \in U_j$. The condition clearly does not depend on the system $(U_i, f_i)_{i \in I}$ but only on the curve ξ. If p lies on ξ we will equivalently say that ξ goes through p and write $p \in \xi$. The set of points of ξ, also called the *locus* of ξ, will be denoted by $|\xi|$. It clearly is an analytic subvariety and a nowhere dense subset of U. We will say that a curve ξ is compact or connected if its locus $|\xi|$ is so.

Given two curves ξ and ζ defined in the same open set U in a surface S, it is not restrictive to assume that they are defined by systems $\{U_i, f_i\}_{i \in I}$ and $\{U_i, g_i\}_{i \in I}$ with the same family of domains $\{U_i\}_{i \in I}$. Then the curve defined by $\{U_i, f_i g_i\}_{i \in I}$ will be called the curve *composed of* ξ and ζ, or the *sum* of ξ and ζ, and denoted by $\xi + \zeta$. Such a curve does not depend on the equations of ξ and ζ but only on the curves themselves and, obviously, $|\xi + \zeta| = |\xi| \cup |\zeta|$. Addition of curves is clearly commutative and associative. If there is ζ' defined in the same open set U as ξ and ζ, and $\xi = \zeta + \zeta'$, then we say that ξ *contains*

ζ, or that ζ is a *component* of ξ. In such a case ζ' is determined by ξ and ζ and we often write $\zeta' = \xi - \zeta$: indeed, if ξ contains ζ, local equations of ζ divide local equations of ξ in a neighbourhood of each point of U and the quotients are local equations of ζ'.

Let ξ be a curve defined by $(U_i, f_i)_{i \in I}$ in a domain $U = \bigcup_i U_i$ in S. If U' is a domain in a surface S' and $\varphi : U' \longrightarrow U$ a morphism such that $\varphi(U') \not\subset |\xi|$, then the restrictions of the pull-backs $f_i \circ \varphi$ to the connected components of the $\varphi^{-1}(U_i)$ are the equations of a curve $\varphi^*(\xi)$ in $\varphi^{-1}(U)$ which is called the *pull-back* or the *inverse image* of ξ by φ. Indeed, if one of the f_i identically vanishes in a conected component of $\varphi^{-1}(U_i)$, then an easy argument using the conectness of U' shows that all $\varphi^*(f_i)$ identically vanish and therefore $\varphi(U') \subset |\xi|$. After this the remaining conditions for the restrictions of the $\varphi^*(f_i)$ to define a curve are obviously satisfied. In the case $\varphi(U') \subset |\xi|$, $\varphi^*(\xi)$ remains undefined. Note that if this occurs, then φ has no non-critical point. Once again the definition does not depend on the system of local equations and it is clear that a point $p' \in S'$ lies on $\varphi^*(\xi)$ if and only if $\varphi(p)$ lies on ξ. If ζ is another curve defined in U, $\varphi^*(\xi + \zeta)$ is defined and equals $\varphi^*(\xi) + \varphi^*(\zeta)$ provided both $\varphi^*(\xi)$ and $\varphi^*(\zeta)$ are defined. As for the pull-back of functions or germs of functions, $(\varphi \circ \psi)^*(\xi) = \psi^*(\varphi^*(\xi))$ if $\psi : U'' \longrightarrow U'$ is another morphism and $\psi(U'') \not\subset |\varphi^*(\xi)|$. If in particular φ is an isomorphism, then one may transform curves in both senses and we will just write $\varphi(\xi')$ for $(\varphi^{-1})^*(\xi')$, ξ' being any curve defined in U'.

0.8   Germs of curves

Curves are restricted to smaller open sets just by restricting their equations. Two curves that restrict to the same one in a suitable open neighbourhood of a closed set K are said to have the same germ at K. We will often use germs of curves at points instead of the curves themselves, as the germs carry all the local information on the curves: a *germ of curve* at a point O is an equivalence class of curves defined in some neighbourhood of O, modulo the equivalence relation of having the same restriction to an open neighbourhood of O. The point O is then called the origin of the germ. If ξ is a curve, we will write ξ_O to denote its germ at O. We will often say just germ instead of germ of curve if no confusion may result.

It is clear that functions giving the same germ at O define in a neighbourhood of O curves that have the same germ at O: then we take as *equations of a germ of curve* ξ at O the germs at O of all equations of all representatives of ξ. Each germ of curve is then determined by any of its equations. If f is a non-zero germ of holomorphic function at O we will write $\xi : f = 0$ or just $f = 0$ to denote the germ of curve ξ at O with equation f.

Assume that ξ and ζ are curves defined by equations f and g, respectively, in a certain open neighbourhood of O. By the definition of curve, the restrictions of ξ and ζ to some open neighbourhood of O agree if and only if f/g is holomorphic and has no zeros in a neighbourhood of O, that is, if and only if the germ of

f/g is an invertible element of $\mathcal{O}_{S,O}$. It turns out that if f and g are now germs of holomorphic functions at O, $f, g \neq 0$, they are equations of the same germ if and only if the germ f/g is an invertible element of $\mathcal{O}_{S,O}$. Thus we get a one to one correspondence between germs of curve at O and non-zero principal ideals of $\mathcal{O}_{S,O}$: a germ ξ corresponds to the ideal generated by any of its equations. In the sequel we will call this ideal the *ideal of* ξ.

Let $\xi : f = 0$ be a germ of curve at O and $I = (f)$ its ideal. We can associate with ξ (or with I) the germ of set at O represented by the set of points $|\xi'|$ of any representative ξ' of ξ. We call it the *locus of* ξ and denote it by $|\xi|$. It may be also written $|\xi| = \mathbf{V}(I) = \mathbf{V}(f)$, according to the customary notations for the loci of zeros of ideals.

Notice that we are considering in particular the *empty germ of curve*, which is the germ at O of any curve going not through O and corresponds to the ideal (1): since no confusion may result, we will denote it by the symbol of the empty set, $\emptyset$. Of course the locus of the empty germ is the germ of the empty set.

Addition of germs at O may be equivalently defined either by adding suitable representatives or by multiplying their equations or ideals: we write $\xi + \zeta$ for the *sum* of the germs ξ and ζ, which we also call the germ *composed of* ξ and ζ: it has equation fg if f and g are equations of ξ and ζ. As for curves, if $\xi = \zeta + \zeta'$, we still write $\zeta = \xi - \zeta'$ and we say that ζ is *contained* in, or is a *component* of, ξ.

A non-empty germ γ is said to be *irreducible* if and only if it cannot be obtained as the sum of two non-empty germs. This is clearly the same as saying that the equations of γ are irreducible elements of the ring $\mathcal{O}_{S,O}$ or that the ideal of γ is a prime ideal. Unique decomposition of a germ as a sum of irreducible ones will be obtained in 2.1.1, as an easy consequence of the factoriality of $\mathcal{O}_{S,O}$.

A curve γ defined in a neighbourhood of a point O is said to be *irreducible at* O if and only if its germ γ_O at O is irreducible.

Assume that O' is a point on a surface S' and that $\varphi : U' \longrightarrow U$ is a morphism between neighbourhoods of O' and O, $\varphi(O') = O$. If $\xi : f = 0$ is a germ of curve at O and $\varphi^*(f) \neq 0$, then $\varphi^*(f)$ is the equation of a germ of curve at O' which is called the *inverse image* or *pull-back* of ξ at O'. It will be denoted by $\varphi^*(\xi)$ (or $\varphi^*_{O'}(\xi)$ if reference to the point O' is needed). Obviously both the condition $\varphi^*(f) \neq 0$ and the germ $\varphi^*(\xi)$ are independent of the choice of the equation f. As the reader may easily check, $\varphi^*(\xi)$ is defined if and only if a representative ξ' of ξ has its pull-back $\varphi^*(\xi')$ defined, in which case all representatives of ξ in neighbourhoods of O contained in U have their pull-backs defined and representing $\varphi^*(\xi)$. Since φ^* is a local morphism, $\varphi^*(\xi) = \emptyset$ if and only if $\xi = \emptyset$. As for curves, $\varphi^*(\xi + \zeta)$ is defined and equals $\varphi^*(\xi) + \varphi^*(\zeta)$ if ζ is another germ at O and both $\varphi^*(\xi)$ and $\varphi^*(\zeta)$ are defined. Also $(\varphi \circ \psi)^*(\xi)$ is defined and equals $\psi^*(\varphi^*(\xi))$ if ψ is a morphism from a neighbourhood of a third point $O'' \in S''$ in a neighbourhood of O', $\psi(O'') = O'$ and both $\varphi^*(\xi)$ and $\psi^*(\varphi^*(\xi))$ are defined.

If φ is biholomorphic, then germs of curve can be transformed in both senses by taking direct or inverse images of their equations by the isomorphism of local rings $\varphi^* : \mathcal{O}_{S,O} \longrightarrow \mathcal{O}_{S',O'}$. In such a case the germs ξ and $\varphi^*(\xi)$ are called *isomorphic* or *analytically isomorphic*. An *analytic invariant* (of germs of curve) is any map defined on germs of curve that maps any two isomorphic germs to the same image. Usually these invariants are referred to by the image of a germ rather than by the map itself. For instance, the reader may easily see, after its definition in section 0.9 below, that the multiplicity of a germ (or the map that sends each germ to its multiplicity) is an analytic invariant.

0.9   Multiplicity and tangent cone

Assume that ξ is a germ of curve at O and f and f' are equations of ξ. Then (0.8), one has $f' = uf$ with u invertible and hence $[f'] = [u][f]$, where the initial form $[u]$ of u is a non-zero complex number, just because u is invertible. It follows in particular that all equations of ξ have the same order: it is called the *multiplicity of the germ* ξ, and sometimes also the *multiplicity of O on* ξ. We will denote it by $e(\xi)$, or even by $e_O(\xi)$ if an explicit reference to the origin O of ξ is needed. Equivalently the multiplicity of $\xi : f = 0$ may be defined by the relation $f \in \mathcal{M}^{e(\xi)} - \mathcal{M}^{e(\xi)+1}$.

It is clear that $e(\xi) = 0$ if and only if ξ is empty, and also that $e(\xi + \zeta) = e(\xi) + e(\zeta)$ for any two germs ξ and ζ at O (*additivity of multiplicity*). One way for getting germs of arbitrary multiplicity is to take them as composed of germs of multiplicity one. Nevertheless it is worth noting that the multiplicity of a germ is not merely a consequence of the number of germs it is composed of, as there are irreducible germs of arbitrary multiplicity (see example 0.11.5 below).

We have just seen above that different equations of a germ ξ have proportional initial forms: all these initial forms define thus the same algebraic cone in the tangent space $T_{S,O}$: it is called the *tangent cone* to ξ. If x, y are local coordinates at O and $f = \sum_{i+j \geq e} a_{i,j} x^i y^j$ is an equation of ξ, $e = e(\xi)$, then the tangent cone to ξ has equation $\sum_{i+j=e} a_{i,j} x^i y^j = 0$. Of course, a initial form being a homogeneous polynomial in two variables, it is a product of linear factors and the tangent cone is composed of lines. Each line composing the tangent cone to ξ is called a *tangent line* or a *principal tangent* to ξ. Two germs are said to be *tangent* if and only if they share a principal tangent. Notice that the multiplicity of the germ equals the order of its tangent cone, that is, the number of principal tangents provided each tangent is counted according to its multiplicity as a component of the tangent cone. Notice also that the tangent cone to $\xi + \zeta$ is composed of the tangent cones to the germs ξ and ζ. We will see in 2.2.6 that the tangent cone to an irreducible germ γ is composed of a single line (necessarily counted with multiplicity $e(\gamma)$).

The notions introduced above for a germ will be applied to its representatives with explicit mention of the point O: if ξ is now a curve, the *multiplicity*, the *tangent cone* and the *principal tangents* of ξ at O are those of its germ ξ_O. In particular we will write $e_O(\xi)$ for the multiplicity of ξ at O, $e_O(\xi) = e(\xi_O)$. Two curves are said to be *tangent* at O if their germs at O are so. If $e = e_O(\xi)$, it is often said that O is an e-fold point of the curve ξ.

0.10 Smooth germs

A germ of curve ξ at O is called a *singular germ* or a *curve singularity* if and only if it has $e = e(\xi) > 1$. Then its representatives are said to have a *singular point* or a *multiple point* at O.

Non-singular non-empty germs of curve at O are called *smooth* germs. Their representatives are, by definition, *smooth* at O. It is also said that they have a *simple point* or a *non-singular point* at O. A curve is *smooth* or *non-singular* if and only if it is smooth at all its points.

Notice that a smooth germ is necessarily irreducible, by the additivity of multiplicities, and that he tangent cone to a smooth germ is just a single line counted once, as it has order one.

Assume that x, y are coordinates in a neighbourhood V of O. Assume that the point O has coordinates (a, b) so that $\bar{x} = x - a$, $\bar{y} = y - b$ are local coordinates at O. Let g be holomorphic in U, say $g = \sum_{i,j \geq 0} a_{i,j} \bar{x}^i \bar{y}^j$ represented as a convergent series in a neighbourhood of O. Assume furthermore that g defines a curve ζ through O, that is, that $a_{0,0} = g(O) = 0$. By the definition, the curve ζ is smooth at O if and only if either $a_{1,0} \neq 0$ or $a_{0,1} \neq 0$ and in such a case the line $a_{1,0}\bar{x} + a_{0,1}\bar{y} = 0$ in $T_{S,O}$ is its tangent line at O.

Since

$$a_{1,0} = (\partial g/\partial \bar{x})_{(0,0)} = (\partial g/\partial x)_{(a,b)}$$
$$a_{0,1} = (\partial g/\partial \bar{y})_{(0,0)} = (\partial g/\partial y)_{(a,b)},$$

the condition for the curve $g = 0$ to be smooth at O may be equivalently stated by saying that one of the derivatives $\partial g/\partial x$, $\partial g/\partial y$ is non-zero at O. Thus, the singular points of ζ in V are the points that satisfy the system of equations

$$g = 0, \quad \frac{\partial g}{\partial x} = 0, \quad \frac{\partial g}{\partial y} = 0.$$

We see in particular that the set of singular points of ζ is closed in V and so if a curve is smooth at a point O, then it is also smooth at all its points in a certain neighbourhood W of O.

Assume that ζ is a smooth curve on the surface S. Then its locus $|\zeta|$ is an one-dimensional analytic submanifold of S, as for each $O \in |\zeta|$ there is an open neighbourhood V of O in S so that $|\zeta| \cap V$ is the locus of zeros of an analytic function regular (i.e., with a non-zero derivative) at O (see, for instance, [41]).

Let us have a closer look at a non-singular point. Assume that ζ is smooth at O (and hence at all its points in a neighbourhood of O) and choose local coordinates x, y at O so that ζ is not tangent to the y-axis, that is, $(\partial g/\partial y)$ is not zero at O. Then, by the implicit function theorem, there is a neighbourhood U of 0 in $\mathbb{C}$, a function $s(x)$ analytic in U and a neighbourhood W of O so that the points on ζ in W are just those of the form $(x, s(x))$ for $x \in U$. We have thus a nice 'explicit equation' $y = s(x)$ for $|\zeta| \cap W$. The (obviously bijective and regular) map

$$U \longrightarrow |\zeta| \cap W$$
$$x \longmapsto (x, s(x))$$

defines the manifold structure of $|\zeta|$ near O and so its inverse, the restriction of x, may be taken as a local coordinate on $|\zeta|$ at O.

In chapter 2 we will also find local representations of the form $y = s(x)$ for the non-smooth irreducible curves, but then the functions s will be no longer uniform, the sort of local parameterizations they will give rise to (uniformizing maps, see section 2.5) will be no longer regular maps, and thus will induce no local manifold structure.

Two germs at O are called *transverse* if and only if they are smooth and have different tangents. Two curves whose germs at O are transverse are said to be *transverse* (or to meet transversally) at O. The reader may easily see that the conditions $(\partial(g, h)/\partial(x, y))(O) \neq 0$, $g(O) = h(O) = 0$ for functions g, h to be local coordinates at O (0.3) may be equivalently stated by saying that the curves $g = 0$ and $h = 0$ are transverse at O. In particular any equation g of a curve ζ smooth at O may be taken as part of a system of local coordinates at O, ζ being then one of the coordinate axes: it is enough to complete g with any equation of a second curve transverse to ζ at O.

0.11   Examples of singular germs

We will describe some of the easiest singularities of plane curve. Assume that x, y are local coordinates at a point O or even, if we want to have affine representatives of the germs, affine coordinates in $\mathbb{C}^2$.

0.11.1 *Nodes:* A *node* is a germ of multiplicity two with two different principal tangents, for instance $x^2 - y^2 + y^3 = 0$, whose principal tangents are $x \pm y = 0$. Let us admit, just for this example, that irreducible germs have a single principal tangent (2.2.6). Then a node is not irreducible and therefore, by the additivity of multiplicities, it is composed of two smooth germs with different tangents.

0.11.2 *Cusps:* Irreducible germs are sometimes called *cusps*. Among the simplest ones there is $\xi : f = x^3 - y^2 = 0$, which clearly has a double point and a single principal tangent (counted twice, thus). Irreducibility comes from the fact that $o(f(x, 0)) = 3$: indeed, if $f = f_1 f_2$, no one of them invertible, then necessarily $o(f_i(x, 0)) = 1$ for one of the factors, say for $i = 1$. Then the tangent line of $f_1 = 0$ is not tangent to ξ as it should be. Notice that we have an 'explicit equation' for ξ, namely $y = x^{3/2}$. The two determinations of the square root give rise to two points on the germ for each small enough and non-zero value of x, but these points do not lie on different germs composing ξ, as ξ is irreducible. The reader may see that these two points exchange their positions when x completes a turn around $x = 0$. This would be not the case if the germ were composed of two non-empty germs.

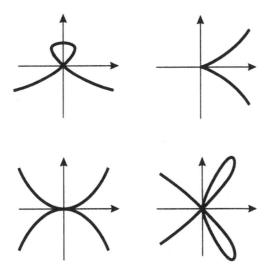

Figure 0.1: The (real parts of) affine representatives of some of the singularities described in section 0.11: from top to bottom and from left to right, the node $x^2 - y^2 + y^3 = 0$, the cusp $x^3 - y^2 = 0$, the tacnode $y^2 - x^4 = 0$ and the ordinary singularity $x^3 - xy^2 + y^4 = 0$.

0.11.3 *Tacnodes:* Singularities composed of two smooth germs that have the same tangent line are called *tacnodes*, they have thus multiplicity two. One of the simplest examples is $y^2 - x^4 = 0$ whose decomposition into two smooth germs is clear. Notice that its multiplicity and tangent cone are the same as those of the cusp above, but this time the 'explicit equation' $y = x^{4/2}$ clearly breaks into the uniform ones $y = x^2$ and $y = -x^2$.

0.11.4 *Ordinary singularities:* A singularity is called *ordinary* if its multiplicity equals the number of its (different) principal tangents. Ordinary singularities of multiplicity two are thus nodes. The same argument we have already used for nodes shows that the ordinary singularities of multiplicity e are composed of e different smooth germs. An easy example is $x^e - y^e + y^{e+1} = 0$.

0.11.5 *Higher-order cusps:* An example of a cusp of multiplicity e is $\xi : y^e - x^{e+1} = 0$. Irreducibility comes by an argument similar to that in 0.11.2: in the case of a non-trivial factorization $f = f_1 f_2$, one of the factors, say f_1, needs to have $o(f_1) = o(f_1(x, 0))$. Then the tangent cone to the germ it defines has no component $y = 0$ which is just the only component of the tangent cone to ξ. The reader may see that any determination of $y = x^{(e+1)/e}$ changes into all the others by successive turns of x around the origin and comes back to its original value after the e-th turn.

Chapter 1

Newton–Puiseux algorithm

In this chapter, throughout sections 1.1 to 1.5, we shall present a purely algebraic method for factorizing the elements of $\mathbb{C}\{x,y\}$ which is essentially due to Newton, though it was later completed and precised by the work of Cramer and Puiseux. Since this method may be applied to any formal power series $f \in \mathbb{C}[[x,y]]$ as well, we will deal with formal power series for a while, and come back to convergent series in section 1.7. As consequences we will obtain the Weierstrass theorems for series in two variables and the main algebraic properties of $\mathbb{C}\{x,y\}$. All these results will be applied to any ring $\mathcal{O}_{S,O}$, O a point on a smooth surface S, in the sequel.

1.1 Newton polygon

Fix on a plane $\pi = \mathbb{R}^2$ a system of orthogonal coordinates α, β. Let us conventionally assume that the first axis is horizontal and oriented from left to right, and the second axis is vertical, oriented from bottom to top.

Assume that $f = \sum_{\alpha,\beta \geq 0} A_{\alpha,\beta} x^\alpha y^\beta$ is an element of $\mathbb{C}[[x,y]]$. For each pair (α, β) with $A_{\alpha,\beta} \neq 0$, we plot on π the point of coordinates (α, β), so that we obtain a discrete set of points with non-negative integral coordinates

$$\Delta(f) = \{(\alpha,\beta) | A_{\alpha,\beta} \neq 0\}$$

which is called the *Newton diagram* of f.

We are interested in a polygonal line whose vertices are points of $\Delta(f)$ and whose sides leave the origin of coordinates and the whole of $\Delta(f)$ in different half-planes: in order to get it, we first translate $\Delta(f)$ by the action of all vectors with non-negative components, so that we obtain the set

$$\Delta'(f) = \Delta(f) + (\mathbb{R}^+)^2.$$

Then we consider the convex hull $\bar{\Delta}(f)$ of $\Delta'(f)$ (i.e., the minimal convex set containing $\Delta'(f)$); the border of $\bar{\Delta}(f)$ consists of two half-lines parallel to the

15

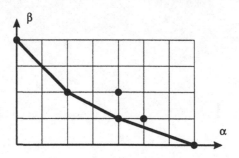

Figure 1.1: $\Delta(f)$ and $N(f)$ for $f = y^4 - x^2y^2 - 2x^4y^2 + x^4y + x^5y + x^7$.

axis and a polygonal line (maybe reduced to a single vertex) joining them: this polygonal line is, by definition, the *Newton polygon* of f. We shall denote it by $N(f)$.

In the sequel we will consider Newton polygons and their sides oriented from left to right and from top to bottom. Thus, if the vertices of a Newton polygon N, taken according to orientation, are $P_i = (\alpha_i, \beta_i)$, $i = 0, \ldots, k$, then $\alpha_{i-1} < \alpha_i$ and $\beta_{i-1} > \beta_i$, $i = 1, \ldots, k$. In such a case we will say that N begins at P_0 and ends at P_k, P_0 and P_k being called, respectively, the first and last end of N. The *height* $h(N)$ and the *width* $w(N)$ of the Newton polygon N are defined, respectively, as the maximal ordinate and the maximal abscissa of its vertices, that is, $h(N) = \beta_0$ and $w(N) = \alpha_0$. Also, if Γ_i is the side of N of ends $(\alpha_{i-1}, \beta_{i-1})$, (α_i, β_i) we will call $h(\Gamma_i) = \beta_i - \beta_{i-1}$ and $w(\Gamma_i) = \alpha_i - \alpha_{i-1}$ the *height* and *width* of Γ_i

The next three lemmas being elementary, their proofs are left to the reader:

Lemma 1.1.1 *The Newton polygon of f begins (resp. ends) on the β-axis (resp. α-axis) if and only if f has no factor x (resp. factor y).*

Lemma 1.1.2 *The Newton polygon of f is reduced to a single vertex if and only if $f = ux^\alpha y^\beta$, where u is an invertible series.*

Lemma 1.1.3 *If $u \in \mathbb{C}[[x,y]]$ is invertible, then $\Delta'(f) = \Delta'(uf)$ and hence also $N(f) = N(uf)$.*

Let O be a point on a smooth surface S. Assume that local coordinates x, y at O have been chosen, after which we identify the local ring at O and the ring of convergent series in x, y: $\mathcal{O}_{S;O} = \mathbb{C}\{x, y\}$. Let ξ be a germ of curve at O. It is clear from 1.1.3 above that all equations of ξ have the same Newton polygon. Therefore one may define the *Newton polygon* $N(\xi)$ of the germ ξ, relative to the local coordinates x, y, as being the Newton polygon of any one of its equations.

The easy proofs of next two lemmas are also left to the reader.

Lemma 1.1.4 *If x does not divide f, then the height of $N(f)$ is $o_y(f(0, y))$. Similarly, if y does not divide f, then the width of $N(f)$ equals $o_x(f(x, 0))$.*

Lemma 1.1.5 *Height and width of Newton polygons are additive, i.e.,*

$$h(\mathbf{N}(f_1 f_2)) = h(\mathbf{N}(f_1)) + h(\mathbf{N}(f_2))$$
$$w(\mathbf{N}(f_1 f_2)) = w(\mathbf{N}(f_1)) + w(\mathbf{N}(f_2)).$$

1.2 Fractionary power series

Let $f \in \mathbb{C}[[x, y]]$: we are interested in solving for y the equation $f(x, y) = 0$. To be a bit more precise, this means we want to find some sort of series in x, $y(x)$, such that $f(x, y(x)) = 0$, $f(x, y(x))$ being the series in x obtained by substituting $y(x)$ for y in f. It will become clear in section 1.3 that the wanted solutions $y(x)$ do not need to belong to $\mathbb{C}[[x]]$. Because of this, we need to deal with series in fractionary powers of x. In this section we shall explain some elementary facts about them.

If z is any free variable over $\mathbb{C}$, we denote by $\mathbb{C}((z))$ the field of fractions of $\mathbb{C}[[z]]$. Its elements are thus the formal Laurent series $\sum_{i=d}^{\infty} a_i z^i$, $d \in \mathbb{Z}$, $a_i \in \mathbb{C}$.

If $n > 1$ is an integer, we choose a free variable over $\mathbb{C}$, t_n, and consider the monomorphism of fields

$$\varphi_{1,n} : \mathbb{C}((x)) \longrightarrow \mathbb{C}((t_n))$$

defined by substituting t_n^n for x in the elements of $\mathbb{C}((x))$. We identify the elements of $\mathbb{C}((x))$ and their images by $\varphi_{1,n}$. In particular, after such identification, we have $x = t_n^n$, so that in the sequel we shall write $x^{1/n}$ instead of t_n and denote by $x^{i/n} = (x^{1/n})^i$ the powers of $x^{1/n}$. An element of $\mathbb{C}((x^{1/n})) = \mathbb{C}((t_n))$ has thus the form

$$s = \sum_{i \geq r} a_i x^{i/n}.$$

Assume now that n' is a second positive integer and that n divides n', say $n' = nd$. Then we have a monomorphism

$$\varphi_{n,n'} : \mathbb{C}((x^{1/n})) \longrightarrow \mathbb{C}((x^{1/n'}))$$

mapping $\sum a_i x^{i/n}$ to $\sum a_i x^{di/dn}$. The monomorphisms $\varphi_{n,n'}$ make commutative diagrams $\varphi_{n,n''} = \varphi_{n',n''} \circ \varphi_{n,n'}$ for any $n \geq 1$, $n' \in (n)$ and $n'' \in (n')$. Thus one may identify each $\mathbb{C}((x^{1/n}))$ with its images by the $\varphi_{n,n'}$ and take $\mathbb{C}\langle\langle x \rangle\rangle$ as being the union of all $\mathbb{C}((x^{1/n}))$ after such identifications (or, equivalently, define $\mathbb{C}\langle\langle x \rangle\rangle$ to be the direct limit of the system $\{\mathbb{C}((x^{1/n})), \varphi_{n,n'}\}$).

In the sequel $\mathbb{C}\langle\langle x \rangle\rangle$ will be taken as being the set of all formal Laurent series

$$\sum_{i \geq r} a_i x^{i/n}$$

with $r, n \in \mathbb{Z}, n \geq 1$. Notice that each such series belongs to all fields $\mathbb{C}((x^{1/n'}))$ for $n' \in (n)$. In particular any two elements of $\mathbb{C}\langle\langle x \rangle\rangle$ may be added or multiplied in a suitable $\mathbb{C}((x^{1/n}))$ containing both of them. The reader may easily prove that in such a way we define an addition and a product making $\mathbb{C}\langle\langle x \rangle\rangle$ a field.

Let

$$s = \sum_{i \geq r} a_i x^{i/n} \qquad\qquad (1.1)$$

be a fractionary power series. As for ordinary series, we define the *order* in x of s to be $o_x(s) = \infty$ if $s = 0$ and

$$o_x(s) = \frac{\min\{i | a_i \neq 0\}}{n}$$

otherwise. Obviously o_x does not depend on the denominator n. Fractionary power series s with $o_x(s) > 0$ will be called *Puiseux series*.

After a simultaneous reduction of all fractionary exponents effectively appearing in s, we may assume that n and $\gcd\{i | a_i \neq 0\}$ have no common factor. Then we say that n is the *polydromy order* of s, usually denoted in the sequel by $\nu(s)$.

Fix n and consider $\mathbb{C}((x^{1/n}))$ as an extension of $\mathbb{C}((x))$: for each n-th root of unity ε, the substitution of $\varepsilon x^{1/n}$ for $x^{1/n}$ induces an automorphism σ_ε of $\mathbb{C}((x^{1/n}))$ over $\mathbb{C}((x))$. If $s \in \mathbb{C}((x^{1/n}))$, the series $\sigma_\varepsilon(s)$, $\varepsilon^n = 1$, will be called the *conjugates* of s. Notice that if s is as in 1.1 above, then

$$\sigma_\varepsilon(s) = \sum_{i \geq r} \varepsilon^i a_i x^{i/n}$$

from which it is clear that the set of conjugates of a given s does not depend on the field $\mathbb{C}((x^{1/n}))$ the series s is belonging to. As is clear from its definition, conjugation of series is an equivalence relation. The set of all conjugates of s will be called the *conjugacy class* of s. Clearly, conjugate series have the same order in x and the same polydromy order, hence in particular, the conjugates of a Puiseux series are Puiseux series too.

Lemma 1.2.1 *The number of different conjugates of s is $\nu(s)$. In particular a series $s \in \mathbb{C}((x^{1/n}))$ belongs to $\mathbb{C}((x))$ if and only if $\sigma_\varepsilon(s) = s$ for all n-th roots of unity ε.*

PROOF: Write

$$s = \sum_{i \geq r} a_i x^{i/n}$$

and assume that $n = \nu(s)$. Then for each prime divisor m of n there is $i \geq r$ so that $a_i \neq 0$ and m does not divide i. One may thus select indices $i_1, \ldots, i_k$ so that $a_{i_j} \neq 0$, $j = 1, \ldots, k$ and $\gcd(n, i_1, \ldots, i_k) = 1$. Now, from an equality

$\sigma_\eta(s) = \sigma_\varepsilon(s)$, $\eta^n = \varepsilon^n = 1$ we get $\eta^{ij}a_{i_j} = \varepsilon^{ij}a_{i_j}$ and hence $\eta^{ij} = \varepsilon^{ij}$ for $j = 1, \ldots, k$. Since $\gcd(n, i_1, \ldots, i_k) = 1$, it follows that $\eta = \varepsilon$ and so s has $n = \nu(s)$ different conjugates as claimed. The second part of the claim easily follows from the first one. ◇

Assume that $s \in \mathbb{C}[[x^{1/n}]]$ is a Puiseux series, that is, $o_x(s) > 0$. One may substitute s for y in the elements f of $\mathbb{C}[[x, y]]$: the resulting series $f(x, s)$ is an element of $\mathbb{C}[[x^{1/n}]]$ and the substitution induces a morphism of $\mathbb{C}$-algebras $\mathbb{C}[[x, y]] \longrightarrow \mathbb{C}[[x^{1/n}]]$.

We say that a Puiseux series s is a *y-root* of f if and only if $f(x, s) = 0$ (as element of $\mathbb{C}[[x^{1/n}]]$, i.e., identically in x). Obviously the series of the form $x^d u$, u invertible, have no y-roots. Note that these are just the series whose Newton polygon has height zero. Of course if f is a polynomial, $f \in \mathbb{C}[[x]][y]$, then its y-roots are ordinary roots as a polynomial. The following lemmas will show the close relationship between y-roots and divisors of a series f.

Lemma 1.2.2 *A Puiseux series $s \in \mathbb{C}[[x^{1/n}]]$ is a y-root of $f \in \mathbb{C}[[x, y]]$ if and only if $y - s$ divides f in $\mathbb{C}[[x^{1/n}, y]]$.*

PROOF: Consider the automorphism of $\mathbb{C}[[x^{1/n}, y]]$ obtained by substituting $y + s$ for y, and write $g^\sharp = g(x, y + s)$ for the image of g. Since $(y - s)^\sharp = y$ and $f^\sharp(x, 0) = f(x, s)$, it is enough to deal with the case $s = 0$ which in turn is clear. ◇

Lemma 1.2.3 *If s is a y-root of f, then all conjugates of s are y-roots of f too.*

PROOF: By 1.2.2 we have $f = (y - s)g$, $g \in \mathbb{C}[[x^{1/n}, y]]$, so, if we still denote by σ_ε the obvious extension of the conjugation automorphism σ_ε to $\mathbb{C}[[x^{1/n}, y]]$, then $f = \sigma_\varepsilon(f) = (y - \sigma_\varepsilon(s))\sigma_\varepsilon(g)$, from which we obtain the claim. ◇

If $s \in \mathbb{C}[[x^{1/n}]]$, let us write $s = s^1, \ldots, s^\nu$ for its different conjugates and $g_s = \prod_1^\nu (y - s^i)$. Note that $g_s \in \mathbb{C}[[x]][y]$ as all its coefficients are invariant by conjugation and therefore 1.2.1 applies to it. It is also clear that $\nu(s) = \nu = h(\mathbf{N}(g_s))$.

Lemma 1.2.4 *A Puiseux series $s \in \mathbb{C}[[x^{1/n}]]$ is a y-root of f if and only if g_s divides f in $\mathbb{C}[[x, y]]$.*

PROOF: The converse being obvious, let us assume that $f(x, s) = 0$. Then, since the case $i = 1$ is guaranteed by 1.2.2, we may assume by induction that $f = (y - s^1) \ldots (y - s^i) f_i$, $f_i \in \mathbb{C}[[x^{1/n}, y]]$. If $i < \nu$, s^{i+1} needs to be a y-root of f_i so that 1.2.2 gives a new equality $f = (y - s^1) \ldots (y - s^{i+1}) f_{i+1}$, and eventually the equality $f = (y - s^1) \ldots (y - s^\nu) f_\nu = g_s f_\nu$. To close, both f and g_s being invariant by conjugation, so is f_ν and therefore it belongs to $\mathbb{C}[[x, y]]$ as claimed. ◇

Lemma 1.2.5 *The series g_s is irreducible in $\mathbb{C}[[x, y]]$.*

PROOF: If $g_s = f_1.f_2$, s needs to be a y-root of at least one of the factors, say of f_1. Then, by 1.2.4, g_s divides f_1 and therefore f_2 is invertible. ⋄

Lemma 1.2.6 *If $h(N(f)) = 0$, then f has no y-roots.*

PROOF: Direct from 1.2.4 and 1.1.5. ⋄

1.3 Search for y-roots of $f(x, y)$

Fix a series
$$f(x, y) = \sum_{\alpha,\beta \geq 0} A_{\alpha,\beta} x^\alpha y^\beta \in \mathbb{C}[[x, y]].$$

In order to determine a non-zero y-root of f by means of an inductive procedure, we start by testing solutions of the form
$$s = ax^{m/n} + \cdots \tag{1.2}$$

where m, n are positive integers, $\gcd(m, n) = 1$, a is assumed to be non-zero to avoid trivialities and the dots indicate higher order terms. It is clear that the initial term of $f(x, s(x))$ is also the initial term of
$$\sum_{\alpha,\beta \geq 0} A_{\alpha,\beta} a^\beta x^{\alpha+\beta m/n} \tag{1.3}$$

the non-explicit terms in 1.2 giving rise to higher order terms. The last series may be written in the form
$$\sum_k \left(\sum_{n\alpha+m\beta=k} A_{\alpha,\beta} a^\beta \right) x^{k/n}$$

after which it is clear that the pairs (α, β) giving rise to a term of degree k/n in 1.3 are just the points of $\Delta(f)$ lying on the line $n\alpha + m\beta = k$. Then, consider all lines $n\alpha + m\beta = k$, $k \in \mathbb{Z}$, that have a non-empty intersection with $\Delta(f)$. From them, pick the line ℓ closest to the origin (i.e., that with minimal k): the terms of lowest degree in 1.3 are then
$$\sum_{(\alpha,\beta)\in\ell\cap\Delta(f)} A_{\alpha,\beta} a^\beta x^{\alpha+\beta m/n}.$$

We shall consider two cases, namely:

Case (a): There is no side of $N(f)$ with slope $-n/m$. Then there is a single point (α_0, β_0) of $\Delta(f)$, in fact a vertex of $N(f)$, on ℓ and thus $f(x, s(x))$ has a single term of minimal degree, namely
$$A_{\alpha_0,\beta_0} a^{\beta_0} x^{(n\alpha_0+m\beta_0)/n},$$

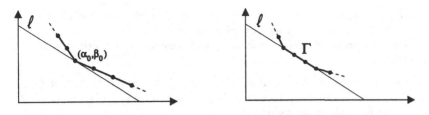

Figure 1.2: Cases (a) (left) and (b) (right) in section 1.3.

which therefore cannot be cancelled by other terms. It follows that $f(x, s(x)) \neq 0$, and so in such a case there is no y-root of $f(x,y)$ with initial term of degree m/n.

Case (b): There is a side of $\mathbf{N}(f)$, say Γ, with slope $-n/m$ and hence on ℓ. The terms of lowest degree in $f(x, s(x))$ are thus

$$\left(\sum_{(\alpha,\beta)\in\Gamma} A_{\alpha,\beta} a^\beta \right) x^{(n\alpha_0 + m\beta_0)/n},$$

where (α_0, β_0) is any point on Γ. Assume that the first and last end of Γ are (α_1, β_1) and (α_0, β_0), respectively. Then, the former expression may be written in the form

$$a^{\beta_0} \left(\sum_{(\alpha,\beta)\in\Gamma} A_{\alpha,\beta} a^{\beta-\beta_0} \right) x^{(n\alpha_0 + m\beta_0)/n} = a^{\beta_0} F_\Gamma(a) x^{(n\alpha_0+m\beta_0)/n}$$

where

$$F_\Gamma(Z) = \sum_{(\alpha,\beta)\in\Gamma} A_{\alpha,\beta} Z^{\beta-\beta_0} \in \mathbb{C}[Z]$$

is a polynomial with non-zero constant term and positive degree equal to the height $\beta_1 - \beta_0$ of Γ. The polynomial $F_\Gamma(Z)$ (or $F_\Gamma(Z) = 0$) is usually called the *equation associated with* Γ. If $s(x)$ is a y-root of f, we must have $F_\Gamma(a) = 0$, hence we have found necessary conditions for a monomial to be the initial term of a solution:

Lemma 1.3.1 *If a y-root of f has initial term $ax^{m/n}$, then there is a side Γ of $\mathbf{N}(f)$ with slope $-n/m$ and furthermore a is a root of the equation associated with Γ.*

Of course f may also have 0 as y-root: this occurs if and only if y is a factor of f, that is, if and only if $\mathbf{N}(f)$ ends above the α-axis.

Notice that if $\mathbf{N}(f)$ has no side with slope $-1/m$, $m \in \mathbb{N}$, then, by 1.3.1, f has no non-zero y-root in $\mathbb{C}[[x]]$. So, if f is chosen so that $\mathbf{N}(f)$ has positive height, no side with slope $-1/m$, $m \in \mathbb{N}$ and last end on the α-axis (in order to avoid the y-root 0), we have a non-trivial example of a series which has no y-roots in $\mathbb{C}[[x]]$, as already announced in section 1.2.

A special property of the equation F_Γ will become very important later on, namely:

Lemma 1.3.2 *If the slope of Γ is $-n/m$, $(n, m) = 1$ and $n > 0$, then $F_\Gamma \in \mathbb{C}[Z^n]$. If a is a root of F_Γ, then so is εa for any n-th root of unity ε, and both roots $a, \varepsilon a$ have the same multiplicity.*

PROOF: The points of $\Delta(f)$ on Γ have the form

$$(\alpha, \beta) = (\alpha_0, \beta_0) + t(-m, n)$$

with $t \in \mathbb{N} \cup \{0\}$. All such points have in particular $\beta \in \beta_0 + (n)$, and so $F_\Gamma \in \mathbb{C}[Z^n]$. Then, to complete the proof, just write $F_\Gamma(z) = P(z^n)$ and decompose P into irreducible factors. $\diamond$

Roots a and εa, $\varepsilon^n = 1$, of F_Γ will be called *conjugate*.

1.4   The Newton–Puiseux algorithm

Let f be, as before, a series in x, y. We assume that its Newton polygon $\mathbf{N}(f)$ has positive height, since otherwise it has no y-roots (1.2.6). In this section we build-up an inductive algorithm, named the *Newton–Puiseux algorithm*, which gives all y-roots of f. We describe its first step next:

Step (0): Since $h(\mathbf{N}(f)) > 0$, either $\mathbf{N}(f)$ ends above the α-axis or it has at least one side. We start determining a y-root $s = s^{(0)}$ of f by performing any of the two steps (0.a), (0.b) below, at least one of them being always possible:

(0.a): If $\mathbf{N}(f)$ ends above the α-axis, take $s^{(0)} = 0$ and stop the algorithm here, or

(0.b): choose a side Γ of $\mathbf{N}(f)$, if any, and a root a of F_Γ. Assume that Γ has equation $n\alpha + m\beta = k$, $(n, m) = (1)$. Introduce new free variables x_1, y_1 and identify $\mathbb{C}\langle\langle x \rangle\rangle = \mathbb{C}\langle\langle x_1 \rangle\rangle$ and $\mathbb{C}((x, y))$ with a subfield of $\mathbb{C}((x_1, y_1))$ by the rules

$$x = x_1^n$$
$$y = x_1^m(a + y_1).$$

Then we have

$$f = \sum_{n\alpha+m\beta \geq k} A_{\alpha,\beta} x^\alpha y^\beta = x_1^k \left(\sum_{n\alpha+m\beta \geq k} A_{\alpha,\beta} x_1^{n\alpha+m\beta-k} (a + y_1)^\beta \right)$$

so that we may define $f_1 = x_1^{-k} f \in \mathbb{C}[[x_1, y_1]]$. Take, as y-root of f,

$$s^{(0)} = x^{m/n}(a + s^{(1)})$$

where $s^{(1)}$ is an element of $\mathbb{C}\langle\langle x_1 \rangle\rangle$ to be determined.

If step (0.a) has been chosen we end by taking $s = s^{(0)} = 0$. If we have performed step (0.b) we perform step (1) by determining, just in the same way, the leading term of $s^{(1)}$ from x_1, y_1 and f_1, and so on. To be clear we describe step (i) assuming by induction that step $((i-1).b)$ has been performed. We assume furthermore that $h(\mathbf{N}(f_i)) > 0$, which will be proved to be always true in Lemma 1.4.1 below. Then we have

$$s^{(0)} = x^{m/n}(a + x_1^{m_1/n_1}(a_1 + \cdots + x_{i-1}^{m_{i-1}/n_{i-1}}(a_{i-1} + s^{(i)}) \cdots))$$
$$= x^{m/n}(a + x^{m_1/nn_1}(a_1 + \cdots + x^{m_{i-1}/n \cdots n_{i-1}}(a_{i-1} + s^{(i)}) \cdots))$$

and perform step (i) as follows:

Step (i): As in step (0), since $h(\mathbf{N}(f_i)) > 0$, either:

    $(i.a)$: take $s^{(i)} = 0$, only in the case $\mathbf{N}(f_i)$ ends above the α-axis and stop the algorithm here, or

    $(i.b)$: take

$$s^{(i)} = x^{m_i/n_i}(a_i + s^{(i+1)})$$

where $-n_i/m_i$ is the slope of a side Γ_i of $\mathbf{N}(f_i)$, say with equation $n_i\alpha + m_i\beta = k_i$, $(n_i, m_i) = (1)$, and a_i is a root of F_{Γ_i}. The series $s^{(i+1)}$ is to be determined in the next step from x_{i+1}, y_{i+1} and f_{i+1}, the new variables x_{i+1}, y_{i+1} being related to x_i, y_i by the rules

$$x_i = x_{i+1}^{n_i}$$
$$y_i = x_{i+1}^{m_i}(a_i + y_{i+1}),$$

and $f_{i+1} = x_{i+1}^{-k_i} f_i \in \mathbb{C}[[x_{i+1}, y_{i+1}]]$, by a computation quite similar to that in step (0.b).

It is worth remarking that there may be many ways for continuing the Newton–Puiseux algorithm at each step, either because both possibilities $(i.a)$ and $(i.b)$ may be chosen, or because after choosing $(i.b)$ one may still choose the side Γ of the Newton polygon and the root a of F_Γ. Notice also that after performing step $(i.b)$ all terms up to the degree $m/n + \cdots + m_i/n \cdots n_i$ in x of the claimed solution $s^{(0)}$ have been determined, while after step $(i.a)$ the whole series $s^{(0)}$ has been determined.

The next lemma shows in particular that always $h(\mathbf{N}(f_i)) > 0$, so that either the Newton–Puiseux algorithm ends by an step $(i.a)$, or it runs indefinitely.

Lemma 1.4.1 *All notations being as above, for any $i \geq 0$ $h(\mathbf{N}(f_{i+1}))$ equals the multiplicity of a_i as a root of F_{Γ_i}.*

PROOF: All cases being identical, we make the proof for the case $i = 0$ in order to simplify the notations. By definition we have

$$f_1 = \sum_{n\alpha + m\beta \geq k} A_{\alpha,\beta} x_1^{n\alpha + m\beta - k}(a + y_1)^\beta,$$

so that

$$f_1(0, y_1) = \sum_{(\alpha, \beta) \in \Gamma} A_{\alpha, \beta}(a + y_1)^{\beta} = (a + y_1)^{\bar{\beta}} F_{\Gamma}(a + y_1)$$

where $\bar{\beta}$ is the ordinate of the end of Γ. It follows that

$$h(\mathbf{N}(f_1)) = o_{y_1}(f_1(0, y_1)) = o_{y_1}(F_{\Gamma}(a + y_1))$$

and hence the claim. ◇

We will see in the next section that all outputs of the Newton–Puiseux algorithm are y-roots of f. We close this section by proving just the converse:

Lemma 1.4.2 *If a Puiseux series s is a y-root of f, then s is one of the series the Newton–Puiseux algorithm gives rise to when applied to f.*

PROOF: If $s = 0$, then f has factor y and so $\mathbf{N}(f)$ ends above the α-axis. The step (0.a) is thus allowed and gives $s = 0$.

If $s \neq 0$, write $s = x^{m/n}(a + s^{(1)})$ with $a \neq 0$. It follows from 1.3.1 that one may perform a step (0.b) with just these m/n and a. Then an easy computation shows that since $f(x, s) = 0$, also $f_1(x_1, s^{(1)}) = 0$ and so the claim follows by iteration. ◇

1.5   Puiseux theorem

We have described in the preceding section a recursive procedure that determines the successive partial sums of certain series s. We will show here that these series are the y-roots of f. First of all we need to see that the polydromy orders of the successive partial sums do not increase indefinitely, so that $s = s^{(0)}$ actually belongs to $\mathbb{C}\langle\langle x \rangle\rangle$. This is clear if a step $(i.a)$ is performed. Then, assume that option (b) is chosen at each step. Using the same notations as in section 1.4, we have:

Lemma 1.5.1 *There exists an integer i_0 such that $n_i = 1$ if $i > i_0$.*

PROOF: Fix any $i \geq 0$. We have $\deg F_{\Gamma_i} = h(\Gamma_i) \leq h(\mathbf{N}(f_i))$. On the other hand, by 1.3.2, if a_i is a root of multiplicity μ_i of F_{Γ_i}, then so are all its conjugates, εa_i, $\varepsilon^{n_i} = 1$, and therefore F_{Γ_i} has n_i different roots of multiplicity μ_i. This gives

$$h(\mathbf{N}(f_i)) \geq \deg F_{\Gamma_i} \geq n_i \mu_i = n_i h(\mathbf{N}(f_{i+1})),$$

the last equality coming from 1.4.1. It follows that, for any i,

$$h(\mathbf{N}(f)) \geq n n_1 \cdots n_i h(\mathbf{N}(f_{i+1}))$$

and so all but finitely many of the n_i must be equal to 1. ◇

Corollary 1.5.2 *The series given rise to by the Newton–Puiseux algorithm are Puiseux series.*

PROOF: We have seen in 1.5.1 that all these series belong to $\mathbb{C}\langle\langle x\rangle\rangle$. It is clear from the first step of the algorithm that they have positive order. ⋄

With the same hypothesis and notations as above, let us assume that $s = s^{(0)} \in \mathbb{C}\langle\langle x\rangle\rangle$ is a Puiseux series obtained from f by the Newton–Puiseux algorithm. We show next that s is actually a y-root of f.

Lemma 1.5.3 $f(x, s) = 0$.

PROOF: An easy computation from the definitions of f_{i+1} and $s^{(i+1)}$ gives

$$f_i(x_i, s^{(i)}) = x_{i+1}^{k_i} f_{i+1}(x_{i+1}, s^{(i+1)}),$$

from which it follows that

$$o_{x_i}(f_i(x_i, s^{(i)})) = \frac{1}{n_i}[k_i + o_{x_{i+1}}(f_{i+1}(x_{i+1}, s^{(i+1)}))],$$

and thus, for any i,

$$o_x(f(x, s)) = \frac{k}{n} + \frac{k_1}{nn_1} + \cdots + \frac{k_{i-1}}{nn_1 \cdots n_{i-1}} + \frac{o_{x_i}(f_i(x_i, s^{(i)}))}{nn_1 \cdots n_{i-1}}.$$

If for some i we perform step $(i.a)$, then f_i is a multiple of y_i and we take $s^{(i)} = 0$, so that we have $f_i(x_i, s^{(i)}) = 0$ and thus, by the former equality, also $f(x, s) = 0$ as wanted.

Assume thus that no step $(i.a)$ is performed. In such a case it is clear from the equality above, using 1.5.1, that $o_x(f(x, s))$ may be arbitrarily increased by continuing the algorithm (at least by $1/nn_1 \cdots n_{i_0}$ at each step), so $o_x(f(x, s)) = \infty$ as wanted. ⋄

We have given a constructive proof of the Puiseux theorem, namely

Theorem 1.5.4 (Puiseux) *If $f \in \mathbb{C}[[x, y]]$ and $h(\mathrm{N}(f)) > 0$, then there is a Puiseux series s which is a y-root of f, namely $f(x, s(x)) = 0$.*

As in section 1.2, if s is a Puiseux series, write $g_s = \prod_1^\nu (y - s^i)$, the s^i, $i = 1, \ldots, \nu = \nu(s)$ being the conjugates of s. Then we have

Corollary 1.5.5 *For any $f \in \mathbb{C}[[x, y]]$,*

(i) *There are Puiseux series $s_1, \ldots, s_m$, $m \geq 0$, so that f decomposes in the form*

$$f = ux^r g_{s_1} \cdots g_{s_m},$$

where $r \in \mathbb{Z}$, $r \geq 0$, and u is an invertible series in $\mathbb{C}[[x, y]]$.

(ii) *Such a decomposition is uniquely determined by f.*

(iii) $h(\mathbf{N}(f)) = \nu(s_1) + \cdots + \nu(s_m)$ *and the y-roots of f are the conjugates of the s_j, $j = 1, \ldots, m$.*

PROOF: We prove claim (i) by induction on $h(\mathbf{N}(f))$: if $h(\mathbf{N}(f)) = 0$, then the claim is obviously satisfied with no Puiseux series at all (i.e., $m = 0$). Otherwise, there is a y-root s of f, by 1.5.4, and therefore (1.2.4) $f = g_s f'$, $f' \in \mathbb{C}[[x, y]]$. Then it is enough to apply the induction hypothesis to f', as $h(\mathbf{N}(f')) < h(\mathbf{N}(f))$ by 1.1.5.

Claim (iii) obviously follows from claim (i) and the definition of the series g_{s_i}, as ux^r has no y-root by 1.2.6.

To prove (ii), notice first that r is the first coordinate of the first vertex of $\mathbf{N}(f)$ and so it does not depend on the decomposition. If g_s does appear as a factor in a second decomposition, then s is a y-root of f and hence it is a conjugate of some s_j, by the part of the claim already proved. Therefore $g_s = g_{s_j}$, after which it is enough to divide both sides by g_s and use induction on m. ◇

Of course, f may also be written as a product of y-linear factors:

Corollary 1.5.6 *If $f \in \mathbb{C}[[x, y]]$, then it has a unique decomposition of the form*

$$f = ux^r \prod_{j=1}^{\ell} (y - z_j),$$

where the z_j are Puiseux series, and, as above, $r \in \mathbb{Z}$, $r \geq 0$, and $u \in \mathbb{C}[[x, y]]$ is invertible. Furthermore, $\ell = h(\mathbf{N}(f))$ and the series z_j are the y-roots of f.

PROOF: A decomposition as claimed, with $\ell = h(\mathbf{N}(f))$, comes from the one in 1.5.5 using the definition of the series g_s. For the uniqueness, assume we have

$$f = ux^r \prod_{j=1}^{\ell} (y - z_j) = u'x^{r'} \prod_{j=1}^{\ell'} (y - z_j').$$

The equality $r = r'$ follows as in the proof of 1.5.5. From the decompositions above it is clear that the set of y-roots of f equals both $\{z_j\}_{j=1,\ldots,\ell}$ and $\{z_j'\}_{j=1,\ldots,\ell'}$. Then uniqueness is clear if $h(\mathbf{N}(f)) = 0$ as in this case there are no y-roots. Assume $h(\mathbf{N}(f)) > 0$ and let s be a y-root of f. Since all conjugates of s are y-roots of f (1.2.3), all of them appear among the z_j and also among the z_j'. Furthermore, g_s divides f in $\mathbb{C}[[x, y]]$ (1.2.4), so that one may divide the former equalities by g_s. After this the proof ends by induction on $h(\mathbf{N}(f))$, the induction hypothesis being applied to f/g_s. The last claim is clear, as already noticed. ◇

Our constructive proof of the Puiseux theorem provides further information that is worth a separate claim:

Proposition 1.5.7 *A Puiseux series s is a y-root of f if and only if it is obtained from f by the Newton–Puiseux algorithm.*

PROOF: Follows from 1.4.2 and 1.5.3. ⋄

Corollary 1.5.8 *Let* $f \in \mathbb{C}[[x,y]]$ *with no factor* x. *Then* f *is irreducible if and only if* $f = ug_s$ *with* $u \in \mathbb{C}[[x,y]]$ *invertible and* s *a Puiseux series. In such a case* $h(\mathbf{N}(f)) = \nu(s)$ *and the set of* y-*roots of* f *is the conjugacy class of* s.

PROOF: Obvious from 1.2.5 and 1.5.5. ⋄

Corollary 1.5.9 *The ring* $\mathbb{C}[[x,y]]$ *is a unique factorization ring.*

PROOF: Follows from 1.5.5 and 1.5.8. ⋄

The formal Weierstrass preparation theorem for two variables is also an easy consequence of the Puiseux theorem. Let $n \in \mathbb{Z}$, $n \geq 0$. Recall that a series $f \in \mathbb{C}[[x,y]]$ is called *regular* of order n in y if and only if $n = o_y(f(0,y))$, i.e., if and only if its Newton polygon has height n and begins on the β-axis. A polynomial $\sum_0^n a_i(x)y^i \in \mathbb{C}[[x]][y]$ is called a *formal Weierstrass polynomial* in y of degree n if and only if $a_n(x) = 1$ and $a_i(0) = 0$ for $i < n$.

Corollary 1.5.10 (formal Weierstrass preparation theorem) *If a series* $f \in \mathbb{C}[[x,y]]$ *is regular of order* n *in* y, *then there is a unique formal Weierstrass polynomial* g *in* y *such that* $f = ug$, *where* $u \in \mathbb{C}[[x,y]]$ *is invertible. Furthermore* $\deg g = n$.

PROOF: The existence follows from 1.5.5, as the regularity of f forces $r = 0$ and, as it is clear from the definition of the g_{s_i}, $g_{s_1} \dots g_{s_m}$ is a formal Weierstrass polynomial in y. Uniqueness being obvious if $n = 0$, assume $n > 0$ and hence $m \geq 1$ by 1.5.4. Then if $f = ug$ in the conditions of the claim, s_1 needs to be a y-root of g. Therefore g_{s_1} divides both f and g (by 1.2.4), after which the claimed uniqueness follows by induction on n. To close $\deg g = o_y g(0,y) = o_y f(0,y) = n$. ⋄

Further algebraic properties of the ring $\mathbb{C}[[x,y]]$ may be easily obtained from the Puiseux theorem and its corollaries proved above. Nevertheless, we will not pursue this way here as we are far more concerned with the ring $\mathbb{C}\{x,y\}$ and for it all these properties will be presented in the forthcoming section 1.8. The interested reader may easily translate claims and proofs from the convergent case to the formal one. Before dealing with multiplicities of y-roots in the next section we close this one with an algebraic corollary of the Puiseux theorem which is often named the Puiseux theorem too. The reader may notice that the Puiseux theorem 1.5.4 easily follows from it if the formal Weierstrass preparation theorem is assumed to be already known.

Corollary 1.5.11 *The field* $\mathbb{C}\langle\langle x \rangle\rangle$ *is algebraically closed.*

PROOF: Let $P \in \mathbb{C}\langle\langle x \rangle\rangle[y]$ be a polynomial with positive degree: we will prove that P has a root in $\mathbb{C}\langle\langle x \rangle\rangle$. We have in fact $P \in \mathbb{C}((x^{1/n}))[y]$ for a suitable n. Since obviously $\mathbb{C}\langle\langle x^{1/n} \rangle\rangle = \mathbb{C}\langle\langle x \rangle\rangle$, we may assume without restriction that $P \in \mathbb{C}((x))[y]$. After cleaning denominators one may even assume that $P \in \mathbb{C}[[x]][y]$, its coefficients having no common factor x. Write $P = P(x, y) = a_m(x)y^m + \cdots + a_0(x)$, $a_i(x) \in \mathbb{C}[[x]]$, $a_m(x) \neq 0$.

If $a_i(0) \neq 0$ for some $i > 0$, then the $\mathbb{C}$-polynomial $P(0, y)$ has at least one root $c \in \mathbb{C}$. After taking $y' = y - c$ and $P'(x, y') = P(x, y' + c)$, we have $h(\mathbf{N}(P')) > 0$, as $P'(0, 0) = 0$ and $P'(0, y') \neq 0$. By the Puiseux theorem 1.5.4, P' has then a y'-root $s \in \mathbb{C}\langle\langle x \rangle\rangle$ and clearly $c + s$ is a root of P.

Otherwise $a_0(0) \neq 0$ and $a_i(0) = 0$ for $i > 0$. In such a case we take $\bar{y} = 1/y$ and $\bar{P}(x, \bar{y}) = \bar{y}^m P(x, 1/\bar{y}) = a_m(x) + \cdots + a_0(x)\bar{y}^m$. Since $\bar{P}(0, 0) = a_m(0) = 0$ and $\bar{P}(0, \bar{y}) = a_0(0)\bar{y}^m \neq 0$, we get $h(\mathbf{N}(\bar{P})) > 0$ and again by 1.5.4, $\bar{P}$ has a $\bar{y}$-root $s \in \mathbb{C}\langle\langle x \rangle\rangle$. Furthermore, since $a_m(x) \neq 0$, necessarily $s \neq 0$ and then $1/s$ is a root of P as wanted. $\diamond$

1.6 Separation of y-roots

As in the preceding section, let $f \in \mathbb{C}[[x, y]]$. It is clear that in practice only finitely many terms of each y-root of f can be determined by means of the Newton–Puiseux algorithm. Usually, the algorithm for a root s is continued till one finds a partial sum of s that is a partial sum of no other y-root of f. In such a case we say that s has been *separated* from the other y-roots. In this section we will give a sufficient condition (1.6.4) for the separation of y-roots. We will also get (1.6.6) some information on higher steps of the Newton–Puiseux algorithm that will be useful in the next section 1.7.

First we need to introduce the multiplicities of y-roots. Let us rewrite the decomposition of f obtained in 1.5.6 as

$$f = ux^r \prod_{j=1}^{h(\mathbf{N}(f))} (y - z_j).$$

If s is a Puiseux series, the *multiplicity of s as y-root of f* is, by definition, the number of factors $y - s$ appearing in the above decomposition. The y-roots of multiplicity one are called *simple y-roots*.

The reader may easily check that one may equivalently define the multiplicity of s as the number of times that the irreducible series g_s appears as a factor of f (1.5.5). In particular the next lemma, which is stated for further reference, needs no proof.

Lemma 1.6.1 *Conjugate y-roots of f have equal multiplicities. The sum of the multiplicities of all y-roots of f equals $h(\mathbf{N}(f))$.*

The next lemma will provide an easier handling of multiplicities of y-roots.

Lemma 1.6.2 *A Puiseux series s is a y-root of f of multiplicity r if and only if, in some ring $\mathbb{C}[[x^{1/n}, y]]$, $f = (y - s)^r f'$ where f' has no root s.*

PROOF: For the if part, decompose f' (as a series in $x^{1/n}, y$) according to 1.5.6. Then no factor in this decomposition equals $y - s$ and after multiplying by $(y - s)^r$ we get a decomposition for f (still as a series in $x^{1/n}, y$) in the conditions of 1.5.6, in which the factor $y - s$ appears just r times. By its own uniqueness, this decomposition must agree with the decomposition of f (now as a series in x, y) we already used for defining the multiplicity of s and hence this multiplicity equals r. The only if part is clear from the definition of multiplicity. $\diamond$

Take the notations as in section 1.4.

Lemma 1.6.3 *For any $i > 0$, the multiplicity of s as y-root of f equals the multiplicity of $s^{(i)}$ as y_i-root of f_i.*

PROOF: It is enough to make the proof for $i = 1$, as the remaining cases follow by an obvious induction. We have

$$s = x^{m/n}(a + s^{(1)})$$

and

$$f(x_1^n, x_1^m(a + y_1)) = x_1^k f_1(x_1, y_1).$$

It is clear that if $(y - s)^\mu$ divides f, then

$$(x_1^m(a + y_1) - s)^\mu = x_1^{m\mu}(y_1 - s^{(1)})^\mu$$

divides $x_1^k f_1$ and thus $(y_1 - s^{(1)})^\mu$ divides f_1. Put $f = (y - s)^\mu f'$ and $f_1 = (y_1 - s^{(1)})^\mu f_1'$. We have

$$x_1^{m\mu} f'(x_1^n, x_1^m(a + y_1)) = x_1^k f_1'(x_1, y_1),$$

from which it is clear that if s is not a y-root of f', then $s^{(1)}$ is not a y_1-root of f_1 either, just as wanted. $\diamond$

Proposition 1.6.4 *If for a given i, a_i is a simple y-root of F_{Γ_i}, then s is a simple root of f and no other y-root of f has the same partial sum of degree $m/n + \cdots + m_i/n \cdots n_i$.*

PROOF: By 1.4.1, $N(f_{i+1})$ has height one and so, using 1.6.1, f_{i+1} has a single y_{i+1}-root and this root is simple: since $s^{(i+1)}$ is a y_{i+1}-root of f_{i+1} by construction, necessarily $s^{(i+1)}$ has multiplicity one and hence, by 1.6.3, s is a simple y-root of f. Moreover, once we know $N(f_{i+1})$ to have height one, it is clear that the Newton–Puiseux algorithm may be continued from step $(i + 1)$ onwards in a single way, and gives thus a single y-root of f of the form

$$x^{m/n}(a + x_1^{m_1/n_1}(a_1 + \cdots + x_i^{m_i/n_i}(a_i + \cdots) \cdots)).$$

All y-roots of f being given by the Newton–Puiseux algorithm (1.5.7), the claim follows. $\diamond$

The next proposition has a similar proof which is left to the reader.

Proposition 1.6.5 *If for some i, the ordinate of the last end of* $\mathbf{N}(f_i)$ *is one and we perform the step (i.a), then the corresponding y-root s of f is simple.*

Notice that in this case all terms of s have been computed, and so s has been separated from the other y-roots.

The converse of 1.6.4 and 1.6.5 together is also true and will be useful later on:

Proposition 1.6.6 *If s is a simple y-root of f, then, for a certain i, either* $\mathbf{N}(f_i)$ *ends at a point of ordinate 1, or* a_i *is a simple root of* F_{Γ_i}.

PROOF: Assume first that the Newton–Puiseux algorithm for the root s ends by an step $(i.a)$ and so $s^{(i)} = 0$. Then 0 is a y_i-root of f_i and it has multiplicity one by 1.6.3. Thus f_i has a single factor y and therefore $\mathbf{N}(f_i)$ ends at a point with ordinate one.

Assume now that no step $(i.a)$ is performed and take i big enough so that no y-root of f other than s has partial sum

$$x^{m/n}\big(a + x_1^{m_1/n_1}(a_1 + \cdots + x_{i-1}^{m_{i-1}/n_{i-1}} a_{i-1})\ldots\big)).$$

Then, since all y-roots of f given rise to by y_i-roots of f_i would have the partial sum displayed above, f_i has $s^{(i)}$ as its only y_i-root. Since $s^{(i)}$ is a simple y_i-root by 1.6.3, we get $h(\mathbf{N}(f_i)) = 1$ by 1.6.1, as wanted. ◇

1.7 The case of convergent series

Let s be a fractionary power series and assume that $o_x(s) \geq 0$, so that s has the form

$$s = \sum_{i \geq 0} a_i x^{i/n}.$$

We say that s is a *convergent fractionary power series* if and only if the ordinary power series

$$s(t^n) = \sum_{i \geq 0} a_i t^i$$

has non-zero convergence radius. It is clear that the condition does not depend on the integer n and that the convergent fractionary power series describe a subring of $\mathbb{C}\langle\langle x \rangle\rangle$.

If s is convergent, one may compose the polydromic function $x \mapsto x^{1/n}$ and the analytic function defined by $s(t^n)$ in a neighbourhood of $t = 0$: we obtain a polydromic function $\bar{s}$, defined in a neighbourhood of $x = 0$, which we call the (polydromic) function *associated* with s.

The reader may easily see that the function $\bar{s}$ does not depend on the integer n used in its definition, it may be assumed in particular that $n = \nu(s)$.

It is clear that if s is convergent, so are all its conjugates, and also that any of them defines the same polydromic function as s.

Lemma 1.7.1 *If s is convergent, the associated function $\bar{s}$ takes $\nu(s)$ different values on each $x_0 \neq 0$ in a suitable neighbourhood U of 0.*

PROOF: Let, as before, $s = \sum a_i x^{i/n}$ and take $n = \nu(s)$. Any two values of $\bar{s}$ on x_0 may be written in the form

$$\sum a_i \varepsilon^i t_0^i, \quad \sum a_i \eta^i t_0^i$$

where $t_0^n = x_0$ and $\varepsilon^n = \eta^n = 1$, $\varepsilon \neq \eta$. Since $n = \nu(s)$, by 1.2.1, the series

$$\sum a_i (\varepsilon^i - \eta^i) t_0^i, \quad \varepsilon^n = \eta^n = 1, \quad \varepsilon \neq \eta$$

are all non-zero. Therefore they define analytic functions which have no zero other than $t = 0$ in a neighbourhood of 0, say for $|t| < \rho$. Then it is enough to take U in the disc $|x| < \rho^n$ and so that $\bar{s}$ is defined in U. ◇

Back to considering convergent series $f \in \mathbb{C}\{x, y\}$, it is clear that the Newton–Puiseux algorithm applies in particular to them to give their y-roots. The relevant fact is that in this case the roots are convergent too:

Theorem 1.7.2 (Convergence of the y-roots) *If $f \in \mathbb{C}\{x, y\}$ is convergent, then so are all y-roots of f.*

PROOF: Let s be a y-root of f. An easy computation shows that if s is a y-root of multiplicity r of f, then s is a simple root of its $(r-1)$-th y-derivative $\partial^{r-1} f / \partial y^{r-1}$. If f is convergent so is this derivative, hence we may take $\partial^{r-1} f / \partial y^{r-1}$ instead of f itself and assume in the sequel that s is actually a simple root of f.

We will use for s the notations of section 1.4 above. Since s is a simple y-root, 1.6.6 applies to it and we will prove that s is convergent using induction on the integer i of 1.6.6: we take $i = i(s)$ to be the level of the first step in the Newton–Puiseux algorithm for s that either is of type (a), or deals with a Newton polygon of height one.

We consider the case $i = 0$ first. If the Newton–Puiseux algorithm begins by a step (0.a), then $s = 0$ and there is nothing to prove. So, assume that we perform a step (0.b). Since $i = 0$, we have $h(\mathbf{N}(f)) = 1$, or, equivalently, $\partial f / \partial y$ is not zero at the origin. Thus, by the implicit function theorem, there is a y-root s' of f in $\mathbb{C}\{x\}$. Since, by 1.6.1, f has a single y-root, then necessarily $s = s'$ and s is convergent as claimed.

Assume now $i > 0$: the Newton–Puiseux algorithm begins by the step (0.b) and one has

$$s = x^{m/n}(a + s_1),$$

s_1 being a y_1-root of

$$f_1 = x_1^{-k} f(x_1^n, x_1^m(a + y_1))$$

which clearly is a convergent series. Since the Newton–Puiseux algorithm for the y_1-root s_1 of f_1 runs on the same steps as that for s, the first one excepted, it is clear that $i(s_1) = i(s) - 1$ and the induction hypothesis may be applied to s_1, f_1: then s_1 is convergent and hence s is convergent too. ◇

1.8   Algebraic properties of $\mathbb{C}\{x, y\}$

In this section we will use the Puiseux theorem and the convergence of y-roots to derive a factorization theorem for convergent series. From it we shall prove the Weierstrass theorems and some algebraic properties of the ring $\mathbb{C}\{x, y\}$ that will be applied to the rings $\mathcal{O}_{S,O}$ in forthcoming chapters. All of them are two-variable versions of well known results on analytic functions in many variables that may be found, for instance, in [40] or [41]. A reader trained in this field may thus skip this part.

First of all we need a convergent version of 1.2.2, namely,

Lemma 1.8.1 *If f is convergent, then a Puiseux series s is a y-root of f if and only if $y - s$ divides f in $\mathbb{C}\{x^{1/n}, y\}$.*

PROOF: If s is a y-root of f, then, by 1.7.2, s is convergent. Thus, substituting $y + s$ for y defines an automorphism of $\mathbb{C}\{x^{1/n}, y\}$ and the same argument used in the proof of 1.2.2 applies here. The converse is clear. ◇

As in preceding sections, if s is a Puiseux series, we write $g_s = \prod(y - s')$, s' running on the conjugates of s. Of course if s is convergent, so is g_s. In such a case, the series g_s being irreducible as an element of $\mathbb{C}[[x, y]]$ (1.2.5), it is also irreducible as an element of $\mathbb{C}\{x, y\}$.

Lemma 1.8.2 *A Puiseux series s is a y-root of f if and only if g_s divides f in $\mathbb{C}\{x, y\}$.*

PROOF: Same as for 1.2.4 but for using 1.8.1 instead of 1.2.2. ◇

Now we have a factorization for f in $\mathbb{C}\{x, y\}$.

Theorem 1.8.3 *If $f \in \mathbb{C}\{x, y\}$, then there are an invertible series $u \in \mathbb{C}\{x, y\}$ and a non-negative integer r, both uniquely determined by f, and convergent Puiseux series $s_1, \ldots, s_m$, uniquely determined by f up to conjugation, so that $f = u x^r g_{s_1} \cdots g_{s_m}$.*

PROOF: Existence follows by induction from 1.8.2, as in the formal case 1.5.5. Uniqueness in 1.5.5 implies uniqueness here. ◇

Remark 1.8.4 Of course, still $h(\mathbf{N}(f)) = \nu(s_1) + \cdots + \nu(s_m)$ and the y-roots of f are the conjugates of the s_j, $j = 1, \ldots, m$, either because 1.5.5 applies or just by direct computation.

The next two corollaries follow from 1.8.3 just as their formal analogues, 1.5.8 and 1.5.9, did from 1.5.5.

Corollary 1.8.5 *Let* $f \in \mathbb{C}\{x,y\}$ *with no factor* x. *Then* f *is irreducible if and only if* $f = ug_s$ *with* $u \in \mathbb{C}\{x,y\}$ *invertible and* s *a convergent Puiseux series. In such a case* $h(\mathbf{N}(f)) = \nu(s)$ *and the set of y-roots of* f *is the conjugacy class of* s.

Corollary 1.8.6 *The ring* $\mathbb{C}\{x,y\}$ *is a unique factorization ring.*

Formal Weierstrass polynomials whose coefficients are convergent are called *Weierstrass polynomials*. Next we will prove the Weierstrass theorems for two variables.

Theorem 1.8.7 (Weierstrass preparation theorem) *If* $f \in \mathbb{C}\{x,y\}$ *is regular of order* n *in* y, *there is a unique Weierstrass polynomial* g *in* y *such that* $f = ug$, *where* u *is invertible in* $\mathbb{C}\{x,y\}$. *Furthermore* $\deg g = n$.

PROOF: Existence follows from 1.8.3, as $r = 0$ by the hypothesis and $g_{s_1} \cdots g_{s_m}$ is a (convergent) Weierstrass polynomial as wanted. Uniqueness and the equality $\deg g = n$ follow from the formal version 1.5.10. ◇

Theorem 1.8.8 (Weierstrass division theorem) *Assume that* g *is a Weierstrass polynomial in* y *of degree* m *and* $f \in \mathbb{C}\{x,y\}$. *Then there exist* $q \in \mathbb{C}\{x,y\}$ *(the quotient) and a polynomial* $r \in \mathbb{C}\{x\}[y]$ *(the remainder), both uniquely determined by* f *and* g, *such that* $f = qg + r$ *and* $\deg r < m$. *Furthermore, if* f *itself is a polynomial in* y, *then* q *is also a polynomial in* y.

PROOF: First of all let us show how to divide by Weierstrass polynomials of degree one. Given any $f \in \mathbb{C}\{x,y\}$ and the Weierstrass polynomial $y - a(x)$, $a(x) \in \mathbb{C}\{x\}$, $a(0) = 0$, it is clear that $f(x,y) - f(x,a(x))$ has y-root $a(x)$. Then, just apply 1.8.1 to get $q \in \mathbb{C}\{x,y\}$ so that $f(x,y) = (y - a(x))q + f(x,a(x))$, as wanted.

Now assume that the Weierstrass polynomial g of degree m is given. Of course the case $m = 0$ is trivial. Assume thus $m > 0$ and, by 1.8.3, decompose g in the form $g = \prod_1^m (y - s_j)$ where the s_j are the $y-roots$ of g. Let n be such that all these y-roots belong to $\mathbb{C}\{x^{1/n}\}$. We will work for a while in $\mathbb{C}\{x^{1/n}, y\}$ in order to profit from the former decomposition of g. We divide first f by $y - s_1$ and get, as above

$$f = q_1(y - s_1) + r_1,$$

with $r_1 \in \mathbb{C}\{x^{1/n}\}$ and $q_1 \in \mathbb{C}\{x^{1/n}, y\}$. Then we divide q_1 by $y - s_2$ and substitute the resulting expression for q_1 in the equality above in order to get

$$f = q_2(y - s_1)(y - s_2) + r_2,$$

where $r_2 \in \mathbb{C}\{x^{1/n}\}[y]$ has degree at most one and $q_2 \in \mathbb{C}\{x^{1/n}, y\}$. We continue by dividing q_2 by $y - s_3$ and so on, till eventually getting, after the m-th division,

$$f = q_m(y - s_1)(y - s_2) \ldots (y - s_m) + r_m,$$

where $r_m \in \mathbb{C}\{x^{1/n}\}[y]$ and has degree at most $m-1$ and still $q_m \in \mathbb{C}\{x^{1/n}, y\}$. We take $q = q_m$ and $r = r_m$, which gives the wanted equality $f = qg + r$.

It remains to prove that actually $q \in \mathbb{C}\{x, y\}$ and $r \in \mathbb{C}\{x\}[y]$. Before this we shall see that they are uniquely determined by f and g. Assume that $f = q'g + r'$, where also $r' \in \mathbb{C}\{x^{1/n}\}[y]$ and has degree at most $m-1$ and $q' \in \mathbb{C}\{x^{1/n}, y\}$. We get $(q - q')g = r' - r$ and so if $q = q'$, then $r = r'$ too. Otherwise the height of the Newton polygon of the series $(q - q')g$ is at least m, against the fact that the polynomial $r - r'$ has degree in y at most $m-1$.

Once uniqueness has been seen, let ε be any n-th root of unity and denote by σ the conjugation automorphism of $\mathbb{C}\{x^{1/n}, y\}$ induced by substitution of $\varepsilon x^{1/n}$ for x in the series. Both f and g being invariant by σ, the equality $f = qg + r$ gives rise to $f = \sigma(q)g + \sigma(r)$ where still $\sigma(r)$ is a polynomial in y of degree at most $m-1$. By the uniqueness of q and r, $\sigma(q) = q$ and $\sigma(r) = r$ and so, by 1.2.1, $q \in \mathbb{C}\{x, y\}$ and $r \in \mathbb{C}\{x\}[y]$ as wanted.

To close, the last claim is just ordinary division of polynomials. ◇

Corollary 1.8.9 *The ring $\mathbb{C}\{x, y\}$ is a Noetherian ring.*

PROOF: As recalled in section 0.2, the ring $\mathbb{C}\{x\}$ is principal and so, in particular, Noetherian. By the Hilbert basis theorem ([48] VI.2, for instance) the ring $\mathbb{C}\{x\}[y]$ is Noetherian too. Assume that I is a non-zero ideal of $\mathbb{C}\{x, y\}$, we will prove that it is finitely generated. Let x^m, $m \geq 0$, be the highest power of x that divides all elements of I: then $I = x^m I'$ where I' is an ideal containing some series without factor x. It is enough to prove that I' is finitely generated, so in the sequel we assume without restriction that there is $f \in I$ with no factor x. Then the Weierstrass preparation theorem (1.8.7) may be applied to f showing that there is a Weierstrass polynomial $g = u^{-1}f$ that belongs to I. Any $f' \in I$ may be divided by g (1.8.8) and so $f' = qg + r$ where $r \in \mathbb{C}\{x\}[y]$ and still $r = f' - qg \in I$. Then, clearly, g and a finite set of generators of $I \cap \mathbb{C}\{x\}[y]$ generate I. ◇

In the next proposition we write $\mathcal{M}$ for the maximal ideal of $\mathbb{C}\{x, y\}$, $\mathcal{M} = (x, y)$.

Proposition 1.8.10 *If I is an ideal of $\mathbb{C}\{x, y\}$, $I \neq (1)$, the following conditions are equivalent:*

　(*i*) *I contains a power of $\mathcal{M}$.*

　(*ii*) *I is $\mathcal{M}$-primary.*

　(*iii*) *$\mathcal{M}$ is the only prime ideal of $\mathbb{C}\{x, y\}$ containing I.*

　(*iv*) *No principal ideal of $\mathbb{C}\{x, y\}$ other than (1) contains I.*

　(*v*) *$\dim_{\mathbb{C}} \mathbb{C}\{x, y\}/I < \infty$.*

PROOF: (i) $\implies$ (ii): If I contains a power of $\mathcal{M}$, then its radical $\mathbf{r}(I)$ contains $\mathcal{M}$, so in fact $\mathbf{r}(I) = \mathcal{M}$ and, its radical being maximal, I is primary.

(ii) $\implies$ (iii): If $\mathcal{P}$ is a prime ideal containing i, then $\mathcal{P}$ contains $\mathbf{r}(I) = \mathcal{M}$ and hence $\mathcal{P} = \mathcal{M}$.

(iii) $\implies$ (iv): If $(f) \supset I$, f non-invertible, for any irreducible factor g of f, (g) is a prime ideal and $(g) \supset I$. This forces $(g) = \mathcal{M}$ against the fact that x, y have no common factor.

(iv) $\implies$ (i): Since $I \not\subset (x)$, choose $f \in I$ not a multiple of x. Using the Weierstrass preparation theorem 1.8.7 on f, one may assume that f itself is a Weierstrass polynomial and still $f \in I$. If f_i is a prime divisor of f, by the hypothesis $I \not\subset (f_i)$ and so $(f_i) \cap I \neq I$. Since the prime divisors of f are finitely many and I cannot be the union of finitely many proper linear subspaces, we may choose $g' \in I$ sharing no factor with f. Dividing g' by f (1.8.8) gives rise to a remainder g which is a polynomial in y that still belongs to I and shares no factor with f. Then f, g are polynomials in y having no common root, so there is a Bezout identity $1 = af + bg$ where a, b are polynomials in y whose coefficients are convergent Laurent series in x. After clearing denominators we get an equality $x^n = a'f + b'g$ with $a', b' \in \mathbb{C}\{x\}[y]$ and therefore $x^n \in I$. Necessarily $n > 0$ as we assume $1 \notin I$. After interchanging the roles of x and y, the same argument shows that $y^m \in I$ for a suitable $m > 0$, from which it easily follows that $\mathcal{M}^{n+m-1} \subset I$.

Lastly, the equivalence (i) $\iff$ (v) is easy using Nakayama's lemma ([48] IX.1, for instance) in $\mathbb{C}\{x, y\}/I$ and is left to the reader. $\diamond$

The reader with some knowledge of commutative algebra may notice that the next corollary says that $\mathbb{C}\{x, y\}$ is a two-dimensional ring, as its maximal chains of embodied prime ideals have three elements. This implies that it is a regular local ring, as it is Noetherian and there are generators for its maximal ideal in number equal to its dimension (see [8], 11.22).

Corollary 1.8.11 *The prime ideals of $\mathbb{C}\{x, y\}$ other than (0) and the maximal one are the principal ideals generated by irreducible elements.*

PROOF: Let $\mathcal{P}$ be a non-zero prime ideal. If no principal proper ideal contains it, by 1.8.10, it is $\mathcal{M}$-primary and so equal to $\mathcal{M}$. Otherwise there is $f \neq 1$ so that $(f) \supset \mathcal{P}$. After taking one of the irreducible factors of f instead f itself, it is not restrictive to assume that f is irreducible. Pick any non-zero $g \in \mathcal{P}$ and write it as $g = f^n g'$ with $g' \notin (f)$ and $n > 0$ because $\mathcal{P} \subset (f)$. Then $g' \notin \mathcal{P}$ which forces $f^n \in \mathcal{P}$. Thus $f \in \mathcal{P}$ and therefore $\mathcal{P} = (f)$. $\diamond$

1.9 Exercises

1.1 Compute the partial sums of the y-roots of

$$y^5 - 5xy^4 - (x^3 - 10x^2)y^3 + (3x^4 - 10x^3)y^2 + (x^8 - 3x^5 + 5x^4)y - x^9 + x^6 - x^5 = 0$$

up to the stage of separating them. Do the same for

$$y^5 - 5xy^4 + 10x^2y^3 - 10x^3y^2 + 5x^4y - 2x^5 = 0$$
$$y^5 - 5xy^4 + 10x^2y^3 - 10x^3y^2 + 5x^4y - x^5 - x^6 = 0$$

1.2 For each $n > 0$ give an example of an irreducible series whose y-roots have been not separated after performing n steps of the Newton–Puiseux algorithm.

1.3 Prove that if Γ is a side of $N(f)$, then $h(\Gamma)$ equals the number of y-roots of f that are computed starting from the side Γ, each y-root counted according to multiplicity.

1.4 Define an addition of Newton polygons by building up $N_1 + N_2$ from the sides of N_1 and N_2 in such a way that $N(fg) = N(f) + N(g)$ for any $f, g \in \mathbb{C}[[x, y]]$.

1.5 Let $f \in \mathbb{C}[[x, y]]$ and s be a y-root of f of multiplicity μ. Using the notations of section 1.4, prove that either

(a) The Newton–Puiseux algorithm for s has a step $(i.\text{a})$ and then $N(f_i)$ ends at a point with ordinate μ or, otherwise,

(b) There is i_0 so that $h(N(f_i)) = \mu$ for $i \geq i_0$.

1.6 Let $f = \sum A_{\alpha,\beta} x^\alpha y^\beta \in \mathbb{C}[[x, y]]$ be irreducible and $s = \sum_{i \geq 1} a_i x^{m_i/n}$ $(n = \nu(s)$, $m_{i+1} > m_i$ and $a_i \neq 0$ for $i \geq 1$) be a y-root of f. Prove that $A_{\alpha,0}$ is not involved in the steps of the Newton–Puiseux algorithm that compute the partial sum of degree m_i/n of s provided

$$\alpha > \frac{m_1 n + (m_2 - m_1)n^1 + \cdots + (m_i - m_{i-1})n^{i-1}}{n},$$

where for all r, $n^r = \gcd(m_1, \ldots, m_r, n)$. *Hint:* use induction on i.

1.7 Let $P = \sum_{j \leq d} a_{i,j} x^i y^j \in \mathbb{C}[[x]][y]$ have degree $d > 0$. Check that the border of the convex hull of $\Delta(P) + \mathbb{R}^+ \times \{0\}$ consists of two horizontal half lines and a polygonal line joining them. This line, oriented from top to bottom, will be denoted by $\tilde{N}(P)$. Check that $N(P)$ consists of the sides of $\tilde{N}(P)$ of negative slope, the latter possibly having a vertical side (slope $= \infty$) and further sides of positive slope. For Γ a side of $\tilde{N}(P)$ define its associated equation F_Γ as in the ordinary case (1.3): prove that if $s = ax^{m/n} + \cdots \in \mathbb{C}\langle\langle x \rangle\rangle$ is a root of P, then there is a side Γ of slope $-n/m$ in $\tilde{N}(P)$ and $F_\Gamma(a) = 0$. (Note that the condition $m/n > 0$ is no longer required in order to substitute s for y in P, as P is now a polynomial in y).

1.8 Let P and $\tilde{N}(P)$ be as above. Describe a Newton–Puiseux-like algorithm that, starting from $\tilde{N}(P)$, gives all roots of P.

1.9 *Continuity of algebraic functions.* Assume that $P \in \mathbb{C}\{x\}[y]$ has degree n and $P(0, y) \neq 0$. For any value $\bar{x}$ of x close enough to $0 \in \mathbb{C}$, denote by $\Lambda(\bar{x})$ the set of roots of $P(\bar{x}, y)$ and write $T = \Lambda(0) \cup \{\infty\}$ where ∞ denotes the improper point of the Riemann sphere $\mathbb{C} \cup \{\infty\}$. Prove that for any real $\varepsilon > 0$ there is $\delta > 0$ so that if $|\bar{x}| < \delta$, then $\Lambda(\bar{x})$ is contained in the union $\bigcup_{p \in T} B(p, \varepsilon)$ of the balls with centres the points in T and radii ε. Prove furthermore that $B(p, \varepsilon) \cap \Lambda(\bar{x}) \neq \emptyset$ if $p \neq \infty$ and also that the same claim is true for $p = \infty$ if and only if the coefficient of y^n in P vanishes for $x = 0$.

1.10 Check that the Newton–Puiseux algorithm runs with no changes if one takes an arbitrary algebraically closed base field K of characteristic zero instead of $\mathbb{C}$, all results of sections 1.1 to 1.6 remaining true over K.

1.11 Try the Newton–Puiseux algorithm on $f = y^p + y^{p+1} + x$ over an algebraically closed field F_p of characteristic p and notice that it gives rise to a sort of fractionary power series $S(x)$ with unbounded denominators. Find the exact point in the proof of 1.5.1 that fails to be true in positive characteristic. Prove that f has no y-root in $F_p\langle\langle x \rangle\rangle$.

Let S_i be the partial sum of S obtained after the i-th step of the Newton–Puiseux algorithm. Prove that the set $\{o_x f(x, S_i(x)) | i \geq 0\}$ has an upper bound, and so that even if series with unbounded denominators were allowed, S would not be a y-root of f. Why does the proof of 1.5.3 not apply here?

1.12 Prove that $\mathbb{C}((x^{1/n}))$ is a Galois extension of $\mathbb{C}((x))$, its Galois group being isomorphic to the group of the n-th roots of unity.

Prove that the same claim is true if convergent series are taken instead of the formal ones.

Chapter 2

First local properties of plane curves

Results from the preceding chapter will be applied in the present one to decompose germs of curve into branches and to parameterize these branches using Puiseux series. Parameterizations of branches will allow us to introduce intersection multiplicity of germs and prove its main properties. The last section contains basic facts about local rings and linear families of germs.

2.1 The branches of a germ

Fix O to be a point on an smooth surface S. The ring $\mathcal{O} = \mathcal{O}_{S,O}$ being isomorphic to a ring of convergent series in two variables, it is a unique factorization ring by 1.8.6. Let us begin by an easy but very important consequence of this fact:

Proposition 2.1.1 *If $\xi : f = 0$ is a germ of curve with origin at O, then it has a uniquely determined decomposition as a sum of irreducible germs: if $f = f_1^{\alpha_1} \ldots f_r^{\alpha_r}$, the f_i being irreducible, then $\xi = \alpha_1 \gamma_1 + \cdots + \alpha_r \gamma_r$, where each γ_i is the irreducible germ $f_i = 0$.*

PROOF: Direct from the unique factorization property applied to f. ⋄

The germs γ_i above are called the *irreducible components* or *branches* of ξ and also the *branches at O* of any representative of ξ. Then irreducible germs are often called *unibranched germs* or just *branches*. Note that the empty germ is the only germ with no branches. For it the above decomposition has no summands, i.e., $r = 0$.

The positive integer α_i is called the *multiplicity of γ_i as a component of ξ*, or the number of times that γ_i counts as a component of ξ. A component γ_i is called *multiple* if $\alpha_i > 1$. This multiplicity has no relation to, and therefore

39

should not be confused with, the multiplicity of γ_i as a germ, as introduced in section 0.9.

Assume that local coordinates x, y at O have been chosen. Let $\xi : f = 0$ be a germ which has not the germ of the y-axis as a branch. Then x does not divide f and, by the Weierstrass preparation theorem (1.8.7), one may write $f = ug$ where $u \in \mathcal{O}_{S,O}$ is invertible and g is a Weierstrass polynomial in y. It is clear from 1.8.7 that g does not depend on the equation f, but only on the germ ξ (provided the coordinates are fixed). The polynomial g is currently called the *Weierstrass equation* of ξ.

The germ ξ is said to be a *reduced germ* if and only if it has no multiple components, that is, $\xi = \gamma_1 + \cdots + \gamma_r$ with all γ_i irreducible and $\gamma_i \neq \gamma_j$ if $i \neq j$. This is the same as saying that the equations of the germ have no multiple factors or that its ideal is radical. Irreducible germs are thus, in particular, reduced. Given a germ ξ, there is one and just one reduced germ ξ_{red} with the same branches as ξ: if $\xi = \alpha_1 \gamma_1 + \cdots + \alpha_r \gamma_r$ with $\alpha_i > 0$, γ_i irreducible for $i = 1, \ldots, r$ and $\gamma_i \neq \gamma_j$ for $i \neq j$, then $\xi_{red} = \gamma_1 + \cdots + \gamma_r$.

Curves whose germs at O are reduced (resp. irreducible) are said to be *reduced* (resp. *irreducible* or *unibranched*) at O. One says that a curve is reduced if and only if it is reduced at all its points.

Reduced curves and their germs are our main field of interest. For them the more elementary idea of defining curves as sets of points is in fact correct, since, as we will see in section 2.3, a reduced curve is determined by its locus and so no information is lost when thinking of it just as the set of its points. However, we will not limit ourselves to considering reduced curves only, as some non-reduced curves will be quite useful in the sequel.

As a direct consequence of the additivity of multiplicities (section 0.9), the multiplicity of a germ ξ is the sum of the multiplicities of its branches, each branch counted according to its multiplicity as a component of ξ: $e(\xi) = \alpha_1 e(\gamma_1) + \cdots + \alpha_r e(\gamma_r)$ if the notations are as in 2.1.1.

2.2 The Puiseux series of a germ

Let $O \in S$ as in the preceding section. Fix local coordinates x, y at O, and so identify $\mathcal{O} = \mathcal{O}_{S,O} = \mathbb{C}\{x, y\}$. Let f be the equation of a germ of curve ξ at O. The y-roots of f will be called the *Puiseux series of the germ* ξ (and also the Puiseux series at O of any representative of ξ). Puiseux series of ξ actually depend on the coordinates, but not on the particular equation f of ξ, since f and uf have the same y-roots for any invertible $u \in \mathcal{O}$. The Puiseux series of a germ are convergent by 1.7.2.

The y-roots and the irreducible factors of f being closely related (1.8.3), so are the Puiseux series and the irreducible components of ξ, as the next proposition summarizes.

Proposition 2.2.1 *By mapping each branch of ξ different from the germ of the y-axis to the set of its Puiseux series, we get a one to one correspondence* *between the branches of ξ other than the germ of the y-axis and the conjugacy classes of the Puiseux series of ξ.*

PROOF: Follows from 1.8.3 applied to an equation f of ξ. ◇

The reader may notice that to each side Γ of $\mathbf{N}(\xi)$ there corresponds a non-empty set of branches of ξ, namely all branches whose Puiseux series are of the form $ax^{m/n} + \cdots$, $-n/m$ being the slope of Γ and a a root of F_Γ. In the sequel these branches will be referred to as the *branches corresponding to* or *associated with* the side Γ.

Remark 2.2.2 If the germ ξ has no branch equal to the germ of the y-axis, the branches of ξ are in one to one correspondence with the conjugacy classes of its Puiseux series. When needed, one usually comes into this situation either by performing a suitable change of coordinates or just by dropping from ξ the germ of the y-axis.

The next two corollaries directly follow from 2.2.1.

Corollary 2.2.3 *The germs ξ and ξ_{red} have the same Puiseux series.*

Corollary 2.2.4 *Let γ be a reduced germ that does not contain the germ of the y-axis: γ is irreducible if and only if all its Puiseux series are in a single conjugacy class.*

The germ of the y-axis is the only irreducible germ with no Puiseux series. In the sequel, when saying that an irreducible germ γ has Puiseux series s we will implicitly assume that γ is not the germ of the y-axis.

Let ξ be a germ of curve that has not the germ of the y-axis as a branch. Assume that ξ decomposes into its branches in the form $\xi = \alpha_1\gamma_1 + \cdots + \alpha_r\gamma_r$, each γ_i irreducible and $\gamma_i \neq \gamma_j$ if $i = j$. By 2.2.1, the Puiseux series of each branch γ_i describe a conjugacy class: if s_i is a Puiseux series of γ_i, then the whole set of Puiseux series of γ_i is

$$\{\sigma_\varepsilon(s_i)\}_{\varepsilon^{\nu_i}=1}$$

where $\nu_i = \nu(s_i)$.

If a single Puiseux series s_i has been chosen for each branch γ_i of ξ, we may take

$$g_{s_i} = \prod_{\varepsilon^{\nu_i}=1} (y - \sigma_\varepsilon(s_i)) \tag{2.1}$$

as an equation of γ_i, and therefore

$$g = \prod_{i=1}^{r} \left(\prod_{\varepsilon^{\nu_i}=1} (y - \sigma_\varepsilon(s_i)) \right)^{\alpha_i} \tag{2.2}$$

as an equation of ξ (its Weierstrass equation in fact). The $\sigma_\varepsilon(s_i)$, $\varepsilon^{\nu_i} = 1$, $i = 1, \ldots, r$, are the Puiseux series of ξ.

It is clear from equation 2.1 that each branch γ_i is determined by any of its Puiseux series. Because of this, and also because conjugate series define the same polydromic function, usually a single Puiseux series is considered for each branch of ξ. Such a series is often called *the* Puiseux series of the branch and should be retained as defined up to conjugacy.

We know from 1.5.7 that the Newton–Puiseux algorithm applied to an equation f of ξ gives all its Puiseux series. In order to avoid unnecessary computation, the Newton–Puiseux algorithm is usually performed taking, at each step and for each side Γ of the corresponding Newton polygon, just one root from each conjugacy class of roots of F_Γ. The reader may check as exercise that in such a way one gets one and only one Puiseux series from each conjugacy class of y-roots of f, that is, one and only one Puiseux series for each branch of ξ.

In practice one performs finitely many steps of the algorithm in order to get partial sums of the Puiseux series of ξ, the degree of these partial sums depending on the number of steps performed. In the most usual case of the germ ξ being reduced, f has no multiple factors and therefore 1.6.6 may be applied to it: then, in order to get a reasonably complete information on ξ, the Newton–Puiseux algorithm for each Puiseux series should be continued up to finding either a step of type a or a coefficient which is a simple root of the corresponding equation F_Γ. Once this is done, each y-root has been separated from the other y-roots, which means that one has computed terms of each Puiseux series enough to distinguish each Puiseux series:

(a) from the Puiseux series of any other branch, we say then that the branches of ξ have been separated, and also

(b) from any of its conjugates, which makes evident the polydromy order of the series as the minimal common denominator of the exponents of the non-zero terms so far computed.

It should be noticed that a multiple root of an equation F_Γ (at any step of the Newton–Puiseux algorithm) may give rise either to many different branches, to be separated at a further step, or to a single branch whose Puiseux series has higher polydromy order. This may be shown by a simple example: for any $\lambda \in \mathbb{C}$, the Newton polygon of the germ

$$\xi_\lambda : x^2 - 2xy + y^2 + \lambda x^5 + x^6 = 0 \quad , \quad \lambda \in \mathbb{C}$$

has a single side Γ which in turn has associated equation $F_\Gamma = (Z - 1)^2$. The double root 1 gives rise to two smooth branches if $\lambda = 0$, and to a single branch if $\lambda \neq 0$, the polydromy orders of the Puiseux series being 1 in the former case and 2 in the latter.

Proposition 2.2.5 *Assume that a Puiseux series s of an irreducible germ γ has initial term $ax^{m/n}$. Assume furthermore that the polydromy order of s is*

$\nu(s) = n$ and write $d = \gcd(m,n)$, $m' = m/d$ and $n' = n/d$. Then, up to a non-zero constant factor, any equation of γ has the form

$$(y^{n'} - a^{n'}x^{m'})^d + \sum_{n\alpha+m\beta>mn} a_{\alpha,\beta}x^\alpha y^\beta.$$

PROOF: Follows from equation 2.1 above by direct computation. One may also argue as follows: y-roots of f, coming either from different sides of $N(f)$ or from non-conjugate roots of the equation of a side, are necessarily non-conjugate y-roots of f and hence they correspond to different branches of $f = 0$. Thus, the germ γ being irreducible, $N(\gamma)$ has a single side, say Γ, and all roots of its associated equation F_Γ are conjugate. By 1.3.1, the slope of Γ is $-n/m$, and the set of roots of F_Γ is the conjugacy class of a, $\{\varepsilon a | \varepsilon^{n'} = 1\}$. Furthermore, $h(N(\gamma)) = n$ by 1.5.6, hence the equation of Γ is $n\alpha + m\beta = nm$ and F_Γ has degree n. Using 1.3.2, it easily turns out that $F_\Gamma = (z^{n'} - a^{n'})^d$ up to a multiplicative constant, and then the claim follows. ◇

Corollary 2.2.6 *An irreducible germ γ has a single principal tangent. Assume that a Puiseux series of γ has initial term $ax^{m/n}$. Then γ is tangent to the x-axis if $m > n$, it is tangent to the y-axis if $m < n$, and its principal tangent is $y - ax = 0$ if $m = n = 1$.*

PROOF: Clear from 2.2.5. ◇

Since the tangent cone to γ has degree $e(\gamma)$, if γ is irreducible its only principal tangent has multiplicity $e(\gamma)$. The tangent cone to a germ of curve ξ being composed of the tangent cones to its branches, each principal tangent to ξ is the principal tangent to at least one of its branches. More precisely, if $\xi = \sum_{i=1}^r \alpha_i \gamma_i$, each γ_i irreducible, and ℓ is a principal tangent to ξ, then its multiplicity as a such equals the sum of the $\alpha_i e(\gamma_i)$ for all branches γ_i of ξ that have principal tangent ℓ.

Corollary 2.2.7 *If ξ is a germ of curve, the sides of $N(\xi)$ with slope bigger than -1 give rise to branches of ξ tangent to the x-axis, those with slope less than -1 give branches tangent to the y-axis, and the branches corresponding to a side with slope -1 are not tangent to either of the axes.*

PROOF: Follows from 1.3.1 and 2.2.6. ◇

Corollary 2.2.8 *If γ is an irreducible germ of curve and one of its Puiseux series s has initial term $ax^{m/n}$, $n = \nu(s)$, then the multiplicity of γ is $e(\gamma) = \min\{n,m\}$.*

PROOF: Follows from 2.2.5. ◇

In particular $e_\gamma = n = \nu(s)$ if and only if $m \geq n$, and so, by 2.2.6,

Corollary 2.2.9 *An irreducible germ γ is not tangent to the y-axis if and only if its multiplicity equals the polydromy order of its Puiseux series.*

It follows in particular that the polydromy order of the Puiseux series of an irreducible germ γ does not depend on the choice of the coordinates as far as the y axis is not tangent to the branch. This polydromy order, which equals the multiplicity of γ by 2.2.9, is often called the *order of* γ. Order-one (i.e. smooth) branches are also called *linear branches*.

2.3   Points on curves around O

In this section we will be concerned with local properties of loci of curves. We will see in particular to what extent curves and germs are determined by their loci.

Let ξ be, as before, a germ of curve with origin at O, assume that it has branches γ_i, counted with multiplicities α_i, for $i = 1, \ldots, r$. Assume furthermore that local coordinates x, y at O have been fixed in such a way that no one of the γ_i is the germ of the y-axis and choose a Puiseux series s_i of each γ_i. Then the polydromic functions $\bar{s}_i$ associated with the Puiseux series s_i (see section 1.7) provide a sort of parametric representation of the locus of a suitable representative of ξ, as shown in the next proposition. Let U be a neighbourhood of O where the local coordinates x, y are defined.

Proposition 2.3.1 *There is an open neighbourhood $V \subset U$ of O, a representative ξ' of ξ in V and a positive real number $\rho > 0$ so that all associated functions $\bar{s}_i$ are defined for $|x| < \rho$ and (x, y) lies on ξ' if and only if $|x| < \rho$ and $y = \bar{s}_i(x)$ for some $i = 1, \ldots, r$.*

PROOF: Fix ρ' so that for all x, y, $|x| < \rho'$, $|y| < \rho'$, the point $(x, y) \in U$ with coordinates x, y does exist. Put $\nu_i = \nu(s_i)$. The functions $t \longmapsto s_i(t^{\nu_i})$ being analytic, they are in particular continuous in a neighbourhood of $t = 0$ and so one may choose ρ, $0 < \rho < \rho'$, so that all values $\bar{s}_i(x)$ are defined and $|\bar{s}_i(x)| < \rho'$ for $|x| < \rho$. Take $V = \{(x, y) \in U \mid |x| < \rho, |y| < \rho'\}$. Let ξ' be the curve

$$ g = \prod_{i=1}^{r} \left(\prod_{\varepsilon^{\nu_i} = 1} (y - \sigma_\varepsilon(s_i)) \right)^{\alpha_i} = 0 $$

in V. As seen above (equation 2.2), $\xi'_O = \xi$ and obviously, $g(x, y) = 0$ if and only if $y = \bar{s}_i(x)$ for some $i = 1, \ldots, r$, as claimed. $\diamond$

The above many-valued local representation of the points on ξ' as $(x, \bar{s}_i(x))$, $i = 1, \ldots, r$, turns into the usual uniform one (see section 0.10) if ξ is smooth and non-tangent to the y-axis, as then there is a single Puiseux series $s = s_1$ and $\nu(s) = 1$ by 2.2.9.

Corollary 2.3.2 *If ξ is a non-empty germ of curve at O, then any representative ξ' of ξ has points other than O in any neighbourhood of O.*

PROOF: Since, by definition, any two representatives of ξ restrict to the same curve in a neighbourhood of O, it is clearly enough to deal with a fixed representative: we take ξ' as in 2.3.1. Since ξ is assumed to be non-empty, $r \geq 1$ and then the claim for ξ' obviously follows from 2.3.1. ◇

The next proposition shows that either an intersection point O of two curves is isolated, or the curves share a branch at O.

Proposition 2.3.3 *Let ξ and ζ be curves defined in a neighbourhood of O. If the germs ξ_O and ζ_O are non-empty and share no branch, then there is a neighbourhood U of O so that $|\xi| \cap |\zeta| \cap U = \{O\}$.*

PROOF: Let f, g be equations of ξ, ζ, respectively, in a neighbourhood of O. By the hypothesis, their germs f_O and g_O are non-invertible and share no factor. Therefore the ideal (f_O, g_O) they generate in the local ring $\mathcal{O}$ of O is not contained in a principal ideal. By 1.8.10, (f_O, g_O) contains a power of the maximal ideal (x, y) and so, say, $x^n, y^n \in (f_O, g_O)$. Then there are functions a, b, c, d, analytic in a neighbourhood U of O, so that $x^n = af + bg$ and $y^n = cf + dg$ in U, from which the claim follows. ◇

Corollary 2.3.4 *Let γ and ξ be curves defined in a neighbourhood of O and assume that γ is irreducible at O. If $|\gamma| \cap U \subset |\xi| \cap U$ for some neighbourhood U of O, then γ_O is a branch of ξ_O.*

PROOF: By 2.3.2, $|\gamma|$ contains points other than O in any neighbourhood of O, so that the claim follows from 2.3.3. ◇

Corollary 2.3.5 *If curves ξ, ζ have the same points in a neighbourhood of O, then they have the same branches at O. If furthermore ξ and ζ are reduced at O, then $\xi_O = \zeta_O$.*

PROOF: Follows from 2.3.4. ◇

Now it turns out that reduced curves are determined by their loci, as already announced, i.e.,

Corollary 2.3.6 *If ξ and ζ are reduced curves defined in the same open set U and $|\xi| = |\zeta|$, then $\xi = \zeta$.*

PROOF: 2.3.5 gives $\xi_O = \zeta_O$ for any $O \in U$, from which the claim. ◇

Remark 2.3.7 From now on we will make no distinction between reduced curves and their sets of points. In particular, in the sequel we will use the same symbol ξ to denote both a reduced curve ξ and its locus $|\xi|$, this causing no confusion by 2.3.6. Identification of reduced curves and their loci applies in particular to smooth curves and allows us to consider them in the sequel as one-dimensional analytic manifolds (see section 0.10). We will say that two smooth curves are *isomorphic* when they are so as analytic manifolds.

The next proposition shows that singular points of reduced curves are isolated:

Proposition 2.3.8 *If a curve ξ is reduced at the point O, then there is a neighbourhood of O that contains no singular point of ξ other than O itself.*

PROOF: Choose local coordinates x, y in a neighbourhood of O so that the germ of the y-axis is not a branch of ξ_O. As seen in section 2.2, if $s_1, \ldots, s_k$ are all Puiseux series of ξ at O, then

$$g = \prod_{j=1}^{k} (y - s_j)$$

is an equation of ξ in a suitable neighbourhood V of O. Note that there are no multiple factors because ξ_O is reduced. Clearly no s_i is a y-root of

$$\frac{\partial g}{\partial y} = \sum_{i=1}^{k} \prod_{j \neq i} (y - s_j)$$

and hence the curves $g = 0$ and $\partial g / \partial y = 0$ share no branch at O. Once we know (from section 0.10) that the singular points of ξ in V need to lie on both $\xi : g = 0$ and $\partial g / \partial y = 0$, the claim is obvious if $k = 1$ and otherwise follows from 2.3.3 above. ◇

Corollary 2.3.9 *If a curve is reduced and compact, then it has finitely many singular points.*

PROOF: Clear from 2.3.8. ◇

2.4   Local rings of germs

Let ξ be a non-empty germ of curve at O and f an equation of ξ. Instead of the ideal (f) of ξ one may consider the quotient ring $\mathcal{O}_{S,O}/(f)$, which is obviously a local ring and does not depend on the equation f: it is called the *local ring* of ξ, and also *the local ring at O* of any representative of ξ. We will use for it the notations $\mathcal{O}_\xi$ or $\mathcal{O}_{\xi,O}$ and write $\mathcal{M}_\xi$ or $\mathcal{M}_{\xi,O}$ for its maximal ideal, $\mathcal{M}_{\xi,O} = \mathcal{M}_{S,O}/(f)$, ξ being either the germ or one of its representatives. In this section we will show some very easy properties of $\mathcal{O}_\xi$ and a deeper result, namely that germs of curve are analytically isomorphic if and only if their local rings are isomorphic as $\mathbb{C}$-algebras. This means that the local ring of a germ is an analytic invariant that encloses all analytically invariant information on the germ. In this section and also in the forthcoming sections 3.10 and 3.11 we will show how geometric characters and properties of a germ ξ may be recovered from its local ring $\mathcal{O}_\xi$.

It is not worthwhile associating any local ring with the empty germ. Thus we will keep the local ring of the empty germ undefined and when speaking about the local ring of a germ ξ we will implicitly assume that $\xi \neq \emptyset$.

When dealing with a fixed germ ξ, we will denote by g the class in $\mathcal{O}_\xi$ of any $g \in \mathcal{O}_{S,O}$. The next two propositions are quite easy and their proofs are left to the reader.

Proposition 2.4.1 *The ring $\mathcal{O}_\xi$ has no nilpotent elements (other than 0) if and only if the germ ξ is reduced and it has no zero-divisors if and only if ξ is irreducible.*

Proposition 2.4.2 *The prime ideals of $\mathcal{O}_\xi$ are the maximal one and the ideals generated by the classes of the irreducible factors of f.*

We see in particular that $\mathcal{O}_\xi$ is a one-dimensional local ring.

Local rings of smooth germs are particularly simple, as they are rings of convergent series in one variable. Indeed, if the curve $\zeta : g = 0$ is smooth at O, we have already seen (in section 0.10) that its equation g may be taken as part of a system of local coordinates, say $x, y = g$. Then $\mathcal{O}_{S;O} = \mathbb{C}\{x, y\}$ and $\mathcal{O}_\zeta = \mathbb{C}\{y\}$. Note in particular that the maximal ideal of $\mathcal{O}_\zeta$ is principal, generated by x, that is, by the class of any equation of any germ transverse to ζ_O (0.10).

Conversely, if the maximal ideal of the local ring of a germ $\zeta : g = 0$ is principal, then the germ is smooth: assume that $\mathcal{M}_\zeta$ is generated by the class of a germ h; then $\mathcal{M}_{S,O} = (g, h)$ and hence the cotangent space $\mathcal{M}_{S,O}/\mathcal{M}_{S,O}^2$ is generated by the linear parts (or differentials at O) of g and h. Since the cotangent space has dimension two, this says in particular that g has non-zero linear part and hence ζ is smooth. Thus we have proved:

Lemma 2.4.3 *A germ of curve ζ is smooth if and only if the maximal ideal of its local ring is principal. In such a case the maximal ideal may be generated by the class of any equation of any germ transverse to ζ, the local ring of ζ is isomorphic to a ring of convergent power series in a single variable and is in particular a principal ring.*

If $\varphi : U' \longrightarrow U$ is a morphism from a neighbourhood U' of a point O' on a surface S' onto a neighbourhood U of O on S, and $\varphi(O') = O$, then the morphism of local rings $\varphi^* : \mathcal{O}_{S,O} \longrightarrow \mathcal{O}_{S',O'}$ obviously induces a monomorphism from the local ring of ξ into the local ring of the inverse image of ξ at O': we will refer to it as the morphism induced by φ.

If φ above is an isomorphism, clearly the morphism it induces between the local rings $\mathcal{O}_{\xi,O}$ and $\mathcal{O}_{\varphi^*(\xi),O'}$ is an isomorphism too. The converse will be a little bit harder to prove. We will make use of next lemma, which is in fact a very particular case of the Krull intersection theorem (see [48] or [58]).

Lemma 2.4.4 *For any germ of curve ξ,*

$$\bigcap_{n \geq 1} \mathcal{M}_\xi^n = (0).$$

PROOF: Put $\mathcal{M} = \mathcal{M}_{S,O}$ and let f be an equation of ξ. We will prove that $\bigcap_{n\geq 1}(f + \mathcal{M}^n) = (f)$. Take local coordinates x, y so that no branch of ξ equals the germ of the y-axis. Then, the claim being obvious if $\xi = \emptyset$, we will make induction on $h(\mathbf{N}(\xi))$. If $h(\mathbf{N}(\xi)) > 0$, take a Puiseux series s of a branch $\gamma : f' = 0$ of ξ. If $g \in \bigcap_{n\geq 1}(f + \mathcal{M}^n)$, then for any n one has an equality $g = v_n f + h_n$ where $h_n \in \mathcal{M}^n$. After substituting s for y in this equality we get $g(x, s(x)) = h_n(x, s(x))$. Since h_n has order n and the order of s is at least $1/\nu(s)$, this implies that $o_x g(x, s(x)) \geq n/\nu(s)$, and so, n being arbitrary, $g(x, s(x)) = 0$. This means that γ is also a branch of $g = 0$, that is, f' divides g. Since by hypothesis f' divides f, f' divides h_n too and all the equalities above may be divided by f' giving the new ones $g/f' = v_n f/f' + h_n/f'$, $n > 0$. Now $h_n/f' \in \mathcal{M}^{n-e(\gamma)}$, so the former equalities prove that $g/f' \in \bigcap_{n\geq 1}(f/f' + \mathcal{M}^n)$ and the claim follows by induction. $\diamond$

Theorem 2.4.5 *Let ξ and ξ' be non-empty germs of curve with origins $O \in S$ and $O' \in S'$ respectively. If $\psi : \mathcal{O}_\xi \longrightarrow \mathcal{O}_{\xi'}$ is an isomorphism of $\mathbb{C}$-algebras, then there is an analytic isomorphism φ between neighbourhoods of O' and O, $\varphi(O') = O$, so that $\varphi^*(\xi) = \xi'$ and ψ is induced by φ.*

PROOF: Take local coordinates x, y at O and x', y' at O' and let us write just $\mathcal{O} = \mathcal{O}_{S,O}$, $\mathcal{O}' = \mathcal{O}_{S',O'}$, $\mathcal{M} = \mathcal{M}_{S,O}$, $\mathcal{M}' = \mathcal{M}_{S',O'}$, $\mathbf{m} = \mathcal{M}_\xi$ and $\mathbf{m}' = \mathcal{M}_{\xi'}$. We denote by $\mathbf{f}$ the class in either $\mathcal{O}_\xi$ or $\mathcal{O}_{\xi'}$ represented by $f \in \mathcal{O}$ or $f \in \mathcal{O}'$ respectively.

First of all notice that if one of the germs is smooth, so is the other, by 2.4.3. Let us assume first that both germs are singular and let $f, g \in \mathcal{O}'$ be representatives of $\psi(\mathbf{x}), \psi(\mathbf{y})$: $\psi(\mathbf{x}) = \mathbf{f}$, $\psi(\mathbf{y}) = \mathbf{g}$. Since $\mathbf{x}, \mathbf{y}$ generate $\mathbf{m}$, $\mathbf{f}, \mathbf{g}$ generate $\mathbf{m}'$ and so their classes modulo $\mathbf{m}'^2$ generate $\mathbf{m}'/\mathbf{m}'^2$. On the other hand, since ξ' is singular, its equation necessarily belongs to $\mathcal{M}'^2$ from which it is clear that the natural morphism $\mathcal{O}' \longrightarrow \mathcal{O}_{\xi'}$ induces an isomorphism $\mathcal{M}'/\mathcal{M}'^2 \simeq \mathbf{m}'/\mathbf{m}'^2$. It turns out that also the classes of f, g in the cotangent space $\mathcal{M}'/\mathcal{M}'^2$ generate it, which is the same as saying that the jacobian determinant $\partial(f, g)/\partial(x', y')$ is non-zero at O'. Thus the map $(x', y') \mapsto (f(x', y'), g(x', y'))$ is an analytic isomorphism φ between suitable neighbourhoods of O' and O.

Now we are going to see that the diagram

$$
\begin{array}{ccc}
\mathcal{O} & \xrightarrow{\varphi^*} & \mathcal{O}' \\
{\scriptstyle \pi}\downarrow & & \downarrow{\scriptstyle \pi'} \\
\mathcal{O}_\xi & \xrightarrow{\psi} & \mathcal{O}_{\xi'}
\end{array}
$$

is commutative, π, π' being the natural morphisms onto the quotient rings. Note that the claim directly follows from the commutativity of this diagram: indeed, if the diagram is commutative, then $\varphi^*(\ker \pi) = \ker \pi'$, thus $\varphi^*(\xi) = \xi'$ and obviously ψ is induced by φ^*, that is, by φ.

Let $h = \sum_{i,j \geq 0} a_{i,j} x^i y^j$ be any element of $\mathcal{O}$ and write $h_n = \sum_{i,j=0}^n a_{i,j} x^i y^j$. On one hand we have

$$\pi'(\varphi^*(h_n)) = \sum_{i,j=0}^n a_{i,j} \mathbf{f}^i \mathbf{g}^j = \psi(\pi(h_n)).$$

On the other hand, since $h - h_n \in \mathcal{M}^{n+1}$ and all morphisms are local,

$$\pi'(\varphi^*(h)) - \pi'(\varphi^*(h_n)) \in \mathbf{m}'^{n+1}$$

and

$$\psi(\pi(h_n)) - \psi(\pi(h)) \in \mathbf{m}'^{n+1}.$$

The last three relations together give

$$\pi'(\varphi^*(h)) - \psi(\pi(h)) \in \mathbf{m}'^{n+1}$$

for all n. Thus, since we know from 2.4.4 that $\bigcap_{n \geq 1} \mathbf{m}'^n = (0)$, necessarily $\pi'(\varphi^*(h)) = \psi(\pi(h))$ as wanted.

In the case of both germs being smooth, it is not restrictive to assume that y and y' are equations of ξ and ξ', respectively (see section 0.10). Then one may choose a representative f of $\psi(\mathbf{x})$ which is a series in x' alone. Furthermore, since $\mathbf{x}$ generates $\mathbf{m}$, $\psi(\mathbf{x}) = \mathbf{f}$ generates $\mathbf{m}' = (\mathbf{x}')$, which implies $o_{x'}(f) = 1$. Then we take $g = y'$. Again $\partial(f,g)/\partial(x',y')$ is not zero at O', $\psi(\mathbf{x}) = \mathbf{f}$ and $\psi(\mathbf{y}) = \psi(0) = 0 = \mathbf{g}$, so that the proof ends as in the former case. $\diamond$

2.5 Parameterizing branches

Clearly, the Puiseux series of a germ of curve ξ depend on the local coordinates. Nevertheless, they may be used for defining holomorphic maps that parameterize the branches of ξ and whose germs at $0 \in \mathbb{C}$ are intrinsically related to ξ. We will deal mostly with irreducible germs. For them we will prove that the germ of the map induced by any Puiseux series is uniquely determined, up to local isomorphism on the source, because it satisfies an universal property of local factorization of holomorphic maps.

Theorem 2.5.1 *Let γ be a curve irreducible at O.*

(a) There exists a neighbourhood U of 0 in $\mathbb{C}$ and a holomorphic map

$$\varphi : U \longrightarrow \gamma \subset S, \quad \varphi(0) = O,$$

that satisfy the following unique factorization property:

UFP: *for any holomorphic map*

$$\psi : V \longrightarrow \gamma, \quad \psi(0) = O,$$

V a neighbourhood of 0 in $\mathbb{C}$, there exists a neighbourhood V' of 0, $V' \subset V$, and a holomorphic map $\psi' : V' \longrightarrow U$ such that $\psi_{|V'} = \varphi \circ \psi'$ and the germ of ψ' at 0 is uniquely determined by this property.

(b) The germ of φ at 0 is uniquely determined, up to an analytic isomorphism between neighbourhoods of 0 in $\mathbb{C}$, by the germ γ_O and the UFP above.

PROOF: Choose the coordinates so that γ is not the y-axis. Assume that f is the Weierstrass equation of γ and $s = \sum_{i>0} a_i x^{i/n}$, $n = \nu(s)$, one of its Puiseux series. Define the holomorphic map φ from a neighbourhood U of 0 in $\mathbb{C}$ into γ by the rule

$$t \longmapsto (t^n, s(t^n))$$

where $s(t^n) = \sum a_i t^i$ (2.3.1).

Assume that ψ is given in a neighbourhood V of 0 by $\psi(z) = (u_1(z), u_2(z))$, $u_i \in \mathbb{C}\{z\}$, $i = 1, 2$. If $u_1(z) = 0$, then from $f(u_1(z), u_2(z)) = 0$ we get $u_2(z) = 0$ and so ψ is constant in a neighbourhood of 0. Since in this case UFP is obviously satisfied, we will assume in the sequel that $u_1(z) \neq 0$.

Write $u_1(z) = z^r \bar{u}_1(z)$, $\bar{u}_1(0) \neq 0$. Then there exists a series v, convergent in a neighbourhood of 0, such that $v^r = \bar{u}_1$ and still $v(0) \neq 0$. One may thus take $\tilde{z} = zv(z)$ as a new coordinate in a smaller neighbourhood V' of 0 in $\mathbb{C}$ and then ψ is given by the rule $\psi(\tilde{z}) = (\tilde{z}^r, u(\tilde{z}))$, $u \in \mathbb{C}\{\tilde{z}\}$. Since the image of ψ lies in γ, still $f(\tilde{z}^r, u(\tilde{z})) = 0$. It follows that $u(x^{1/r})$ is a y-root of f, and so we have $u(x^{1/r}) = \sigma_\varepsilon(s)$ for a suitable n-th root of unity ε. In particular r needs to be a multiple of $n = \nu(s)$, say $nr' = r$, and then $u(\tilde{z}) = \sum a_i \varepsilon^i \tilde{z}^{r'i}$. After this it is clear that $\psi'(\tilde{z}) = \varepsilon \tilde{z}^{r'}$ gives the wanted factorization.

For the uniqueness of the germ of ψ', assume we have $\varphi(w(\tilde{z})) = \psi(\tilde{z})$ for all $\tilde{z}$ in a neighbourhood of 0 which is equivalent to the equalities $w(\tilde{z})^n = \tilde{z}^r$ and $\sum a_i w(\tilde{z})^i = \sum a_i \varepsilon^i \tilde{z}^{r'i}$. From the first one we get $w(\tilde{z}) = \eta \tilde{z}^{r'}$ with $\eta^n = 1$, and then the second one gives $\sum a_i \eta^i \tilde{z}^{r'i} = \sum a_i \varepsilon^i \tilde{z}^{r'i}$, that is, $\eta^i = \varepsilon^i$ for all i for which $a_i \neq 0$. Since $n = \nu(s)$, n and the integers i for which $a_i \neq 0$ share no factor, so that from the former equalities one easily gets $\eta = \varepsilon$ and thus $w = \psi'$ as wanted.

Lastly, the uniqueness of the germ of φ up to isomorphism claimed in part (b) will follow by an argument which is standard in such a situation. Assume that ϕ also satisfies UFP. Then, after a suitable shrinking of the neighbourhoods of definition, ϕ factors through φ, $\phi = \varphi \circ \phi'$, and φ factors through ϕ, $\varphi = \phi \circ \varphi'$. It follows that $\varphi = \varphi \circ (\phi' \circ \varphi')$ in a neighbourhood of 0; since φ also factors through itself in the obvious way, $\varphi = \varphi \circ Id$, the uniqueness we have proved for ψ', used for $\psi = \varphi$, shows that $\phi' \circ \varphi' = Id$ in a neighbourhood of 0. Then ϕ' has non-zero derivative at 0 and thus gives an analytic isomorphism between neighbourhoods of 0 just as wanted. $\diamond$

Any map φ satisfying UFP above will be called a *uniformizing map* of γ at O. The local coordinate t at $0 \in \mathbb{C}$ is often called a *uniformizing parameter* of γ. We have just proved that different uniformizing maps have the same germ at 0 but for a change of the uniformizing parameters. In fact, as seen in the proof above, uniformizing maps of γ are given by its Puiseux series, i.e.,

Corollary 2.5.2 (of the proof of 2.5.1) *Assume that x, y are local coordinates at O and $s = \sum_{i>0} a_i x^{i/n}$, $n = \nu(s)$, is a Puiseux series of the germ of γ*

at O. Then the map φ defined by the rule

$$\varphi(t) = (t^n, s(t^n)) = (t^n, \sum a_i t^i),$$

for t in a suitable neighbourhood of 0 in $\mathbb{C}$, is a uniformizing map of γ.

Let γ be now an irreducible germ with origin at O, γ' a representative of γ and φ a uniformizing map of γ' at O. The map φ induces, via pull-back, the morphism of local rings

$$\varphi^* : \mathcal{O}_{S,O} \longrightarrow \mathbb{C}\{t\}$$

where t is a uniformizing parameter and, for any $g \in \mathcal{O}_{S,O}$, $\varphi^*(g)$ is the germ of $g' \circ \varphi$, g' being any representative of g. Since $\varphi(0) = O$, clearly φ^* is a local morphism, that is, the images of the non-invertible elements of $\mathcal{O}_{S,O}$ are non-invertible in $\mathbb{C}\{t\}$. Notice that, by 2.5.1, φ^* is determined by γ up to an isomorphism $\mathbb{C}\{t\} \simeq \mathbb{C}\{t'\}$ of rings of convergent power series in one variable. If, as above, s denotes a Puiseux series of γ, one may take $\varphi^*(g)$ to be the series $g(t^n, s(t^n))$. Let f be an equation of γ. It is clear that $\varphi^*(g) = 0$ if and only if g has y-root s, that is, by 1.2.4, if and only if g belongs to the ideal (f) of γ. In other words, $\ker \varphi^* = (f)$ and thus φ^* induces a local monomorphism of local rings

$$\bar{\varphi} : \mathcal{O}_\gamma = \mathcal{O}_{S,O}/(f) \longrightarrow \mathbb{C}\{t\},$$

still determined by γ up to an isomorphism $\mathbb{C}\{t\} \simeq \mathbb{C}\{t'\}$. In the sequel the morphism $\bar{\varphi}$ will be called a *uniformizing morphism* of γ: it is to be understood as the algebraic counterpart of a uniformizing map.

Still let g be a germ of analytic function at O: we define the *order of g along* γ, $o_\gamma(g)$, as being the order of $\varphi^*(g)$ as a series in t, that is, the order at 0 of $g' \circ \varphi$, g' being any representative of g. Notice that $o_\gamma(g)$ does not depend on either γ' or φ, but only on the germs γ and g. Since $\ker \varphi^* = (f)$, one has $o_\gamma(g) = \infty$ if and only if g belongs to the ideal (f) of γ, while $o_\gamma(g) = 0$ if and only if g is invertible, as φ^* is a local morphism. The next propositions, which we state for future reference, directly follow:

Proposition 2.5.3 *Assume that x, y are local coordinates at O, s is a Puiseux series of an irreducible germ γ with origin at O and n is the polydromy order of s. Then we have:*

(a) *For any germ of function g at O,*

$$o_\gamma(g) = o_t \varphi^*(g) = o_t g(t^n, s(t^n)) = n o_x g(x, s(x)) = \sum_{\varepsilon^n = 1} o_x g(x, \sigma_\varepsilon(s(x))),$$

(b) *$o_\gamma(g) = \infty$ if and only if g belongs to the ideal of γ,*

(c) *$o_\gamma(g) = 0$ if and only if g is invertible in $\mathcal{O}_{S,O}$.*

Proposition 2.5.4 *If f and g are germs of analytic functions at O, then*

$$o_\gamma(fg) = o_\gamma(f) + o_\gamma(g)$$

and

$$o_\gamma(f + g) \geq \min\{o_\gamma(f), o_\gamma(g)\},$$

the inequality being strict only in the case $o_\gamma(f) = o_\gamma(g)$.

The reader familiar with valuations may see that o_γ induces a discrete valuation of the local ring $\mathcal{O}_\gamma$.

Let ξ be a curve through O and assume that it is reduced but not necessarily irreducible at O. One may easily put together uniformizing maps of representatives of the different branches of its germ ξ_O in order to obtain a sort of uniformizing map for ξ at O. Let $\gamma_1, \ldots, \gamma_r$ be representatives of the branches of ξ_O, all contained in ξ, and let $\varphi_i : U_i \longrightarrow \gamma_i$, be a uniformizing map of γ_i at O, $i = 1, \ldots, r$. Then these maps induce a map

$$\varphi : \coprod_{i=1}^{r} U_i \longrightarrow \xi,$$

$\coprod$ meaning disjoint union, which is often said to be a uniformizing map of ξ at O.

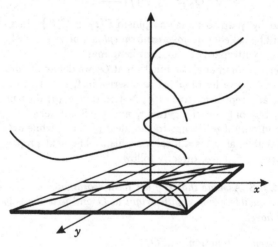

Figure 2.1: A uniformizing map φ is pictured as a vertical projection, and the open sets U_i as arcs of smooth space curves: the germ ξ is $(y^2 - x^3)(y - x^2)(y^2 - x) = 0$.

The reader will have no difficulty in proving that if $\psi : U \longrightarrow \xi$ is a holomorphic and non-constant map defined in an open neighbourhood of 0 in $\mathbb{C}$, such that $\psi(0) = O$, then ψ factors through φ in an essentially unique way: in fact, after shrinking U if needed, ψ is a map into just one of the γ_i and then it factors through the corresponding φ_i.

Let us take all pull-back morphisms φ_i^* to define a new morphism

$$\varphi^* : \mathcal{O}_{S,O} \longrightarrow \prod_{i=1}^{r} \mathbb{C}\{t_i\}$$

by taking $\varphi^*(g) = (\varphi_1^*(g), \ldots, \varphi_r^*(g))$. The reader may also consider φ^* as being the pull-back morphism induced by φ. If s_i is a Puiseux series of γ_i and n_i the polydromy order of s_i, $i = 1, \ldots, r$, one may take φ^* to be the map

$$\varphi^*(g) = (g(t_1^{n_1}, s_1(t_1^{n_1})) \cdots g(t_r^{n_r}, s_r(t_r^{n_r}))),$$

and any other such morphism, obtained from other uniformizing maps, is the composition of φ^* above and a product $\psi_1 \times \cdots \times \psi_r$ of isomorphisms ψ_i : $\mathbb{C}\{t_i\} \simeq \mathbb{C}\{t_i'\}$.

Let f_i, $i = 1, \ldots, r$ be equations of the different branches of ξ_O: since ξ_O is reduced, any two f_i, f_j, $i \neq j$ are coprime and $f = f_1 \ldots f_r$ is an equation of ξ_O. Then we have

$$\ker \varphi^* = \bigcap_i \ker \varphi_i^* = (f_1) \cap \cdots \cap (f_r) = (f),$$

and hence φ^* induces a monomorphism

$$\bar{\varphi} : \mathcal{O}_{\xi,O} = \mathcal{O}_{S,O}/(f) \longrightarrow \prod_{i=1}^{r} \mathbb{C}\{t_i\}.$$

We will call $\bar{\varphi}$ a *uniformizing morphism* of ξ_O, thus extending the definition already given for irreducible germs. As for the morphisms φ^* above, if $\bar{\varphi}$ is a uniformizing morphism of ξ then any other uniformizing morphism of ξ is of the form $\psi \circ \bar{\varphi}$, where $\psi = \psi_1 \times \cdots \times \psi_r$ and each ψ_i is an isomorphism $\psi_i : \mathbb{C}\{t_i\} \simeq \mathbb{C}\{t_i'\}$. Uniformizing morphisms will be considered from a different viewpoint in the forthcoming section 3.11 (see 3.11.6).

2.6 Intersection multiplicity

In this section we introduce the intersection multiplicity of two curves or germs of curve at a point O. The intersection multiplicity of two germs of curve is a numerical invariant of the couple of germs that depends on the germs themselves and on their relative position as well, it measures at which degree the branches of one germ approach the branches of the other. Intersection multiplicity will play a central role in the analysis of singularities we shall develop in forthcoming chapters. Let us just quote, in addition, that intersection multiplicity also plays a central role in the (global) geometry of algebraic curves on a smooth projective algebraic surface S, as the global intersection number of a pair of algebraic curves on S equals the sum of their intersection multiplicities at their common points (see [62], lect. 12, for instance). If S is the projective plane, the global intersection number is the product of degrees and one has Bezout's theorem (see for instance [37], 5.3). In any case the global intersection number equals the number of intersection points if all intersections are transverse (due to 2.6.8 below).

First we define the intersection multiplicity of a couple of irreducible germs:

Definition: If γ and γ' are irreducible germs at O and f is any equation of γ, the *intersection multiplicity* of γ and γ' is

$$[\gamma.\gamma'] = o_{\gamma'}(f).$$

Notice that the definition does not depend on the equation f of γ and also that it makes no use of coordinates, $o_{\gamma'}$ being already intrinsically defined in section 2.5. Now, for the general case, assume that ξ and ζ are germs of curve and $\xi = \gamma_1 + \cdots + \gamma_\ell$ and $\zeta = \gamma'_1 + \cdots + \gamma'_{\ell'}$ are their decompositions in irreducible germs, these decompositions being uniquely determined by ξ and ζ themselves (2.1.1):

Definition: The *intersection multiplicity* of ξ and ζ is

$$[\xi.\zeta] = \sum_{\substack{1 \le i \le \ell \\ 1 \le j \le \ell'}} [\gamma_i.\gamma'_j].$$

If ξ and ζ are analytic curves, we define their *intersection multiplicity at O*, $[\xi.\zeta]_O$, as being that of their corresponding germs at O, that is,

$$[\xi.\zeta]_O = [\xi_O.\zeta_O].$$

Remark 2.6.1 As customary, in the above definition an empty sum is taken to be zero. The intersection multiplicity of two irreducible germs being always positive (by 2.5.3), the intersection multiplicity of two germs equals zero if and only if one of the germs is empty. It is also clear from its definition that intersection multiplicity is biadditive, that is, for any germs $\xi, \xi_1, \xi_2, \zeta, \zeta_1, \zeta_2$ at O, we have

$$[\xi_1 + \xi_2, \zeta] = [\xi_1, \zeta] + [\xi_2, \zeta]$$
$$[\xi, \zeta_1 + \zeta_2] = [\xi, \zeta_1] + [\xi, \zeta_2].$$

Remark 2.6.2 If f is any equation of ξ, then it may be written $f = f_1 \ldots f_\ell$ where each f_i is an equation of γ_i and so, by 2.5.4, we have

$$[\xi.\zeta] = \sum_{j=1}^{\ell'} o_{\gamma'_j}(f).$$

Fix any system of local coordinates x, y so that the germ of the y-axis is not a branch of either ξ or ζ. Denote by $s_1, \ldots, s_k$ the Puiseux series of the branches γ_i of ξ, all series of each branch included and (if ξ is non-reduced) each series repeated according to the number of times its corresponding branch appears in the summation $\xi = \gamma_1 + \cdots + \gamma_\ell$. Under the same conventions, denote by $s'_1, \ldots, s'_{k'}$ the Puiseux series of the branches of ζ. Then we have

Proposition 2.6.3 *The intersection multiplicity of the germs ξ and ζ is*

$$[\xi.\zeta] = \sum_{\substack{1 \le i \le k \\ 1 \le j \le k'}} o_x(s_i - s'_j).$$

PROOF: We know from section 2.2 that

$$f = \prod_{i=1}^{k}(y - s_i(x))$$

is an equation of ξ. Then it is enough to use 2.6.2 and 2.5.3. ◇

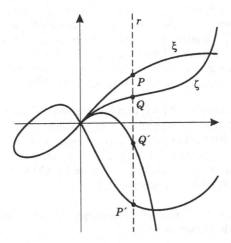

Figure 2.2: Halphen formula: the intersection multiplicity of ξ and ζ is obtained by adding up the infinitesimal orders of the segments PQ, $P'Q$, PQ', $P'Q'$ when the vertical line r approaches the origin.

The formula in 2.6.3 is due to Halphen and so it is called the *Halphen formula*. Since a particular case of it was already used by Zeuthen (to determine multiplicities of fixed points of one-dimensional algebraic correspondences), sometimes it is also called the *Zeuthen formula* (see [96], sections 2 and 14). The reader may realize that the Halphen formula computes the intersection multiplicity as the sum the infinitesimal orders, for $x_0 \to 0$, of all arcs on the curve $x = x_0$ with one end on a representative of ξ and the other on a representative of ζ.

Since for any given germs ξ and ζ there are local coordinates for which the Halphen formula 2.6.3 holds, the next corollaries directly follow. The first one claims the symmetry of the intersection multiplicity, which was not obvious from its definition.

Corollary 2.6.4 *For any two germs ξ, ζ, we have*

$$[\xi.\zeta] = [\zeta.\xi].$$

Corollary 2.6.5 *We have $[\xi.\zeta] = \infty$ if and only if the germs ξ and ζ have a common branch.*

Remark 2.6.6 Directly from the Halphen formula we get the equalities

$$[\xi.\zeta] = \sum_{j=1}^{k'} o_x f(x, s_j') = \sum_{i=1}^{k} o_x g(x, s_i),$$

where f and g are equations of ξ and ζ respectively. They show that the intersection multiplicity may also be computed by substituting the Puiseux series of one germ in the equation of the other. If, in particular, γ is irreducible and has Puiseux series $s = ax^{m/n} + \cdots$, $a \neq 0$, $n = \nu(s)$, then its intersection multiplicities with the x-axis and the y-axis are, respectively, $\nu(s)o_x(s) = m$ and $\nu(s) = n$. The reader may compare with 2.2.8 and 2.6.9 below.

An easy consequence of 2.6.6 is that the intersection multiplicities of $\xi : f = 0$ with the x-axis (resp. the y-axis) is $o_x f(x, 0)$ (resp. $o_y f(0, y)$), just the width (resp. height) of $\mathbf{N}(\xi)$ if the axis is not a component of ξ (1.1.4).

It is also easy to see that if both ξ and ζ are smooth and no one is tangent to the y-axis, then $[\xi, \zeta]_O$ equals the least order for which the analytic functions of x implicitly defined by the equations of ξ and ζ have different derivatives at zero.

As one can expect, the intersection multiplicity is in some way related to the multiplicities of the curves at O:

Proposition 2.6.7 *If ξ and ζ are germs of curve at O,*

$$[\xi.\zeta] \geq e(\zeta)e(\xi)$$

and the equality holds if and only if ξ and ζ share no principal tangent.

PROOF: Since the tangents to a given germ are the tangents to its branches (sect. 0.9) and the intersection multiplicity is biadditive, it is enough to prove the claim for ξ and ζ irreducible. In such a case, choose the coordinates so that both germs are not tangent to either of the axes. Then the Puiseux series of ξ is, by 2.7.1, $s = ax + \cdots$, where $a \neq 0$ is the slope of the tangent to ξ.

Similarly the Puiseux series of ζ is $s' = bx + \cdots$ where b is the slope of the tangent to ζ. On the other hand $\nu(s) = e_O(\xi)$ and $\nu(s') = e_O(\zeta)$ by 2.2.9, so that, according to the Halphen formula 2.6.3, we have to add up the orders in x of the $e_O(\xi)e_O(\zeta)$ differences between the different conjugates of s and s': since any such difference has the form

$$\sigma_\varepsilon(s) - \sigma_\eta(s') = (a - b)x + \cdots$$

the claim follows. ⋄

Recall from section 0.10 that two germs are transverse if and only if they are smooth and have different principal tangents. Then an easy corollary of 2.6.7 is:

Corollary 2.6.8 *Two germs ξ and ζ are transverse if and only if $[\xi.\zeta] = 1$.*

Assume that ξ is smooth: the reader may easily see that any positive integer may be obtained as the intersection multiplicity at O of ξ and another germ (which one may even choose smooth). That this is not the case if ξ is singular is clear from proposition 2.6.7 above. Singular germs ξ behave in a far more complex way when intersected with other germs: describing such behaviour makes a deep description of the singularity of ξ. Infinitely near points will be introduced in chapter 3 to help in this description and our main viewpoint will be to classify singularities according to their intersection behaviour. The next proposition shows that the multiplicity of a germ may be easily recovered from its intersection multiplicities with other germs.

Proposition 2.6.9 *For any germ ξ at O,*

$$e(\xi) = \min_\zeta\{[\xi.\zeta]\},$$

ζ *ranging on all non-empty germs of curve at O.*

PROOF: It follows from 2.6.7 that $e(\xi) \leq [\xi.\zeta]$ for any non-empty ζ. The equality is obtained by taking ζ smooth and non-tangent to ξ, again by 2.6.7. ⋄

2.7 Pencils and linear systems

In this section we shall just introduce pencils and linear systems of germs of curve and prove some of their easiest properties. The whole of the forthcoming chapter 7 being devoted to studying these families of germs, our main goal here is to define a Zariski topology in $\mathcal{O}_{S,O}$ and use it to give a precise meaning to the phrase *a generic germ in a linear system...*, which will be quite useful in the sequel.

A *linear family* of germs of curve is a set of germs of curve at a point O of the form $\{\xi : f = 0 | f \in F - \{0\}\}$, F a linear subspace of $\mathcal{O}_{S,O}$. The reader may notice that the same linear family is defined by the linear subspace uF if u is any invertible element of $\mathcal{O}_{S,O}$.

We are more interested in the linear families defined by ideals, we call them *linear systems*. A linear system has thus the form $L = \{\xi : f = 0 | f \in I - \{0\}\}$, I an ideal of $\mathcal{O}_{S,O}$. It is clear in such a case that I contains all equations of each member of the family and it is determined by L just by this property. Any ideal of $\mathcal{O}_{S,O}$ being finitely generated (1.8.9), one may take a system of generators $f_1, \ldots, f_r$ of I and then $\mathcal{L}$ is the set of all germs of curve that have an equation $\sum_{i=1}^{r} a_i f_i = 0$ for some $a_i \in \mathcal{O}_{S,O}$.

We are in particular considering the linear system defined by the ideal (1), clearly the only linear system containing the empty germ. It will be called the *irrelevant linear system*.

Note that there is a well determined minimal linear system containing given germs $\xi_i : f_i = 0$, $i = 1, \ldots, r$, namely the linear system defined by the ideal $(f_1, \ldots, f_r)$: we will call it the linear system *generated* by $\xi_1, \ldots, \xi_r$. It is worth noting that, by contrast, the linear family defined by the linear subspace generated by $f_1, \ldots, f_r$ depends on the equations f_i and not only on the germs ξ_i.

A linear family of germs of the form

$$\mathcal{P} = \{\xi : f = 0 | f = \lambda_1 f_1 + \lambda_2 f_2, \ (\lambda_1, \lambda_2) \in \mathbb{C}^2 - \{(0,0)\}\},$$

where $f_1 = 0$ and $f_2 = 0$ are different non-empty germs will be called a *pencil*. The germs $f_1 = 0$ and $f_2 = 0$ being assumed to be different, the $\mathbb{C}$-linear subspace L of $\mathcal{O}_{S,O}$ generated by f_1, f_2 has dimension two. The map

$$L - \{0\} \longrightarrow \mathcal{P}$$
$$f \longmapsto \xi : f = 0$$

defines on $\mathcal{P}$ a structure of one-dimensional projective space for which λ_1, λ_2 may be taken as the homogeneous coordinates of $\lambda_1 f_1 + \lambda_2 f_2 = 0$. Indeed, the converse being obvious, we need just to see that different equations, $g_1 = \lambda_1 f_1 + \lambda_2 f_2$ and $g_2 = \mu_1 f_1 + \mu_2 f_2$, of the same germ need to be proportional. To get a contradiction, assume that g_1, g_2 are linearly independent and define the same germ: then f_1 and f_2, which can be written as linear combinations of g_1, g_2, also define the same germ, against the hypothesis.

If g is the greatest common divisor of all elements in (or just generators of) a linear subspace F of $\mathcal{O}_{S,O}$, then one may write $F = gF'$ where F' is a linear subspace and the greatest common divisor of all elements in F' is 1. The germ $\zeta : g = 0$ is called the *fixed part* of the linear family L defined by F: all germs in L are composed of ζ and a germ in the linear family L' (the *variable part* of L) defined by F'. If $g = 1$ we say that L has no fixed part. Notice that this is always the case for the variable part L' above.

As the reader may easily check, the variable part of a pencil is a pencil. In the case of F being an ideal so is F' and hence both L and L' are linear systems. In such a case the condition of having no fixed part is equivalent to saying that there is no principal proper ideal containing I, that is, either $I = (1)$ and so the linear system is the irrelevant one, or I satisfies all the equivalent conditions of 1.8.10. In particular I is then an $\mathcal{M}_{S,O}$-primary ideal.

It is clear that the linear family L has no fixed part if and only if there is no non-empty germ contained in all elements of L. We may say a little bit more, namely:

Lemma 2.7.1 *If a linear family L has no fixed part and $\xi \in L$, then there is a germ in L which shares no component with ξ.*

PROOF: Let F be the subspace defining L. If $\gamma : g = 0$ is a branch of ξ, the equations of the elements of L which contain γ obviously describe the subspace $(g) \cap F$ that cannot be the whole of F as L is assumed not to have a fixed part. Then since ξ has finitely many components and F cannot be the union of finitely many strictly smaller linear subspaces, the claim follows. ◇

The next three lemmas state some elementary properties of pencils that are essentially due to their one-dimensional character. The proof of the first one is left as an exercise.

Lemma 2.7.2 *If two germs in a pencil have the same multiplicity but different tangent cones, then all germs in the pencil have the same multiplicity and no two have the same tangent cone. If two different germs in a pencil either have the same tangent cone or have different multiplicities, then all germs in the pencil have the same multiplicity and the same tangent cone, but for a single germ whose multiplicity is higher.*

Lemma 2.7.3 *Let $\mathcal{P}$ be a pencil without fixed part and γ an irreducible germ, both with the same origin O. There is a $\xi_0 \in \mathcal{P}$ such that $[\gamma.\xi] = [\gamma.\xi'] < [\gamma.\xi_0]$ for any $\xi, \xi' \in \mathcal{P}$, $\xi \neq \xi_0$, $\xi' \neq \xi_0$.*

PROOF: Take local coordinates x, y at O, assume that the germs in $\mathcal{P}$ have equations $\lambda_1 f_1 + \lambda_2 f_2 = 0$, $(\lambda_1, \lambda_2) \in \mathbb{C}^2 - \{(0,0)\}$, and write $t \mapsto (x(t), y(t))$ a uniformizing map of a representative of γ. If $r = \min_{i=1,2}\{o_t f_i(x(t), y(t))\}$, $r \neq \infty$ because $\mathcal{P}$ has no fixed part. Then the initial term of $\lambda_1 f_1(x(t), y(t)) + \lambda_2 f_2(x(t), y(t))$ has the form $(\lambda_1 a + \lambda_2 b)t^r$ with either a or b different from zero, from which the claim obviously follows after taking ξ_0 to be $bf_1 - af_2 = 0$. ◇

Lemma 2.7.4 *If $\mathcal{P}$ is a pencil, all intersection multiplicities $[\xi.\zeta]$, $\xi, \zeta \in \mathcal{P}$, $\xi \neq \zeta$, are equal.*

PROOF: We assume that $\mathcal{P}$ has no fixed part, as otherwise the claim is obvious. It is clearly enough to see that for any fixed $\xi \in \mathcal{P}$, all intersection multiplicities $[\xi.\zeta]$, $\zeta \in \mathcal{P}$, $\zeta \neq \xi$ are equal. This in turn is clear, as for any branch γ of ξ, $\infty = [\gamma.\xi] > [\gamma.\zeta]$ because $\mathcal{P}$ has no fixed part, and so, by 2.7.3, all $[\gamma.\zeta]$ are equal provided $\zeta \neq \xi$. ◇

If O is any point on a smooth surface S, we will define a topology on the local ring $\mathcal{O}_{S,O}$. Fix local coordinates x, y at O, so that the local ring $\mathcal{O}_{S;O}$ is identified with the series ring $\mathbb{C}\{x, y\}$. Then define the closed sets as the sets of the form

$$\{f = \sum_{i,j} a_{i,j} x^i y^j \mid P(a_{i,j}) = 0, \, P \in \mathcal{P}\}$$

where $\mathcal{P}$ is a subset of the ring of the polynomials in the (infinitely many) variables $X_{i,j}$, $i, j \geq 0$. This easily defines a topology which will be called the *Zariski topology* on $\mathcal{O}_{S,O}$. A Zariski-closed set is thus the set of all the series whose coefficients satisfy certain given algebraic relations. The next lemmas will be useful in the sequel:

Lemma 2.7.5 *The $\mathcal{M}_{S,O}$-primary ideals are closed subsets of $\mathcal{O}_{S,O}$ defined by finitely-many homogeneous linear equations in the coefficients of the series.*

PROOF: Write $\mathcal{M} = \mathcal{M}_{S,O}$. If I is $\mathcal{M}$-primary there is an n so that $\mathcal{M}^n \subset I$. Identify $\mathcal{O}_{S,O}/\mathcal{M}^n$ with the finite-dimensional space of the polynomials in x, y of degree less than n. Since $\mathcal{M}^n \subset I$, $f = \sum_{i,j=0}^{\infty} a_{i,j} x^i y^j \in I$ if and only if $\sum_{i+j<n} a_{i,j} x^i y^j \in I/\mathcal{M}^n$. The latter being a linear subspace of $\mathcal{O}_{S,O}/\mathcal{M}^n$, it is defined by finitely many homogeneous linear equations in the $a_{i,j}$, hence the claim. $\diamond$

Lemma 2.7.6 *If $\varphi : \mathcal{O}_{S,O} \longrightarrow \mathcal{O}_{S',O'}$ is a local morphism (i.e., such that $\varphi(\mathcal{M}_{S,O}) \subset \mathcal{M}_{S',O'}$), then it is continuous for the Zariski topologies.*

PROOF: Assume that Zariski topologies are defined from local coordinates x, y at O and x', y' at O'. Put $g_1 = \varphi(x)$, $g_2 = \varphi(y)$.

Let us see first that for any $f \in \mathcal{O}_{S,O}$, $\varphi(f)$ is obtained by substituting g_1, g_2 for x, y in f, that is,

$$\varphi(f) = f(g_1, g_2).$$

Note that this makes sense as both g_1, g_2 are non-invertible series, just because the morphism is assumed to be local. In fact the equality is clear if f is a polynomial, and then the general case follows by using this equality for the partial sums f_n of f: the morphism being local, one has $\varphi(f) - \varphi(f_n) \in \mathcal{M}_{S',O'}^{n+1}$, and since $\varphi(f_n) = f_n(g_1, g_2)$, it is clear in turn that $\varphi(f_n) - f(g_1, g_2) \in \mathcal{M}_{S',O'}^{n+1}$. This gives $\varphi(f) - f(g_1, g_2) \in \mathcal{M}_{S',O'}^{n+1}$ for all n. After this the claim follows, as $\bigcap_n \mathcal{M}_{S',O'}^n = (0)$.

Now it follows from the equality displayed above that the coefficients of the series $\varphi(f)$ may be written as polynomials in the coefficients of f, which in turn clearly shows that the inverse image of a closed set is a closed set. $\diamond$

Notice that 2.7.6 shows in particular that the identity map of $\mathcal{O}_{S,O}$ is a homeomorphism no matter which coordinates are used to define the Zariski topology on each copy of $\mathcal{O}_{S,O}$, and hence that the Zariski topology on $\mathcal{O}_{S,O}$ does not depend on the choice of the local coordinates.

If F is any subset of $\mathcal{O}_{S,O}$, we take the relative topology on F, and call F irreducible if and only if it cannot be decomposed into the union of two relatively closed strictly smaller subsets. Equivalently, F is irreducible if and only if any two non-empty relatively open subsets of F have non-empty intersection.

Lemma 2.7.7 *The ideals of $\mathcal{O}_{S,O}$ are irreducible subsets.*

PROOF: Let us write $\mathcal{O} = \mathcal{O}_{S,O}$ and $\mathcal{M} = \mathcal{M}_{S,O}$. Assume that F is an ideal and write it as $F = gI$, where g is the greatest common divisor of all elements in F and I is then an ideal that contains a power of $\mathcal{M}$, say $I \supset \mathcal{M}^n$ (I is either $\mathcal{M}$-primary or (1)). The reader may easily check that multiplying by g is a continuous map $\times g : \mathcal{O} \longrightarrow \mathcal{O}$, after which, since continuous images of irreducible sets are irreducible, it is enough to prove that I is irreducible. Write $f = \sum a_{i,j} x^i y^j$. Since $I \supset \mathcal{M}^n$, $f \in I$ if and only if its class mod $\mathcal{M}^n$ belongs to $I/\mathcal{M}^n$ which is a finite-dimensional linear space. After choosing a basis of it, one may determine linear forms $A_{i,j} = A_{i,j}(Y_1, \ldots, Y_k)$ for $i + j < n$, so that $f \in I$ if and only if $a_{i,j} = A_{i,j}(\lambda_1, \ldots, \lambda_k)$, $i+j < n$, for some $\lambda_1, \ldots, \lambda_k \in \mathbb{C}$. If P is any polynomial in the variables $X_{i,j}$, $i, j \geq 0$, denote by $\bar{P}$ the polynomial in the variables $Y_1, \ldots, Y_k$ and $X_{i,j}$, $i + j \geq n$, obtained from P by substituting $A_{i,j}$ for $X_{i,j}$ for $i + j < n$. It is clear that P takes the value zero on all series in I if and only if $\bar{P}$ is the polynomial zero and so it follows that if the product of two polynomials $P, P' \in \mathbb{C}[X_{i,j}]$ is zero on all series in I, then at least one of the polynomials is zero on all series in I. From this fact the irreducibility will easily follow. Indeed, assume $I \subset T \cup T'$, T and T' being closed sets in $\mathcal{O}$ defined by sets of polynomials $\mathcal{P}$ and $\mathcal{P}'$ respectively. If T' does not contain I, there is a $P' \in \mathcal{P}'$ which is not zero on some series in I. Since $I \subset T \cup T'$, for all $P \in \mathcal{P}$, PP' takes the value zero on the whole of I. Then the former fact says that all $P \in \mathcal{P}$ are zero on all series in I and so $I \subset T$. ⋄

Assume that L is a linear system of germs, I being its corresponding ideal. It will be useful to give a precise meaning to the somewhat old-fashioned phrasing *a generic germ in L satisfies the property* Π: this will mean, by definition, that there is a non-empty Zariski-open subset of I the elements of which define germs which satisfy the property Π. In other words, if a germ in L does not satisfy the property Π, then the coefficients of any of its equations satisfy a certain set of algebraic relations which are in turn not satisfied by the coefficients of some equation in I. Usually the whole set of all equations of all germs satisfying Π will be a non-empty Zariski-open subset.

The reader may easily see, as an easy example, that the set of all non-empty germs at O is a linear system without fixed part, and that once a tangent direction d at O is fixed, a generic non-empty germ is smooth and non-tangent to d.

We will often make use without further reference of the following consequence of the irreducibility of any ideal of $\mathcal{O}_{S,O}$ seen above: if a generic germ in L satisfies the property Π and also a generic germ in L satisfies the property Π′, then a generic germ in L satisfies both properties Π and Π′.

2.8 Exercises

2.1 A germ ξ at $O = (0,0)$ is called *algebraic* if it has an equation in $\mathbb{C}[x, y]$. The notion obviously depends on the coordinates and is often used for germs in the affine plane $\mathbb{C}^2$, relative to the affine coordinates. Check that the polynomial $x^2 - y^2 + y^3$ is irreducible in $\mathbb{C}[x, y]$ while $\xi : x^2 - y^2 + y^3 = 0$ has two branches at the origin. Prove that these branches are not algebraic.

2.2 Let

$$y^9 - x^5 y^7 - x^4 y^6 - x^6 y^5 + x^9 y^4 + x^{11} y^3 + x^{10} y^2 - x^{15} = 0$$
$$y^4 + 4x^2 y^3 + 5x^4 y^2 - 2x^3 y^2 + 2x^6 y - 4x^5 y - x^7 + x^6 = 0$$
$$y^4 - 2x^2 y^3 - 4x^5 y - x^7 + x^6 = 0$$

define germs of curve at the origin of $\mathbb{C}^2$. For each of them, compute partial sums of its Puiseux series up to the stage of separating the branches and getting the order of each branch.

2.3 Let ξ be a reduced germ of curve and Γ a side of $N(\xi)$. Prove that if the slope of Γ is non-less (resp. non-greater) than -1, then the height (resp. width) of Γ equals the sum of the orders of the branches associated with Γ.

2.4 Let ξ be a reduced germ of curve, Γ_1, Γ_2 two different sides of its Newton polygon and ξ_i the germ composed of the branches of ξ corresponding to the side Γ_i, $i = 1, 2$. Assume that Γ_1 is above Γ_2. Draw two vertical lines through the ends of Γ_1 and two horizontal lines through the ends of Γ_2, thus limiting a rectangle $\mathbf{R}$. Prove that $[\xi_1.\xi_2]$ equals the area of $\mathbf{R}$.

2.5 Let C be the algebraic curve in $\mathbb{P}^2$

$$F = \sum_{\alpha,\beta=0}^{d} a_{\alpha,\beta} X_0^{d-\alpha-\beta} X_1^{\alpha} X_2^{\beta} = 0$$

and assume for simplicity that no one of the X_i divides F. The Newton polygon of C, $\mathbf{N}$, is, by definition, the border of the convex hull of the set $\{(\alpha, \beta)|a_{\alpha,\beta} \neq 0\}$. We will consider three polygonal lines contained in $\mathbf{N}$, namely $\mathbf{N}_0$, joining the α-axis and the β-axis, $\mathbf{N}_1$, joining the β axis and the line $\alpha + \beta = d$, and $\mathbf{N}_2$ joining this line and the α-axis. Take, in $\mathbb{P}^2$, $O = [1, 0, 0]$, $O' = [0, 0, 1]$ and $O'' = [0, 1, 0]$.

(a) Prove that $\mathbf{N}_0$ is the Newton polygon of the germ C_O if local coordinates $x = X_1/X_0$, $y = X_2/X_0$ at O are used.

(b) Take local coordinates $x' = X_1/X_2$, $y' = X_0/X_2$ at O'. Prove that the affine transformation $(\alpha, \beta) \longmapsto (\alpha, d-\alpha-\beta)$ maps $\mathbf{N}_1$ onto $N(C_{O'})$ and therefore that to each side of $\mathbf{N}_1$ corresponds a non-empty set of branches of $C_{O'}$. Furthermore, check that the branches corresponding to a side of $\mathbf{N}_1$ of positive slope are tangent to $X_1 = 0$ while those corresponding to a side of negative slope are tangent to $X_0 = 0$.

(c) State and prove a claim similar to (b) above, relative to $\mathbf{N}_2$ and $C_{O''}$.

2.6 Let $f_1, \ldots, f_r$ be analytic functions defined in a neighbourhood U of O. Define $\mathbf{V}(f_1, \ldots, f_r)$ as being the germ at O of the set $\{p \in U | f_i(p) = 0, i = 1, \ldots, r\}$.

(a) Prove that $\mathbf{V}(f_1, \ldots, f_r)$ does not depend on $f_1, \ldots, f_r$, but only on the ideal that their germs generate in $\mathcal{O}_{S,O}$. Thus for any ideal I of $\mathcal{O}_{S,O}$ define $\mathbf{V}(I) = \mathbf{V}(f_1, \ldots, f_r)$, $f_1, \ldots, f_r$ being representatives of a system of generators of I.

(b) Prove that $\mathbf{V}(I) = \mathbf{V}(\mathbf{r}(I))$, $\mathbf{r}(I)$ being the radical of the ideal I.

(c) Prove that $\mathbf{V}(I)$ is the germ of $\{O\}$ if and only if I is $\mathcal{M}_{S,O}$-primary and that $\mathbf{V}(I) = \emptyset$ if and only if $I = (1)$.

(d) If $I \neq (1)$ is not $\mathcal{M}_{S,O}$-primary, write it as $I = gI'$, where $g \in \mathcal{M}_{S,O}$ and I' is an ideal which is either $\mathcal{M}_{S,O}$-primary or equal to (1). Prove that $\mathbf{V}(I) = \mathbf{V}(g)$.

(e) Prove the following two-dimensional version of the *Nullstellensatz* (see [40] or [41] for the general case): for any two ideals I, J of $\mathcal{O}_{S,O}$, $\mathbf{V}(I) = \mathbf{V}(J)$ if and only if $\mathbf{r}(I) = \mathbf{r}(J)$.

2.7 Give a different proof of 2.3.3 by substituting the Puiseux series of one of the germs in the equation of the other.

2.8 Give examples of

(a) two curves in $\mathbb{C}^2$ sharing infinitely many points but no branch in either of these points, and

(b) a reduced curve in $\mathbb{C}^2$ having infinitely many singular points.

2.9 Let γ be a curve on a surface S and assume that the germ of γ at a point O is irreducible. Prove that there are neighbourhoods U of 0 in $\mathbb{C}$ and V of O in S and a uniformizing map of γ at O which is a homeomorphism between U and $V \cap |\gamma|$.

2.10 Assume that ξ is a curve which is reduced at a point O. Let x, y be local coordinates at O and assume that $\eta : x = 0$, the germ of the y-axis, is not a branch of ξ at O. Prove that there is a neighbourhood U of O where x, y are defined and such that for any other neighbourhood U' of O, $U' \subset U$, there is a positive real number δ so that for any λ, $0 < |\lambda| < \delta$, there are just $[\eta.\xi_O]$ points in $U \cap \xi$ with first coordinate λ, all these points belong to U' and are transverse intersections of ξ and $x - \lambda = 0$. (*Hint:* consider the Weierstrass equation g of ξ and the germ $\partial g / \partial y = 0$, arguing from exercise 1.9 and 2.3.3).

2.11 (See [35], IV, I.1) Notations and conventions being as in 2.10, put $n = [\eta.\xi_O]$ and for a fixed $x \neq 0$, $\rho = |x| < \delta$, let $\{(x, y_j)\}_{j=1,\ldots,n}$ be the points in $U \cap |\xi|$ with first coordinate x.

(a) Prove that for each i there is a uniquely determined continuous map $\alpha_j : [0, 2\pi] \longrightarrow \mathbb{C}$ so that $\alpha_j(0) = y_j$ and $(\rho e^{i\theta}, \alpha_i(\theta)) \in |\xi|$.

(b) Check that $\alpha_j(2\pi) = y_{j'}$ for some $j' = 1, \ldots, n$. Prove that the map $\sigma : y_j \longmapsto y_{j'}$ is a permutation of $\{y_j\}_{j=1,\ldots,n}$.

(c) Prove that the cycles $\{\sigma^h(y_j)\}_{h \in \mathbb{Z}}$ of σ are in one to one correspondence with the branches of ξ_O.

2.12 Let U be a neighbourhood of zero in $\mathbb{C}$ and $\psi : U \longrightarrow S$ a non-constant holomorphic map into a surface S, $O = \psi(0)$.

(a) Prove that there are a smaller neighbourhood of 0, $U' \subset U$ and a curve γ irreducible at O, so that $\psi(U') \subset |\gamma|$. Prove that the germ γ_O is uniquely determined by the germ of ψ at 0.

(b) After a suitable shrinking of U', factorize $\psi_{|U'} = \varphi \circ \psi'$ where φ is a uniformizing map of γ at O and $\psi' : U' \longrightarrow \mathbb{C}$ is analytic. Then define the *degree* of ψ, $d(\psi)$, as being the order of zero of ψ' at 0 and prove that it is determined by the germ of ψ at 0. If, after taking local coordinates at O, $\psi(z) = (\psi_1(z), \psi_2(z))$, prove that $d(\psi) e_O(\gamma) = \min\{o(\psi_1), o(\psi_2)\}$.

(c) After shrinking U' once again if needed, prove that for any smaller neighbourhood $U'' \subset U'$ of 0, there is a neighbourhood V of O in S so that for all $p \in |\gamma| \cap V$, $p \neq O$, $\psi^{-1}(p) \cap U'$ has $d(\psi)$ elements and all of them belong to $U'' - \{O\}$.

2.13 Let $\Psi : U \longrightarrow U'$ be a morphism between open neighbourhoods U, U' of points O, O' in surfaces S, S', respectively, $\Psi(O) = O'$, and γ a curve in U irreducible at O. Let φ be a uniformizing map of γ and put $\psi = \Psi \circ \varphi$. If ψ is non-constant, take a curve ζ, irreducible at O' and containing the image by ψ of a suitable neighbourhood of 0 (see exercise 2.12).

(a) Prove that, once Ψ is given, the germ of ψ at 0 and $\zeta_{O'}$ are uniquely determined by γ_O.

(b) Define $\Psi_*(\gamma_O)$, the *direct image* of γ_O by Ψ, as being $d(\psi)\zeta_{O'}$ (notations from exercise 2.12) if ψ is non-constant, and the empty germ otherwise. Prove that there is a neighbourhood V_0 of O in U so that for any smaller neighbourhood $V \subset V_0$ there is a neighbourhood V' of O' in U' so that for all $p \in V' \cap \zeta$, $p \neq O'$, $\Psi^{-1}(p) \cap |\gamma| \cap V_0$ has $d(\psi)$ elements and all of them belong to $V - \{O\}$.

(c) Take $[\gamma_O.\Psi^*(\xi)] = \infty$ if $\Psi^*(\xi)$ is undefined and prove that for any germ of curve ξ at O',
$$[\gamma_O.\Psi^*(\xi)] = [\Psi_*(\gamma_O).\xi]$$
if $\Psi_*(\gamma_O) \neq \emptyset$, $[\gamma_O.\Psi^*(\xi)] = \infty$ otherwise (*projection formula*).

(d) Extend the definition of Ψ_* to arbitrary germs at O by linearity and prove that the projection formula still holds.

2.14 Take Ψ as in exercise 2.13 and, after taking local coordinates at O and O', assume that it is given by $\Psi(x, y) = (f(x, y), g(x, y))$, where both f and g are holomorphic and no one identically vanishes in a neighbourhood of the origin. Let ξ be a germ of curve at O. Prove that the slopes of the sides of the Newton polygon of $\Psi_*(\xi)$ (see exercise 2.13) are the ratios $-[\gamma.\zeta_2]/[\gamma.\zeta_1]$ for all branches γ of ξ such that $\Psi_*(\gamma) \neq \emptyset$, ζ_1 and ζ_2 being $f = 0$ and $g = 0$, respectively.

2.15 *A dynamic approach to intersection multiplicity.* Let $\xi : f = 0$ and $\zeta : g = 0$ be curves defined in a neighbourhood U of O where local coordinates x, y are also defined. Assume that ξ is reduced at O and that $n = [\xi.\zeta] < \infty$. The goal is to prove that there are n intersections of ξ and $\zeta_\lambda : g - \lambda = 0$ that approach O when λ approaches zero.

(a) Assume ξ irreducible at O and different from the y-axis. Define $\Phi : U \longrightarrow \mathbb{C}^2$ by the rule $x' = g(x, y)$, $y' = y$ and take η_λ to be the line $x' = \lambda$. Prove that $\Phi_*(\xi_O) \neq \emptyset$ and $\Phi^*(\eta_\lambda) = \zeta_\lambda$.

(b) Still assume ξ_O irreducible and take γ so that $\Phi_*(\xi_O) = d\gamma_{(0,0)}$. Apply exercise 2.10 to γ and then use exercise 2.13 to prove that there is a neighbourhood V_0 of O in U such that for any neighbourhood V of O, $V \subset V_0$, there is $\delta \in \mathbb{R}^+$ so that for all $\lambda \in \mathbb{C}$, $0 < |\lambda| < \delta$, the curves ξ and ζ_λ have n different intersection points in V_0, all of them belong to V, and are transverse intersections of ξ and ζ_λ.

(c) Prove that the claim of part (b) above still holds true if ξ is just assumed to be reduced at O.

2.16 Prove that if the linear subspace $F \subset \mathcal{O}_{S,O}$ defining a linear family L contains all equations of all germs in L, then necessarily F is an ideal of $\mathcal{O}_{S,O}$ and hence the linear family L is a linear system.

2.17 Let $\mathbf{N}$ be a Newton polygon (of some germ of curve) with both ends on the axes. Prove that the germs whose Newton polygons have no vertex below $\mathbf{N}$ describe a linear system $\mathcal{L}(\mathbf{N})$ without fixed part. A germ ξ is said to be *non-degenerate with respect to* $\mathbf{N}$ if and only if $\mathbf{N}(\xi) = \mathbf{N}$ and no one of the equations associated with the sides has a multiple root. Prove that generic germs in $\mathcal{L}(\mathbf{N})$ are non-degenerate with respect to $\mathbf{N}$.

2.18 Let ξ be a germ of curve with origin at O.

(a) Prove that g divides zero in $\mathcal{O}_{\xi,O}$ if and only if the germs ξ and $g = 0$ share a branch. Describe the elements $g' \in \mathcal{O}_{S,O}$ so that $gg' = 0$.

(b) Prove that $g \in \mathcal{O}_{\xi,O}$ is nilpotent if and only if all branches of ξ are branches of $g = 0$. Prove that $\mathcal{O}_{\xi,O}$ has no nilpotent elements other than zero if and only if ξ is reduced.

2.19 Let C be an irreducible algebraic curve in $\mathbb{P}_2$ and $\pi : X \longrightarrow C$ a birational (necessarily regular) map from a projective, smooth and irreducible algebraic curve X onto C (this determines X up to isomorphism of algebraic varieties, X is then called the non-singular model of C, see [37], Chap. 7). Prove that for each $O \in C$ there is a one to one correspondence between $\pi^{-1}(O)$ and the set of branches of C at O, so that for each $p \in \pi^{-1}(O)$, the restriction of π to a suitable neighbourhood of p in X is a uniformizing map of the corresponding branch at O.

Chapter 3

Infinitely near points

The classical algebro-geometric way of studying the singularity of a germ of plane curve is to analyze its behaviour when intersected with other germs. In this chapter we begin the study of the intersection behaviour of (singular) plane germs: this performs a deep geometrical analysis of singularities which will lead to their classification. Our main tool in this study will be the points *infinitely near* to a point O on a smooth surface S. Infinitely near points were first introduced by M. Noether [63] and extensively studied by Enriques ([35], book IV). In spite of the fact that infinitely near points lie on different surfaces, the whole set of them provides a sort of infinitesimal space which displays the local geometry at O of the curves on S.

Classical authors used to define infinitely near points by means of different kinds of birational transformations dependent on many arbitrary choices, which nevertheless led them to basically correct and consistent results. The modern definition we shall present in this chapter follows the same idea but uses a single type of transformation, named *blowing-up*. This makes a definition free of arbitrary choices and avoids some undesired effects of the classical transformations.

3.1 Blowing up

Once a point O on a (smooth) surface S has been fixed, choose local coordinates x, y at O and assume that they are analytic in an open neighbourhood U of O. Take a complex projective line $\mathbb{P}_1$, with homogeneous coordinates z_0, z_1, and let $\bar{U}$ be the subvariety of $U \times \mathbb{P}_1$ defined by the equation $xz_1 - yz_0 = 0$. Let the affine lines $\mathbb{A}_1 : z_0 \neq 0$ and $\mathbb{A}_1' : z_1 \neq 0$, with coordinates $z = z_1/z_0$ and $z' = z_0/z_1$, respectively, be an open covering of $\mathbb{P}_1$; then we have $U \times \mathbb{P}_1 = U \times \mathbb{A}_1 \cup U \times \mathbb{A}_1'$ and the trace of $\bar{U}$ on each piece $U \times \mathbb{A}_1$ and $U \times \mathbb{A}_1'$ is defined by the equations

$$xz - y = 0 \quad \text{and} \quad x - yz' = 0,$$

respectively. We may state

Lemma 3.1.1 $\bar{U}$ *is a (smooth and connected) surface. The projection* $U \times \mathbb{P}_1 \longrightarrow U$ *induces an analytic morphism* $\pi : \bar{U} \longrightarrow U$ *whose restriction to* $\bar{U} - \pi^{-1}(O)$ *is an isomorphism onto* $U - \{O\}$.

PROOF: Put $V = \bar{U} \cap U \times \mathbb{A}_1$ and $V' = \bar{U} \cap U \times \mathbb{A}_1'$. That $\bar{U}$ is a smooth and connected surface is clear from the equations for V and V' given above. For the remaining part of the claim it is enough to say that the morphisms

$$U - \{x = 0\} \longrightarrow V$$
$$(x, y) \longmapsto (x, y, y/x)$$

and

$$U - \{y = 0\} \longrightarrow V'$$
$$(x, y) \longmapsto (x, y, x/y)$$

patch together to give the inverse of the claimed isomorphism. ◇

Remark 3.1.2 Since in the open set V we have $y = xz$, it is clear that x, z may be taken as analytic coordinates in the whole of V. Similarly, y, z' are analytic coordinates in the whole of V'. Using such coordinates the restrictions of π to V and V' are given by $(x, z) \mapsto (x, zx)$ and $(y, z') \mapsto (yz', y)$, and the coordinates of a point in both V and V' are related by $y = zx$, $z' = 1/z$.

The morphism $\pi : \bar{U} \longrightarrow U$ we have just defined, or rather its inverse correspondence, has at O the local effect we want. All we need is to extend it to the whole of S, which is not difficult because $\pi_{|\bar{U}-\pi^{-1}(O)}$ is an isomorphism. Let us note first that the graph of $\pi_{|\bar{U}-\pi^{-1}(O)}$ clearly equals the trace on $\bar{U} \times (S - \{O\})$ of $\{(p, q, p)|p \in S, q \in \mathbb{P}_1\}$, clearly a closed subset of $S \times \mathbb{P}_1 \times S$. Then the graph of $\pi_{|\bar{U}-\pi^{-1}(O)}$ is closed in $\bar{U} \times (S - \{O\})$ and we may define (see section 0.4):

Definition: Let $\bar{S}$ be the surface obtained by patching together $\bar{U}$ and $S - \{O\}$ by means of the isomorphism

$$\pi_{|\bar{U}-\pi^{-1}(O)} : \bar{U} - \pi^{-1}(O) \simeq U - \{O\}$$

and still denote by π the morphism π extended by the identity on $S - \{O\}$: then $\pi : \bar{S} \longrightarrow S$ is called the *blowing-up of* O *on* S. We also say that $\bar{S}$ is obtained from S *by blowing up the point* O and that O is the *centre* of the blowing-up.

It is clear that, after blowing up O, $S - \{O\}$ remains essentially unmodified while to the point O it corresponds on $\bar{S}$ just a projective line, say $E = \pi^{-1}(O) = \{O\} \times \mathbb{P}_1$. Such a line E is called the *exceptional divisor of* π. The restriction of π,

$$\pi_{|\bar{S}-E} : \bar{S} - E \longrightarrow S - \{O\}$$

is an isomorphism. It may be noticed that the exceptional divisor E is contained in $V \cup V'$ and has equations $x = 0$ in V and $y = 0$ in V'. This in particular shows

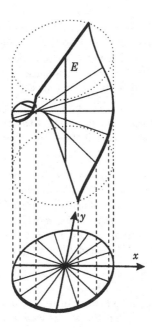

Figure 3.1: Blowing up.

that E is smooth at all its points and hence is a one-dimensional submanifold of $\bar{S}$, clearly isomorphic to a complex projective line and hence compact and connected. Note also that $\bar{S} - E$ is dense in $\bar{S}$, which will be often useful in the sequel.

The next proposition shows that blowing up O on S does not depend on the open subset U and the local coordinates x, y we have used in the definition:

Proposition 3.1.3 *Let* $\pi' : \bar{S}' \longrightarrow S$ *be a second blowing-up of O on S, obtained from an open neighbourhood U' of O and coordinates x', y' on it. Then there is a unique S-isomorphism $\varphi : \bar{S} \longrightarrow \bar{S}'$. Furthermore φ restricts to a linear projectivity between E and $E' = \pi'^{-1}(O)$.*

PROOF: Since both x, y and x', y' are local coordinates at O, we may determine a third open neighbourhood of O, $U'' \subset U \cap U'$, and analytic functions defined in U'', $a_{i,j}$, $i, j = 0, 1$, such that

$$x' = a_{0,0}x + a_{0,1}y$$
$$y' = a_{1,0}x + a_{1,1}y$$

and $\det(a_{i,j}(p)) \neq 0$ for all $p \in U''$. Then it is easy to see that

$$U'' \times \mathbb{P}_1 \longrightarrow U'' \times \mathbb{P}_1$$
$$(p, z_0, z_1) \mapsto (p, a_{0,0}z_0 + a_{0,1}z_1, a_{1,0}z_0 + a_{1,1}z_1)$$

is a U''-isomorphism that restricts to a U''-isomorphism ψ from $\pi^{-1}(U'')$ onto $\pi'^{-1}(U'')$. Then φ is obtained by extending ψ to $\bar{S}$ using $(\pi'_{|\bar{S}'-E'})^{-1} \circ \pi_{|\bar{S}-E}$.

For the uniqueness of φ it is enough to remark that the condition of S-morphism determines φ on $\bar{S} - E$, which is a dense subset of $\bar{S}$.

Lastly it is clear from the definition of φ that $\varphi_{|E}$ is a linear projectivity . ◇

We have thus seen not only that the blowing-up of O is intrinsically defined by S and O up to an S-isomorphism, but also that the projective structure of the exceptional divisor is intrinsic. We will show next that analytic isomorphisms lift to blown up surfaces.

Corollary 3.1.4 *Assume that $\pi : \bar{S} \longrightarrow S$ and $\hat{\pi} : \bar{S}' \longrightarrow S'$ are the blowing-ups of points O and O' on S and S', respectively. If $\varphi : W \longrightarrow W'$ is an analytic isomorphism between open neighbourhoods of O and O' and $\varphi(O) = O'$, then there is a unique isomorphism $\bar{\varphi} : \pi^{-1}(W) \longrightarrow \hat{\pi}^{-1}(W')$ such that $\varphi \circ \pi_{|\pi^{-1}(W)} = \hat{\pi} \circ \bar{\varphi}$. Furthermore $\bar{\varphi}$ restricts to a linear projectivity between the exceptional divisors $E = \pi^{-1}(O)$ and $E' = \hat{\pi}^{-1}(O')$.*

PROOF: The condition $\varphi \circ \pi_{|\pi^{-1}(W)} = \hat{\pi} \circ \bar{\varphi}$ determines $\bar{\varphi}$ on $\pi^{-1}(W) - E$, and hence also on the whole of $\pi^{-1}(W)$ because $\pi^{-1}(W) - E$ is dense in $\pi^{-1}(W)$. For the existence of $\bar{\varphi}$, let x, y be local coordinates in an open neighbourhood $U \subset W$ of O, and take $x \circ \varphi^{-1}$ and $y \circ \varphi^{-1}$ as local coordinates in $U' = \varphi(U)$. Using such coordinates to construct the blowing-ups, it is clear that $\varphi \times Id : U \times \mathbb{P}_1 \longrightarrow U' \times \mathbb{P}_1$ restricts to an isomorphism $\pi^{-1}(U) \simeq \hat{\pi}^{-1}(U')$ whose obvious extension to $\pi^{-1}(W)$ is $\bar{\varphi}$. The last part of the claim follows from the definition of $\bar{\varphi}$ above. ◇

Remark 3.1.5 It follows in particular from 3.1.4 that for any open neighbourhood W of O in S, $\pi^{-1}(W)$ may be canonically identified with the surface $\overline{W}$ obtained from W by blowing up O.

3.2   Transforming curves and germs

Let $\pi : \bar{S} \longrightarrow S$ be the blowing-up of O on S and let x, y be, as before, a system of local coordinates at O. From now on, in order to avoid confusions, if f is an analytic function defined in some open set $U \subset S$, we denote by $\bar{f}$ the function $f \circ \pi$ which is analytic in $\pi^{-1}(U)$. This applies in particular to the functions $x \circ \pi$ and $y \circ \pi$ that we have so far denoted by x and y.

Let ξ be a curve in an open subset W of S: we will denote by $\bar{\xi}$ the pull-back (see section 0.7) of ξ by π, $\bar{\xi} = \pi^*(\xi)$, and will call it the *total transform* of ξ (after blowing up O). As one may expect, if $O \in \xi$, then $E = \pi^{-1}(O)$ needs to be contained in $\bar{\xi} = \pi^*(\xi)$. More precisely:

Lemma 3.2.1 *The total transform of ξ has the form*

$$\bar{\xi} = \tilde{\xi} + e_O(\xi)E$$

where $\tilde{\xi}$ is a curve in $\pi^{-1}(W)$ with finitely many intersections with E.

PROOF: Assume that x, y are local coordinates at O and f an equation of ξ, all defined in a neighbourhood U of O. Consider, as in the preceding section, $\pi^{-1}(U)$ covered by the open subsets $V : \bar{x} \neq 0$ and $V' : \bar{y} \neq 0$. Write $f = f_e + f_{e+1} + \cdots + f_i + \dots$, where $e = e_O(\xi)$ and the f_i are forms of degree i in the local coordinates x, y. Using in V the coordinates $\bar{x}$ and $z = \bar{y}/\bar{x}$ (3.1.2), we have as equation of $\bar{\xi}$ in V:

$$\bar{f} = f_e(\bar{x}, \bar{x}z) + f_{e+1}(\bar{x}, \bar{x}z) + \cdots + f_i(\bar{x}, \bar{x}z) + \cdots$$
$$= \bar{x}^e f_e(1, z) + \bar{x}^{e+1} f_{e+1}(1, z) + \cdots + \bar{x}^i f_i(1, z) + \cdots.$$

Then $\tilde{f} = \bar{f}/x^e$ defines a curve in V with finitely many intersections with E, namely the points $(0, z)$ for which $f_e(1, z) = 0$. A similar computation shows that also y^e divides $\bar{f}$ in V' and $\tilde{f}' = \bar{f}/y^e$ defines in V' a curve with finitely many intersections with E. Since $\tilde{f}/\tilde{f}' = y^e/x^e$ has no zeros in $V \cap V'$, these curves patch together to give a curve in $\pi^{-1}(U)$, which in turn may be trivially extended to the wanted curve $\tilde{\xi}$ in the whole of $\pi^{-1}(W)$ by using the local equations of $\bar{\xi}$ in $\pi^{-1}(W) - E$. ◇

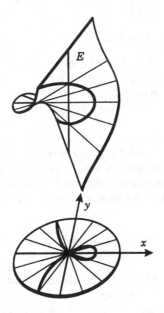

Figure 3.2: A node and its strict transform after blowing up

Definition: The *strict transform* $\tilde{\xi}$ of ξ by blowing up O is the curve obtained from its total transform $\bar{\xi}$ after removing the component $e_O(\xi)E$, that is:

$$\tilde{\xi} = \bar{\xi} - e_O(\xi)E.$$

Notice that it follows from 3.2.1 that the strict transform $\tilde{\xi}$ has no component E. We will see next that the intersection of the strict transform of ξ and the exceptional divisor depends only on the tangent cone to ξ.

Theorem 3.2.2 *There is a linear projectivity τ between the pencil of tangent lines to S at O and the exceptional divisor E of blowing up O, such that for any curve ξ on S, ξ has principal tangent ℓ at O if and only if its strict transform $\tilde{\xi}$ goes through $\tau(\ell)$. Moreover, the multiplicity of ℓ as a component of the tangent cone to ξ equals the intersection multiplicity $[\tilde{\xi}.E]_{\tau(\ell)}$.*

PROOF: Take the same notations as above. Define τ as mapping the tangent line at O of vector (z_0, z_1) to the point $p = (0, 0, (z_0, z_1)) \in E \subset U \times \mathbb{P}_1$, which clearly is a linear projectivity. Let us assume that $p \in V$ so that $z_0 \neq 0$, and p has coordinates $\bar{x} = 0$, and $z = z_1/z_0$. The proof if $p \in V'$ is similar.

Only the claim about multiplicities needs to be verified, as it implies the preceding one. Let us compute thus the intersection multiplicity: as seen in the proof of 3.2.1, if ξ has equation $f = f_e + f_{e+1} + \cdots + f_i + \cdots$ in a neighbourhood of O, $f_e \neq 0$, then $\tilde{\xi}$ has equation

$$f_e(1, z) + \bar{x} f_{e+1}(1, z) + \cdots + \bar{x}^{i-e} f_i(1, z) + \cdots$$

in a neighbourhood of p. Since E has equation $\bar{x}$ in V, the intersection multiplicity $[\tilde{\xi}.E]_p$ is the multiplicity of z_1/z_0 as a root of $f(1, z)$, which equals the multiplicity of the factor $z_1 Z_0 - z_0 Z_1$ in the form $f_e(Z_0, Z_1)$. This gives the claim, because f_e is the equation of the tangent cone. ◇

Remark 3.2.3 Just by the property claimed in 3.2.2, τ does not depend on the coordinates: $\tau(\ell)$ is the point of E the strict transforms of all curves that have principal tangent ℓ are going through. In the sequel we will refer to the point $\tau(\ell) \in E$ as the point corresponding to ℓ, or as the point corresponding to the direction of ξ if ξ is any curve or germ of curve whose only principal tangent at O is ℓ.

Remark 3.2.4 Once the coordinates have been taken as in the proof of 3.2.2, the point corresponding to the direction of $ax - by = 0$ is $(0, 0, (b, a))$.

After 3.2.2 one may understand the blowing-up as a transformation that removes O from S, glues on its place the whole projective line of the tangent directions to S at O and leaves unmodified the remaining points of S. In particular, on each curve ξ through O, the point O is substituted by the points corresponding to the principal tangents to ξ at O.

The reader may use 3.2.2 to show that the strict transform of a curve which has an ordinary singularity at O (0.11.4) is smooth in a neighbourhood of E: the strict transforms of its branches going through different points of E, we say that the branches have been separated by blowing up. That this is not the case for all curves may be seen by transforming a tacnode (0.11.3).

By adding-up the intersection multiplicities of the strict transform and the exceptional divisor at their common points one easily gets an equality which will be very useful later on, namely:

Corollary 3.2.5 *If ξ is an analytic curve in a neighbourhood of O, then*

$$e_O(\xi) = \sum_{p \in E} [\tilde{\xi}.E]_p$$

where E and $\tilde{\xi}$ are the exceptional divisor and the strict transform of ξ after blowing up O, respectively.

PROOF: Follows from 3.2.2 as $e_O(\xi)$ is the order of the tangent cone. ⋄

Remark 3.2.6 If a curve ξ is composed of curves ξ_1, ξ_2, then its total and strict transforms are composed of the total and strict transforms, respectively, of the curves ξ_i:

$$\bar{\xi} = \bar{\xi}_1 + \bar{\xi}_2$$
$$\tilde{\xi} = \tilde{\xi}_1 + \tilde{\xi}_2.$$

Indeed, the first equality was already seen in section 0.7, as the total transforms are just pull-backs, while the second equality easily follows from the first one using the additivity of multiplicity.

Since the coordinates may be arbitrarily chosen, the next theorem shows in particular that if ξ is irreducible at O, then its strict transform $\tilde{\xi}$ meets E at a single point p (as we already know from 2.2.6 and 3.2.2) and is irreducible at p. Assume that local coordinates x, y at O are chosen and use the same notations as before:

Theorem 3.2.7 *If ξ is irreducible at O, non-tangent to the y-axis and has Puiseux series*

$$s = \sum_{i \geq n} a_i x^{i/n}, \quad n = \nu(s),$$

then $\tilde{\xi}$ meets the exceptional divisor at a single point, say p, which belongs to V and has coordinates $\bar{x} = 0, z = a_n$. Furthermore $\tilde{\xi}$ is irreducible at p and has Puiseux series

$$\tilde{s} = \sum_{i > n} a_i \bar{x}^{\frac{i-n}{n}}$$

if $\bar{x}$ and $\tilde{z} = z - a_n$ are used as local coordinates at p.

PROOF: Take

$$f = \prod_{\epsilon^n = 1} \left(y - \sum_{i \geq n} a_i \epsilon^i x^{i/n} \right)$$

as a local equation of ξ at O. After a suitable shrinking of U, it is not restrictive to assume that f is defined in the whole of it. Since the initial form of f is $(y - a_n x)^n$, it is clear from 3.2.2 that $\tilde{\xi}$ meets E at a single point p and, by 3.2.4, p belongs to V and has coordinates $\bar{x} = 0$ and $z = a_n$ as claimed.

Furthermore, in V,

$$\bar{f} = \prod_{\varepsilon^n = 1} \left(\bar{x}z - \sum_{i \geq n} a_i \varepsilon^i \bar{x}^{i/n} \right)$$

$$= \bar{x}^n \prod_{\varepsilon^n = 1} \left(z - a_n - \sum_{i > n} a_i \varepsilon^i \bar{x}^{\frac{i-n}{n}} \right)$$

and thus

$$\tilde{f} = \prod_{\varepsilon^n = 1} \left(z - a_n - \sum_{i > n} a_i \varepsilon^{i-n} \bar{x}^{\frac{i-n}{n}} \right).$$

After taking $\bar{x}$ and $\tilde{z} = z - a_n$ as local coordinates at p, it is clear that the $\tilde{z}$-roots of $\tilde{f}$ are the conjugates of $\bar{s}$ and the proof is complete. $\diamond$

Corollary 3.2.8 *If γ is a curve irreducible at O, then π restricts to a bijection between $|\tilde{\gamma}|$ and $|\gamma|$.*

PROOF: Obvious, as $|\tilde{\gamma}| \cap E$ is a single point by 3.2.7 while π clearly restricts to a bijection between $|\tilde{\gamma}| - E$ and $|\gamma| - \{O\}$. $\diamond$

Corollary 3.2.9 *If ξ is smooth at O, then so is $\tilde{\xi}$ at p. If ξ is smooth, then also $\tilde{\xi}$ is smooth and furthermore π restricts to an isomorphism of one-dimensional complex manifolds between $\tilde{\xi}$ and ξ.*

PROOF: Take local coordinates x, y at O so that ξ is not tangent to the y-axis and notations as in 3.2.7. Then $1 = \nu(s) = n = \nu(\bar{s})$, which proves that $\tilde{\xi}$ is smooth at p and even transverse to the exceptional divisor (the reader may also prove this from 3.2.5). Therefore points corresponding by π in suitable neighbourhoods W' of p in $\tilde{\xi}$ and W of O in ξ have equal local coordinates $\bar{x} = x$ and so π induces, locally at p, an isomorphism of one-dimensional complex manifolds $W' \simeq W$. Once the local situation at p has been dealt with, it is enough to recall that $\pi_{|\bar{s}-E}$ is an isomorphism to complete the proof. $\diamond$

Corollary 3.2.10 *If ξ is a reduced curve on S, its strict transform $\tilde{\xi}$ is reduced too.*

PROOF: Obvious from 3.2.6, 3.2.7 and the fact that a blowing-up induces isomorphism outside of the exceptional divisor. $\diamond$

It is clear that two curves that agree in a neighbourhood of O have total and strict transforms that agree in a neighbourhood of E. Thus, if ξ is now a germ of curve at O, the germs at E of the total and strict transforms of any representative ξ' of ξ do not depend on the representative ξ'. If in particular a point $p \in E$ is chosen, then one may consider the germs at p, $\bar{\xi}'_p$ and $\tilde{\xi}'_p$, of the total and strict transforms of ξ': still they depend only on the germ ξ and not on ξ', so that we call them the *total* and *strict transforms of ξ at p* and we use for them the notations $\bar{\xi}_p$ and $\tilde{\xi}_p$, respectively, with direct reference to the germ ξ.

If E_p denotes the germ of E at p, then, obviously from 3.2.1, $\bar{\xi}_p = \tilde{\xi}_p + e_O(\xi)E_p$, E_p is not a branch of $\tilde{\xi}_p$ and $\tilde{\xi}_p$ is empty but for finitely many points $p \in E$.

Let ξ be a germ of curve and ℓ a tangent line to S at O. By putting together all branches of ξ with tangent line ℓ, we find that ξ has a unique decomposition $\xi = \xi_\ell + \zeta$ where ℓ is the only tangent line to ξ_ℓ and is not a tangent line to ζ. We will say that ξ_ℓ is the ℓ-*component* of ξ. Obviously $\xi_\ell = \emptyset$ if and only if ℓ is not a principal tangent to ξ and $\xi = \sum_\ell \xi_\ell$.

Assume, using the notations from 3.2.2, that $p = \tau(\ell)$. From 3.2.7 we obtain:

Corollary 3.2.11 *The germ $\tilde{\xi}_p$ depends only on the ℓ-component of ξ. Furthermore, the correspondence $\xi \mapsto \tilde{\xi}_p$ induces a bijection between the set of germs at O with only principal tangent ℓ and the set of the germs at p with no branch equal to E_p.*

PROOF: Use the notations as above. The first part of the claim is obvious because ζ_p is empty by 3.2.2. For the remaining part and since the decomposition of a germ into its components is unique, it is enough to prove that $\xi \mapsto \tilde{\xi}_p$ induces a bijection between the set of irreducible germs at O with principal tangent ℓ and the set of the irreducible germs at p different from E_p. For this, choose local coordinates in such a way that ℓ is tangent to the x-axis. If

$$s = \sum_{i > n} a_i x^{i/n}, \quad n = \nu(s),$$

is a Puiseux series of ξ, then, by 3.2.7,

$$\tilde{s} = \sum_{i > n} a_i \bar{x}^{\frac{i-n}{n}} = \bar{x}^{-1} s(\bar{x})$$

is a Puiseux series of $\tilde{\xi}_p$. Since it is clear that $s \mapsto \tilde{s}$ defines a bijection between the set of Puiseux series in x with order greater than one and the set of all Puiseux series in $\bar{x}$, and furthermore such a bijection preserves polydromy orders and conjugation of series, the claim follows. $\diamond$

Corollary 3.2.12 *Let ζ be an analytic curve defined in an open neighbourhood of E. If ζ does not contain E, then there is a curve ξ on S, defined in an open neighbourhood of O, such that $\tilde{\xi}$ and ζ agree in a neighbourhood of E. Furthermore, the germ of ξ at O is uniquely determined by this property.*

PROOF: We will prove first that ζ and E meet at finitely many points. Let K be the subset of E of the points p such that there is an open neighbourhood U of p in $\bar{S}$ and a local equation of ζ in U whose restriction to $U \cap E$ is zero. K is an open set in E by its own definition, we will show that it is also closed. If $p' \in E$, take as U' an open neighbourhood of p' in $\bar{S}$ such that $E \cap U'$ is connected and a local equation f' of ζ is defined in U'. Assume that there is a $p \in K \cap U'$: then there are a neighbourhood U of p in S and an equation f of ζ in U that is identically zero in $U \cap U' \cap E$. Since in $U \cap U'$, $f' = uf$, with u invertible, also f' is identically zero in $U \cap U' \cap E$. Now $E \cap U'$ is a connected one-dimensional analytic manifold and the restriction of f' to $U' \cap E$ is analytic and takes the value zero for all points in the non empty open subset $U \cap U' \cap E$: thus f' identically vanishes in the whole of $E \cap U'$ and $p' \in K$ as wanted. Since E is connected and $K = E$ obviously implies that ζ contains E, necessarily $K = \emptyset$. Let now p be a point on both E and ζ. Since E is smooth and hence irreducible at p, ζ and E cannot share a branch at p, otherwise $p \in K$. Then we know from 2.3.3 that p is an isolated intersection point and hence, by the compactness of E, the number of intersection points is finite.

Write $p_i = \tau(\ell_i)$, $i = 1, \ldots, s$, the points of $E \cap \zeta_{red}$, and denote by ζ_i the germ of ζ at p_i. By 3.2.11, for each $i = 1, \ldots, s$, we may take a curve ξ_i defined in a neighbourhood of O with a single tangent ℓ_i at O and such that $\tilde{\xi}_i$ represents ζ_i: then the curve $\xi = \xi_1 + \cdots + \xi_s$ will do the job, because each $\tilde{\xi}_i$ has no intersection with E other than p_i by 3.2.2. Lastly, for the uniqueness of the germ ξ_O of ξ, it is enough to remark that the condition of the claim determines the germ of $\tilde{\xi}$ at each $p \in E$, and hence, by 3.2.11, also the ℓ-component of ξ_O for each tangent line ℓ at O. $\diamond$

Remark 3.2.13 Since the restriction of the blowing-up to $S - E$ is an isomorphism, one may easily extend the curve ξ of 3.2.12 in such a way that $\tilde{\xi} = \zeta$ in the whole set of definition of ζ.

In the sequel we will say that the germ ξ_O (and sometimes also the curve ξ) has been obtained from ζ by *blowing down* E.

We close this section with a characterization of the curves on $\bar{S}$ that are total transforms of curves on S. The proof, which is easy from 3.2.5 and 3.2.12, is left to the reader. A remark similar to 3.2.13 above may be made.

Corollary 3.2.14 *If ζ is a curve on $\bar{S}$, defined in a neighbourhood of E, put $\zeta = \zeta' + eE$, where $e \geq 0$ and ζ' is assumed to have no component E. Then there exists ξ on S whose total transform agrees with ζ in a neighbourhood of E, if and only if $e = \sum_{p \in E}[\zeta'.E]_p$. In such a case the germ of ξ at O is uniquely determined by the germ of ζ at E.*

3.3   Infinitely near points

Still let O be a point on a smooth surface S. The exceptional divisor E of blowing up O on S will be called the *first infinitesimal neighbourhood of O on*

S. Consequently, its points will be called the *points in the first infinitesimal neighbourhood of O (on S)*.

Since blowing up a point on a smooth surface gives rise to a smooth surface, we may continue defining using induction: if $i > 0$, the *points in the i-th infinitesimal neighbourhood of O (on S)* are, by definition, the points in the first infinitesimal neighbourhood of some point in the $(i-1)$-th infinitesimal neighbourhood of O.

In the sequel we will often drop the word infinitesimal by saying just neighbourhood instead of infinitesimal neighbourhood: this will make no confusion with the ordinary neighbourhoods of O on S in the topological sense, the meaning of the word neighbourhood being clear in each case from the context. Sometimes we will refer to O as being the only point in its own 0-th neighbourhood.

The points which are in the i-th neighbourhood of O, for some $i > 0$, are also called *points infinitely near to O*. Sometimes the points on S will be called *ordinary points* or also *proper points* in order to distinguish them from the infinitely near ones, as the word point will be used for both kinds of point.

From now on, we will put $\mathcal{N}_O^*$ for the set of points infinitely near to O, and $\mathcal{N}_O = \mathcal{N}_O^* \cup \{O\}$. If $p, q \in \mathcal{N}_O$, we will say that p *precedes* q, and write $p < q$, if and only if q is infinitely near p. We will write $p \leq q$ in case of q being equal or infinitely near to p. It is clear that $\leq$ is an ordering: it will be called the *natural ordering* of the infinitely near points.

It should be noticed that a point p infinitely near to O is an ordinary point on a well determined smooth surface S_p which is related to S by the composition π_p of the sequence of blowing-ups giving rise to p:

$$\pi_p : S_p = S^{(i)} \longrightarrow S^{(i-1)} \longrightarrow \cdots \longrightarrow S^{(1)} = \bar{S} \longrightarrow S.$$

In the sequel each local ring $\mathcal{O}_{S_p,p}$ will be considered as an extension of $\mathcal{O}_{S,O}$, and hence also as an $\mathcal{O}_{S,O}$-module, via the pull-back monomorphism

$$\pi_p^* : \mathcal{O}_{S,O} \longrightarrow \mathcal{O}_{S_p,p}.$$

Let us denote by p_j the centre of $S^{j+1} \longrightarrow S^j$ above, so that $p_0 = O$ and $p_1, \ldots, p_{i-1}$ are the infinitely near points preceding p. Assume that $\varphi : S \longrightarrow T$ is an analytic isomorphism from S onto on an open neighbourhood of $O' = \varphi(O)$ on another surface T. By 3.1.4, $\varphi = \varphi_0$ lifts to a uniquely determined sequence of isomorphisms φ_j, $j = 1, \ldots, i$, giving rise to a commutative diagram

$$
\begin{array}{ccccccccc}
S^i & \longrightarrow & S^{i-1} & \longrightarrow & \ldots & \longrightarrow & S^1 & \longrightarrow & S \\
\varphi_i \downarrow & & \varphi_{i-1} \downarrow & & & & \varphi_1 \downarrow & & \varphi \downarrow \\
T^i & \longrightarrow & T^{i-1} & \longrightarrow & \ldots & \longrightarrow & T^1 & \longrightarrow & T
\end{array}
$$

where each $T^j \longrightarrow T^{j-1}$ is the blowing-up of $\varphi_{j-1}(p_{j-1})$. In particular we say that φ_i is the isomorphism induced by φ on S_p. If $\psi : T \longrightarrow T'$ is a second isomorphism, obviously we have, because of the uniqueness of the induced morphisms, $(\psi \circ \varphi)_i = \psi_i \circ \varphi_i$. In the sequel we will write $\varphi(p) = \varphi_i(p)$ thus extending the action of φ to infinitely near points. Furthermore, if the ordinary points on S are identified with their images by φ, then we will also identify the points infinitely near to O and those infinitely near to O' through the corresponding liftings of φ. In particular if S is an open neighbourhood of $O' = O$ in T, we will make no difference between the points infinitely near to O on S and the points infinitely near to O on T.

Still let p be a point infinitely near to O. If ξ is an analytic curve defined in an open neighbourhood of O in S, one may consider the successive strict (resp. total) transforms of ξ by the blowing-ups composing π_p until we reach a curve on S_p whose germ at p will be denoted by $\tilde{\xi}_p$ (resp. $\bar{\xi}_p$) and called the *strict* (resp. *total*) *transform of* ξ *with origin at* p. We say that ξ *goes through* or *contains* p (or also that p *lies on* or *belongs to* ξ) if and only if the germ $\tilde{\xi}_p$ is non-empty. The strict transforms of an empty germ being empty, if p belongs to ξ, then all points preceding p belong to ξ too. We put $\mathcal{N}_O(\xi)$ for the set of points equal or infinitely near to O that lie on ξ. The reader may notice that $\mathcal{N}_O(\xi) = \mathcal{N}_O(\xi_{red})$.

We define the *multiplicity* of ξ at p (also called *multiplicity of p on ξ*) as being the multiplicity of $\tilde{\xi}_p$: it we will be written $e_p(\xi)$. Hence, by definition, $e_p(\xi) = e_p(\tilde{\xi}_p)$ and p is called an $e_p(\xi)$-fold (infinitely near) point of ξ. In particular ξ goes through p if and only if $e_p(\xi) > 0$. If $e_p(\xi) > 1$ (resp. $e_p(\xi) = 1$), p is called a *multiple* (resp. *simple*) (infinitely near) point of ξ.

Since, clearly, all the above notions depend only on the germ of the curve at O, in the sequel they will be freely applied to germs of curve as well. If ξ is now a germ of curve with origin at O, we shall write $e_p(\xi)$, $\bar{\xi}_p$ and $\tilde{\xi}_p$ for the *multiplicity*, the *total transform* and the *strict transform* of the germ ξ at p, all of them being defined as those of any representative of ξ. We will also write $\mathcal{N}(\xi)$ for the set all points on the germ ξ equal or infinitely near to its origin.

If φ is an isomorphism defined in a neighbourhood of O, obviously we have $e_p(\xi) = e_{\varphi(p)}(\varphi(\xi))$ and $\varphi(\mathcal{N}_O(\xi)) = \mathcal{N}_O(\varphi(\xi))$.

It is also clear from its definition that the multiplicity at an infinitely near point is additive:

$$e_p(\xi_1 + \xi_2) = e_p(\xi_1) + e_p(\xi_2).$$

Since the strict transform of a smooth curve is also smooth (3.2.9), if ξ is smooth at O, then $e_p = 1$ for all p equal or infinitely near to O on ξ.

If ξ is an irreducible germ at O, it follows from 2.2.6 and 3.2.2 that there is a single point on ξ in each neighbourhood of O. Therefore, the points in $\mathcal{N}(\xi)$ are sequentially ordered by the natural ordering and so its multiplicities on ξ still make a sequence which is currently named the *sequence of multiplicities* of ξ.

The ordered set $\mathcal{N}(\xi)$, ξ a germ of curve, may be represented by an indefinite tree-graph which has a vertex for each point on ξ and an edge joining two vertices if and only if they correspond to points one of which is in the first neighbourhood of the other. The vertex corresponding to O is taken as the basis or root of the tree. This graph is often called the *tree* of the singularity. Currently it is weighted by attaching to each vertex the multiplicity on ξ of the corresponding point. Figure 3.3 shows an example that the reader may check as exercise. We will deal more extensively with graphic representations of singularities in the forthcoming section 3.9.

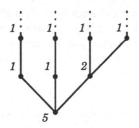

Figure 3.3: The tree of $xy(x - y)(x^3 - y^2) = 0$

The basic property of the infinitely near points is that they 'explain' the intersection multiplicity of two curves or germs by means of a formula due to M. Noether:

Theorem 3.3.1 (Noether's formula) *Let ξ and ζ be are analytic curves defined in a neighbourhood of O. The intersection multiplicity $[\xi.\zeta]_O$ is finite if and only if ξ and ζ share finitely many points infinitely near to O, and in such a case*

$$[\xi.\zeta]_O = \sum_{p \in \mathcal{N}_O(\xi) \cap \mathcal{N}_O(\zeta)} e_p(\xi) e_p(\zeta).$$

Obviously one may leave the summation in the claim to run for $p \in \mathcal{N}_O$ as well. Let us make two remarks before proving 3.3.1:

Remark 3.3.2 If two curves ξ, ζ share no branches at O, then $[\xi.\zeta]_O < \infty$ by 2.6.5, and so they share finitely many points infinitely near to O, by 3.3.1. In particular different branches of a germ share finitely many infinitely near points.

Remark 3.3.3 The Noether formula may be understood as being still true if the curves have a common branch at O, both sides having the value ∞ in such a case.

PROOF OF 3.3.1: If $[\xi.\zeta]_O = \infty$ then ξ and ζ have a common branch at O (2.6.5) and so they share all infinitely near points on such a branch. Otherwise, i.e., if $[\xi.\zeta]_O \neq \infty$, the claim easily follows from the next lemma using induction on $[\xi.\zeta]_O$. ◇

Lemma 3.3.4 *If ξ and ζ are curves in a neighbourhood of O, and $\tilde{\xi}$ and $\tilde{\zeta}$ are their respective strict transforms after blowing up O, we have*

$$[\xi.\zeta]_O = e_O(\xi) e_O(\zeta) + \sum_{p \in E} [\tilde{\xi}.\tilde{\zeta}]_p,$$

the summation running on the points in the first neighbourhood E of O.

PROOF OF 3.3.4: As before, the case $[\xi.\zeta]_O = \infty$ is obvious. Also the case in which one of the curves does not go through O is clear, because such a curve has no points in the first neighbourhood of O (by 3.2.2). We assume thus $0 < [\xi.\zeta]_O < \infty$, and, furthermore, since both sides of the formula in the claim are additive, that one of the curves, say ζ, is irreducible at O.

Assume that local coordinates x, y at O are taken in such a way that ζ is tangent to the x-axis, and therefore the (only) point on ζ in the first neighbourhood of O, say p, corresponds to the direction of the x-axis. Then a Puiseux series of ζ has the form

$$s = a_m x^{m/n} + \cdots, \quad m > n, \quad n = \nu(s) = e_O(\zeta).$$

Let f be an equation of ξ and $e = e_O(\xi)$. We use $\bar{x}$ and $z = \bar{y}/\bar{x}$ as local coordinates at p, so that, by 3.2.7, the Puiseux series of the strict transform of ζ is

$$\tilde{s}(\bar{x}) = \bar{x}^{-1} s(\bar{x}).$$

The equation $\tilde{f}$ of the strict transform of ξ being related to f by the equality

$$\bar{x}^e \tilde{f}(\bar{x}, z) = f(\bar{x}, \bar{x}z),$$

one may compute

$$[\xi.\zeta]_O = no_x(f(x, s(x))) = no_{\bar{x}}(f(\bar{x}, s(\bar{x}))) =$$
$$= no_{\bar{x}}(f(\bar{x}, \bar{x}\tilde{s}(\bar{x}))) = ne + no_{\bar{x}}(\tilde{f}(\bar{x}, \tilde{s}(\bar{x}))) =$$
$$= ne + [\tilde{\xi}.\tilde{\zeta}]_p$$

as wanted. ◇

Sometimes the formula in 3.3.4, which is essentially equivalent to that in 3.3.1, is also called Noether's formula.

It should be noticed that 2.6.7 is an easy consequence of the Noether formula, which gives a far more precise determination of the intersection multiplicity.

Let ξ be a germ of curve with origin at O. Define the *branches* of $\mathcal{N}(\xi)$ as being its maximal totally ordered subsets.

Corollary 3.3.5 *By mapping each branch γ of a germ of curve ξ to $\mathcal{N}(\gamma)$, we get a one to one map between the branches of ξ and those of $\mathcal{N}(\xi)$.*

PROOF: $\mathcal{N}(\gamma)$ is clearly a totally ordered subset of $\mathcal{N}(\xi)$. Furthermore it is a maximal one as, if p in the i-th neighbourhood of O does not belong to γ, then we take q be the point on γ in the i-th neighbourhood of O and no one of the relations $p = q$, $p < q$ and $p > q$ may hold true.

Assume that $T \subset \mathcal{N}(\xi)$ is non-empty and totally ordered. If T is finite let p be the maximal point of T and γ any branch of ξ through p: since any point preceding p still belongs to γ, clearly $T \subset \mathcal{N}(\gamma)$. Otherwise, since the set of points in $\mathcal{N}(\xi)$ that belong to two or more branches is finite (by 3.3.2), we take a point p in T belonging to a single branch γ of ξ: as above all points in T preceding p belong to γ. Since no point in T after p can belong to a branch of ξ other than γ, we get $T \subset \mathcal{N}(\gamma)$. This proves that the map is onto, while its injectivity directly follows from 3.3.2. $\diamond$

The infinitely near points on a reduced germ determine it, as the next corollary shows.

Corollary 3.3.6 *If ξ, ζ are reduced germs of curve, then $\xi = \zeta$ if and only if $\mathcal{N}(\xi) = \mathcal{N}(\zeta)$.*

PROOF: The converse being obvious, assume that $\mathcal{N}(\xi) = \mathcal{N}(\zeta)$. By 3.3.5, for any branch γ of one germ there is a branch γ' of the other such that $\mathcal{N}(\gamma) = \mathcal{N}(\gamma')$. Then, by 3.3.2, $\gamma = \gamma'$, both germs have the same branches and the claim follows, as a reduced germ is the sum of its branches. $\diamond$

3.4 Enriques' definition of infinitely near points

The Noether formula shows the infinitely near points playing the role they are mainly introduced for, namely to give a geometric interpretation of the intersection multiplicity as a sum of contributions of all (ordinary or infinitely near) points shared by the curves or germs.

Enriques ([35], IV.I) gave an alternative definition of infinitely near points (*per abstrazione*) that directly points to the Noether formula. When introduced, such definition had the added advantage of being intrinsic, while other classical definitions make use of global birational transformations, all more or less locally equivalent to a blowing-up but depending on many arbitrary choices (see exercise 3.1 for a description of the most usual of these transformations).

Although Enriques' definition is not really useful today, we will briefly explain it in this section in order to facilitate a deeper understanding of infinitely near points. We refer to [83] for further details.

Let $\mathcal{G}$ be the set of all irreducible germs of curve with origin at O. Define an equivalence relation in $\mathcal{G}$ by the rule: $\gamma \sim \gamma' \iff [\gamma.\gamma'] > e_O(\gamma)e_O(\gamma')$. Each equivalence class is called a point in the first neighbourhood of O, and the quotient set the first neighbourhood of O. We say that an irreducible germ γ goes through a point p in the first neighbourhood of O if and only if γ is a representative of p. The multiplicity of γ at p is then defined by the rule

$$e_p(\gamma) = \min_{\gamma' \in p}([\gamma.\gamma'] - e_O(\gamma)e_O(\gamma')).$$

After this one may iterate to obtain the points in the second neighbourhood of O: for each fixed point p in the first neighbourhood of O we define in p, which is the set of all irreducible germs through itself, an equivalence relation by the rule $\gamma \sim_p \gamma' \iff [\gamma.\gamma'] > e_O(\gamma)e_O(\gamma') + e_p(\gamma)e_p(\gamma')$. The equivalence classes of

$\sim_p$ are now the points in the first neighbourhood of p. The multiplicity of an irreducible germ at a given q in the first neighbourhood of p is then defined as

$$e_q(\gamma) = \min_{\gamma' \in q}([\gamma.\gamma'] - e_O(\gamma)e_O(\gamma') - e_p(\gamma)e_p(\gamma')).$$

Points in the further neighbourhoods of O are defined inductively in the same way. The notions introduced for irreducible germs are easily extended to the non-irreducible ones and the Noether formula becomes then an obvious consequence of the definition. Nevertheless, other geometric facts about infinitely near points (as, for instance, the surfaces they are lying on) become far less evident due to the abstract character of the definition.

3.5   Proximity

A very important notion regarding relative position of infinitely near points is introduced next:

Definition: Let p, q be points equal or infinitely near to O. The point q is said to be *proximate to* p if and only if it belongs, as an ordinary or infinitely near point, to the exceptional divisor E^p of blowing up the point p.

Remark 3.5.1   1. It may be equivalently said that the points proximate to p are all points in the first neighbourhood of p, and all points in the second and successive neighbourhoods of p which lie on one of the successive strict transforms of E^p.

2. It is clear from the definition that a point proximate to p lies either in the first neighbourhood of p or in the first neighbourhood of a point proximate to p.

3. Since E^p is smooth, it is in particular irreducible at any of its points and hence in the first neighbourhood of each point proximate to p there is just a single point proximate to p.

4. The relation of proximity is obviously invariant under isomorphisms: if $\varphi : S \longrightarrow S'$ is an isomorphism of surfaces and $O \in S$, then q is proximate to p if and only if $\varphi(q)$ is proximate to $\varphi(p)$.

Example 3.5.2 The reader may easily see that the points in the first and second neighbourhood of the origin on $y^2 = x^3$ are both proximate to the origin.

The notion of proximity enables us to state the law according to which the multiplicities of the points on a germ may occur:

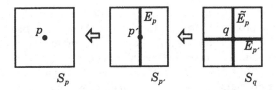

Figure 3.4: Blowing up a point p and then p' in the first neighbourhood of p we get a single point q still proximate to p.

Theorem 3.5.3 (Proximity equalities) *For any (ordinary or infinitely near) point p on a germ of curve ξ,*

$$e_p(\xi) = \sum_q e_q(\xi)$$

where the summation runs on all points q proximate to p on ξ.

PROOF: Using corollary 3.2.5 at the point p, we have

$$e_p(\xi) = e_p(\tilde{\xi}_p) = \sum_{q \in E^p} [E^p . \tilde{\tilde{\xi}}_p]_q$$

where E^p is the exceptional divisor of blowing up p and $\tilde{\tilde{\xi}}_p$ the strict transform, after blowing up p, of any representative of $\tilde{\xi}_p$. Since E^p is smooth, all points on E^p are simple. Then it is enough to compute each intersection multiplicity $[E^p . \tilde{\tilde{\xi}}_p]_q$ using the Noether formula 3.3.1. ⋄

As an obvious consequence of 3.5.3 we have:

Corollary 3.5.4 *For any (ordinary or infinitely near) point p on a germ ξ,*

$$e_p(\xi) \geq \sum_q e_q(\xi),$$

the summation running on the points q on ξ in the first neighbourhood of p. If in particular ξ is an irreducible germ, then its sequence of multiplicities is a non-increasing one.

The proximity equality 3.5.3 gives far more information than 3.5.4, as is shown next. Assume that γ is an irreducible germ and let p, q be points on γ, q in the first neighbourhood of p. We know from 3.5.4 that $e_p(\gamma) \geq e_q(\gamma)$. Assume that this inequality is strict, i.e., $e_p(\gamma) > e_q(\gamma)$: then γ goes through so many points $q_1, \ldots, q_r$ proximate to p in the first, $\ldots$, r-th neighbourhoods of q as for having

$$e_p(\gamma) - e_q(\gamma) = \sum_{i=1}^{r} e_{q_i}(\gamma).$$

Necessarily $r \geq 1$ and the points $q_1, \ldots, q_r$ are determined by p and q: each q_i is the only point proximate to p in the first neighbourhood of q_{i-1} ($q_0 = q$). Thus, because of the inequality $e_p(\gamma) > e_q(\gamma)$, γ needs to go through a certain number of points in the successive neighbourhoods of q which are in a well defined position with respect to p and q. We will come back on this argument in a while, after obtaining more information on proximity.

Assume that we have a finite sequence of blowing-ups

$$S_i \xrightarrow{\pi_i} S_{i-1} \to \ldots \xrightarrow{\pi_2} S_1 \xrightarrow{\pi_1} S_0 = S$$

each π_j being the blowing-up of a point p_{j-1} on S_{j-1}, $p_0 = O$. We do not assume that each p_j belongs to the exceptional divisor of the preceding blowing-up. Let $\pi = \pi_1 \circ \cdots \circ \pi_i$ be the composite morphism and $F = \bigcup_{j=1}^{i} (\pi_j \circ \cdots \circ \pi_i)^{-1}(p_{j-1})$: we call F the *exceptional divisor* of π. Notice that $F = \pi^{-1}(O)$ if all the p_j are infinitely near to O, in particular if $i = 1$, F is the exceptional divisor of a blowing-up as already defined.

Proposition 3.5.5 *The exceptional divisor F is the union of finitely many smooth curves, each isomorphic to a projective line. These curves will be called the components of the exceptional divisor, any two of them either do not meet or meet transversally at a single point and no three have a common point.*

PROOF: We use induction on the number i of blowing-ups, the case $i = 1$ being clear from the definition of blowing up.

Assume that $i > 1$, take $\psi = \pi_1 \circ \cdots \circ \pi_{i-1}$ and let F' be the exceptional divisor of ψ, so that we have $\pi = \psi \circ \pi_i$. Then, since F is composed of the exceptional divisor E of π_i and the strict transform $\tilde{F}'$ of F' by π_i, most of the job is done: E is clearly isomorphic to a projective line and so are all components of $\tilde{F}'$ because each of them is isomorphic to its image by π_i (using either 3.2.9 or the fact that π_i induces isomorphism outside of E) and by induction we know the claim to be true for F'.

Using again the induction hypothesis and that π_i restricts to an isomorphism $S_i - E \simeq S_{i-1} - \{p_{i-1}\}$, we need only take care of the intersections of E with the other components of F. If $p_{i-1} \notin F'$, then $\tilde{F}' \cap E = \emptyset$ (by 3.2.2) and the proof is complete. Otherwise, because of the induction hypothesis, p_{i-1} may belong either to one or two components of F', such components being transverse at p_{i-1} in the latter case. If p_{i-1} belongs to a single component of F', say G, then $\tilde{G}$ is the only component of $\tilde{F}'$ that meets E, and furthermore, since p_{i-1} is simple on G, by 3.2.5, $\tilde{G} \cap E$ is a single point and the intersection is transverse. If p_{i-1} belongs to two components, G_1 and G_2, of F', then $\tilde{G}_1$ and $\tilde{G}_2$ are the only components of $\tilde{F}'$ that meet E. Each intersection consists of a single point and is transverse, again by 3.2.5. Furthermore, by 3.2.2, the intersection points are different because G_1 and G_2 meet transversally at p_{i-1}. ◇

Remark 3.5.6 The proof of 3.5.5 shows that each component of F is the iterated strict transform of (and hence isomorphic to) the exceptional divisor

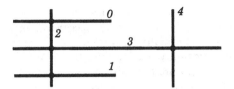

Figure 3.5: The exceptional divisor of successively blowing up $O = (0,0)$ and the points $p_1, \ldots, p_4$ in its first, ... , fourth neighbourhood on $y^2 - x^3 = 0$. The numbers indicate which point each component is corresponding to, according to 3.5.6.

of one of the blowing-ups composing π. Thus, each component of F may be identified with the first neighbourhood of one of the points p_j we have blown up. Nevertheless, since by definition the first neighbourhoods of different points are disjoint, the reader should notice that, in general, the former identifications do not patch together to give an identification of F to the union of the first neighbourhoods of all points we have blown up. See exercise 4.5 for further information.

In particular 3.5.5 and 3.5.6 apply to the sequence of blowing-ups giving rise to a given infinitely near point q. Then $S_i = S_q$ and $\pi = \pi_q$, using the notations of section 3.3, and we obtain:

Corollary 3.5.7 *If q is infinitely near to O, then q is proximate to just one or two points equal or infinitely near to O.*

PROOF: Assume that the points preceding q are $O = p_0, p_1, \ldots, p_{i-1}$, $i \geq 1$. Keep the notations from the proof of 3.5.5 and denote by E_j the exceptional divisor of π_j, $j = 1, \ldots, i$. Then $q \in E = E_i$ and by 3.5.5 it may belong to at most one of the components of $\tilde{F}'$, which are the strict transforms of the E_j, $j < i$. Thus q is proximate to p_{i-1} and to at most one of the p_j, $j < i - 1$, as claimed. ◊

Now, we may be more precise about the sequence of multiplicities of an irreducible germ of curve.

Theorem 3.5.8 *Assume that γ is an irreducible germ and p and q are points on γ, q in the first neighbourhood of p. Write $n = e_p(\gamma)$, $n' = e_q(\gamma)$ and perform the Euclidean division*

$$n = hn' + r, \quad 0 \leq r < n'.$$

Then the points $q_1, \ldots, q_h$ in the first, ... , h-th neighbourhoods of q and proximate to p belong to γ with the multiplicities

$$e_{q_i}(\gamma) = n', \quad i = 1, \ldots, h-1$$
$$e_{q_h}(\gamma) = r.$$

Moreover, no point proximate to p other than $q, q_1, \ldots, q_h$ belongs to γ.

PROOF: It is clear from 3.5.3 that there are finitely many points proximate to p on γ. Write $q = q_0, q_1, \ldots, q_{h-1}$ for the successive points on γ which are proximate to p and have on γ multiplicity n', $h \geq 1$.

It follows from the proximity equality 3.5.3 that $n \geq hn'$. If $n = hn'$, then, still by 3.5.3, $q_0, q_1, \ldots, q_{h-1}$ are the only points proximate to p on γ and the claim is satisfied. Thus, we assume in the sequel $n > hn'$. Let q_h be the point on γ in the first neighbourhood of q_{h-1}. Since $n > hn'$, again by 3.5.3, q_h is proximate to p too. Take $r = e_{q_h}(\gamma)$, $r < n'$ by hypothesis while 3.5.3 now gives $n \geq hn' + r$. If $n = hn' + r$ the claim follows using 3.5.3 once again, so we end the proof by proving that $n > hn' + r$ may not occur.

Indeed, assume that $n > hn' + r$ and denote by q_{h+1} the point on γ in the first neighbourhood of q_h. This inequality and 3.5.3 force q_{h+1} to be proximate to p. Since $n' > r$, then, also by 3.5.3, q_{h+1} is proximate to q_{h-1} and, obviously, q_{h+1} is proximate to q_h too, against 3.5.7. ◇

Remark 3.5.9 Theorem 3.5.8 gives a necessary condition for a non-increasing sequence of integers $\{n_i\}$ to be the sequence of multiplicities of an irreducible curve: if $n_i > n_{i+1}$, then $n_{i+1} = n_{i+2} = \cdots = n_{i+j}$ as far as $n_{i+1} + n_{i+2} + \cdots + n_{i+j} \leq n_i$ and, furthermore, if this inequality is strict and $(j+1)n_{i+1} > n_i$, then, necessarily, $n_{i+j+1} = n_i - n_{i+1} - \cdots - n_{i+j}$. It will follow from results in the forthcoming chapters 4 and 5 that, conversely, if all these conditions are satisfied and, furthermore, $n_i = 1$ for all i large enough, then the sequence (n_i) is the sequence of multiplicities of an irreducible curve (see exercises 4.4 and 5.4).

Note on space curves: The sequence of multiplicities may be similarly defined for an irreducible space curve, using blowing-ups of its ambient space. The main difference with plane curves comes from 3.5.8, which need not be true for space curves. Classical Italian geometricians (Enriques and Chisini [35], Vol. 2, p. 558, for instance) claimed that the sequence of multiplicities of an irreducible germ in a three-space agrees with that of its generic plane projection. It was Zariski ([92] I.3) who first pointed out this statement to be false. The example of Zariski, $\gamma : x = t^4, y = t^6 + t^7, z = t^7$, has sequence of multiplicities $4, 2, 1, 1, \ldots$, so that it follows from 3.5.9 that there is no plane germ (and thus no plane projection of γ) with such a sequence of multiplicities. Rather surprisingly, the same germ γ has its multiplicity sequence computed in Enriques and Chisini's book ([35], Vol. 2, p. 573). In fact the Italian statement fails to be true even for space curves whose sequence of multiplicities satisfies the necessary condition of 3.5.9 (Vicente [84], see also [18] and [17]).

3.6 Free and satellite points

Corollary 3.5.7 allows an infinitely near point p to be proximate to either one or two of its preceding points, that is, in other words, to lie on one or two components of the exceptional divisor of π_p. This makes an intrinsic distinction between infinitely near points:

Definition: A point p infinitely near to O is a *free point* if and only if it is proximate to just one point equal or infinitely near to O. Otherwise, if p is proximate to more than one point (i.e., to exactly two points, by 3.5.7), p is called a *satellite point*.

One may equivalently say that p is a free (resp. satellite) point if and only if it is a simple (resp. double) point of the exceptional divisor of the composition of blowing-ups π_p. For instance the point q in figure 3.4 is a satellite point.

Sometimes we will say that a satellite point is a *satellite of* the last free point that precedes it.

The notions of free and satellite points are obviously invariant under isomorphisms (by 3.5.1), but they depend on the point O: the same point p may be satellite if considered as infinitely near to O and free if considered as infinitely near to another point O', $O < O' < p$.

The next proposition directly follows from the above definition:

Proposition 3.6.1 *All points in the first neighbourhood of O are free points. There is a single satellite point in the first neighbourhood of each free point, while the first neighbourhood of each satellite point contains just two satellite points.*

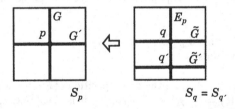

Figure 3.6: The two satellite points q, q' in the first neighbourhood of a satellite point p. G and G' are the components of the exceptional divisor through p.

An easy consequence of the proximity equalities is:

Proposition 3.6.2 *If q is a satellite point on a germ ξ, then q is proximate to some multiple point on ξ. In particular there are no satellite points on a smooth germ.*

PROOF: Assume that q is in the first neighbourhood of p', and that it is proximate to p, $p \neq p'$: then p' is also proximate to p (3.5.1) and by 3.5.3, $e_p(\xi) \geq e_{p'}(\xi) + e_q(\xi) \geq 2$. $\diamond$

The notions of free and satellite point were first introduced by Enriques ([35], IV.I), related to the analysis of the infinitely near points on an irreducible curve from its Puiseux series. According to Enriques, the main difference between free and satellite points is the type of condition a germ of curve must satisfy in order to go through them. Assume that p is infinitely near to O and q is in the first neighbourhood of p. Denote by γ an irreducible germ with origin at O and going through p. We will see later on (section 5.7) that if q is free, then γ goes through q if and only if certain power of one of the coefficients of its Puiseux series has a well determined value. Because of this Enriques says that the position of q depends on a continuous parameter (the above coefficient) and calls q a free point. By contrast, if q is a satellite point, γ goes or goes not through q depending on the multiplicities of γ at the preceding points: to be precise, if q is also proximate to p_1 and $p_2, \ldots, p_s = p$, $s > 1$, are the points between p_1 and q, γ goes through q if and only if $e_{p_1}(\gamma) > e_{p_2}(\gamma) + \cdots + e_{p_s}(\gamma)$ (3.5.3). Since going through satellite points depends only on discrete invariants, Enriques understood these points as rigidly joined to their preceding points and so he called them satellite points of the last free point they are infinitely near to.

3.7   Resolution of singularities

The theorem we will prove in this section implies that one may transform any reduced compact curve into a smooth one by means of a finite sequence of blowing-ups. This is one of the oldest and best known results on singularities of curves (and certainly the most popular with those that do not like to deal with singularities).

Theorem 3.7.1 *A reduced germ of curve contains at most finitely many multiple infinitely near points.*

PROOF: Let $\xi = \gamma_1 + \cdots + \gamma_s$ be the decomposition of ξ into its branches: since for each p on ξ, $e_p(\xi) = e_p(\gamma_1) + \cdots + e_p(\gamma_s)$, it is clear that $e_p(\xi) > 1$ if and only if either $e_p(\gamma_i) > 1$ for some i or p belongs to at least two branches γ_i, γ_j, $i \neq j$. Since ξ is reduced, $\gamma_i \neq \gamma_j$ for $i \neq j$ and so there are finitely many points on ξ belonging to more than one branch (by the Noether formula, 3.3.2). It is thus enough to prove the claim for irreducible germs and hence ξ is assumed to be irreducible from now on.

Take local coordinates x, y at O and let f be an equation of ξ. Define $d_x = o_\xi(\partial f/\partial x)$, $d_y = o_\xi(\partial f/\partial y)$ and then

$$d = d(\xi) = \min(d_x, d_y).$$

But for the case of the derivative $\partial f/\partial x$ being identically zero, we may consider the germ $\xi_x : \partial f/\partial x = 0$ and then $d_x = [\xi.\xi_x]$. Also, in the case $\partial f/\partial y \neq 0$, we may take $\xi_y : \partial f/\partial y = 0$ and then $d_y = [\xi.\xi_y]$. Before continuing the proof of 3.7.1, we will establish three lemmas. The first partial result we need is:

Lemma 3.7.2 $d_x = \infty$ *if and only if ξ is the germ of the x-axis. Also $d_y = \infty$ if and only if ξ is the germ of the y-axis, and hence $d(\xi) < \infty$.*

PROOF OF 3.7.2: Of course only the first claim needs a proof. The *if* part is clear, as if f is a multiple of y, then so is $\partial f/\partial x$. For the converse, assume that ξ is not the germ η_x of the x-axis. Since ξ is irreducible, it is non-empty and does not contain the η_x. Therefore $0 < o_x f(x, 0) = [\xi . \eta_x] < \infty$. It easily follows that $0 \le o_x(\partial f/\partial x)(x, 0) = o_x f(x, 0) - 1 < o_x f(x, 0)$. Now, by 2.5.3, if $d_x = \infty$, $\partial f/\partial x$ needs to be a multiple of f against the above inequalities. $\diamond$

Lemma 3.7.3 *The integer d is independent of the equation of ξ and the local coordinates used in its definition.*

PROOF OF 3.7.3: If f' is another equation of ξ, then $f' = uf$ with u invertible, hence
$$\frac{\partial f'}{\partial x} = u\frac{\partial f}{\partial x} + \frac{\partial u}{\partial x}f.$$
Since $o_\xi f = \infty$, 2.5.4 and the above equality show that $d_x = o_\xi(\partial f'/\partial x)$ and hence it does not depend on the equation f. Since the same argument applies to d_y, the first half of the claim has been proved.

Assume now that x', y' are new local coordinates at O and put
$$d' = \min(d_{x'}, d_{y'}).$$

The equalities
$$\frac{\partial f}{\partial x'} = \frac{\partial f}{\partial x}\frac{\partial x}{\partial x'} + \frac{\partial f}{\partial y}\frac{\partial y}{\partial x'}$$
$$\frac{\partial f}{\partial y'} = \frac{\partial f}{\partial x}\frac{\partial x}{\partial y'} + \frac{\partial f}{\partial y}\frac{\partial y}{\partial y'}$$

together with 2.5.4 easily give
$$d_{x'} \ge d, \quad d_{y'} \ge d$$

and, thus, $d' \ge d$. By swapping the roles of old and new coordinates we get $d \ge d'$ and hence $d = d'$ as wanted. $\diamond$

Lemma 3.7.4 *If the x-axis is tangent to ξ, then $d_y < d_x$ and hence $d = d_y$.*

PROOF OF 3.7.4: Take a uniformizing map of ξ of the form $(t^n, v(t))$, where $s(x) = v(x^{1/n})$ is a Puiseux series of ξ and $\nu(s) = n$ (2.5.2). Since $f(t^n, s(t)) = 0$ for all t in a neighbourhood of 0 in $\mathbb{C}$,
$$n\frac{\partial f}{\partial x}(t^n, v(t))t^{n-1} + \frac{\partial f}{\partial y}(t^n, v(t))\frac{dv}{dt} = 0,$$

from which we get
$$d_x = o_t(dv/dt) - n + 1 + d_y.$$

Furthermore, we have $o_t v > n$ because ξ is tangent to the x-axis (2.2.6). In particular $o_t v > 0$, from which $o_t(dv/dt) = o_t v - 1$ and so $o_t(dv/dt) > n - 1$.

Since on the other hand $d_y \neq \infty$, as clearly ξ is not the germ of the y-axis, the claim follows. ◇

END OF THE PROOF OF 3.7.1: We will use induction on d. If $d = 0$ at least one of the derivatives $\partial f/\partial x, \partial f/\partial y$ is not zero at O, then ξ is smooth at O and the claim is obvious.

Assume thus $d > 0$, which clearly implies $n = e_O(\xi) > 1$. Choose the coordinates so that ξ is tangent to the x-axis. Then, since the y-axis is not tangent to ξ, $n = o_y f(0, y)$ and an easy computation shows that the initial form of $\partial f/\partial y$ has degree $n-1$. Thus, in particular $\partial f/\partial y \neq 0$, the germ ξ_y is defined and has $e_O(\xi_y) = n - 1$. On the other hand, from 3.7.4 we have $d = d_y = [\xi.\xi_y]$.

Let p be the point on ξ in the first neighbourhood of O: p corresponds to the direction of the x-axis and therefore (3.1.2) we may take $\bar{x}$ and $z = \bar{y}/\bar{x}$ as local coordinates at p, $\bar{x}$ being an equation of the exceptional divisor. We have

$$\bar{f} = f(\bar{x}, \bar{y}) = \bar{x}^n \tilde{f}$$

where $\tilde{f}$ is an equation at p of the strict transform $\tilde{\xi}$ of ξ. By taking derivatives,

$$\bar{x}^n \frac{\partial \tilde{f}}{\partial z} = \frac{\partial \bar{f}}{\partial z} = \frac{\partial \bar{f}}{\partial \bar{y}} \bar{x} = \overline{\frac{\partial f}{\partial y}} \bar{x},$$

that is,

$$\bar{x}^{n-1} \frac{\partial \tilde{f}}{\partial z} = \overline{\frac{\partial f}{\partial y}}.$$

We have already seen that $e_O(\xi_y) = n - 1$, so the last equality shows that the strict transform of ξ_y has equation $\partial \tilde{f}/\partial z$ and hence it equals $\tilde{\xi}_z$. The Noether formula (3.3.4 in fact) gives thus

$$d = [\xi.\xi_y] = n(n - 1) + [\tilde{\xi}.\tilde{\xi}_z]_p \geq n(n - 1) + d(\tilde{\xi}) > d(\tilde{\xi}),$$

as we already know that $n > 1$.

Now, by the induction hypothesis, $\tilde{\xi}$ has finitely many multiple points infinitely near to p and the claim follows. ◇

Since we know from 2.3.9 that a compact and reduced curve has finitely many proper singular points, it follows from 3.7.1 that:

Corollary 3.7.5 *On a compact reduced curve there are at most finitely many proper or infinitely near multiple points.*

It is enough to blow up all multiple points for getting a smooth curve; more precisely,

Corollary 3.7.6 *If ξ is a compact reduced curve on S, there is a finite sequence of blowing-ups $\pi_i : S_i \longrightarrow S_{i-1}, i = 1, \ldots, r, S_0 = S$, such that the iterated strict transform of ξ by the π_i is a smooth curve.*

PROOF: If ξ has no multiple point, then the empty sequence of blowing-ups satisfies the claim. If ξ contains $r > 0$ multiple points, at least one of them, say O, needs to be a proper point on S by 3.5.4. Then let π be the blowing-up of O on S and $\tilde{\xi}$ the strict transform of ξ. If p is a proper multiple point on $\tilde{\xi}$, then either p is in the first neighbourhood of O, or π is an isomorphism in a neighbourhood of p. It follows that there are $r - 1$ multiple points on $\tilde{\xi}$ and hence the claim by induction on r. $\diamond$

The composition of a sequence of blowing-ups in the conditions of 3.7.6 is called a *resolution* of the singularities of ξ. The proof of 3.7.6 shows that in order to resolve the singularities of a reduced curve ξ it is enough to blow up one of its proper multiple points, then one of the proper multiple points of its strict transform, and so on until reaching a smooth curve after finitely many blowing-ups. Assume that after each blowing-up in the former sequence, say $\pi_i : S_i \longrightarrow S_{i-1}$ with exceptional divisor E_i, we identify $S_i - E_i$ with its image by π_i, and, consequently, all points infinitely near to a point $q \in S_i - E_i$ with the points infinitely near to $\pi_i(q)$: then the reader may easily see that the points we need to blow up in order to resolve the singularities of ξ are just the points which are equal to or identified with the multiple (proper or infinitely near) points on ξ.

Blowing-ups are a particular type of birational transformation. Transforming an algebraic or analytic variety into a non-singular one by means of birational transformations is an old and still partially unsolved problem. First results, for algebraic curves, are due to Kronecker and M. Noether [63] and the latter ones are the Hironaka theorems [44], [6] that apply to any algebraic variety over a field of characteristic zero and to compact complex analytic varieties as well. The case of algebraic varieties of arbitrary dimension over a field of positive characteristic still remains open. A common characteristic of almost all proofs is the use of certain control functions to assure that the singularities have been, in some sense, simplified after each transformation. The function we have used in the proof of 3.7.1, $d(\xi)$, is classical: it was used by Noether [63], see also [35], IV.II. Notice that the multiplicity of the singular point does not work as a control function because it can remain constant after many blowing-ups: the reader may check that on the curve $y^2 = x^{2s+1}$ there are s successive double points.

The germs ξ_x and ξ_y we have introduced in the proof of 3.7.1 are particular examples of polar germs of ξ. We will deal extensively with polar germs in chapter 6, the proof of 3.7.1 shows that they are deeply related to the geometry of the singularity.

There is also a finiteness result for the satellite points on a germ.

Corollary 3.7.7 *A germ of curve contains at most finitely many satellite points.*

PROOF: Let ξ be a germ of curve: since $\mathcal{N}(\xi) = \mathcal{N}(\xi_{red})$, we may assume ξ to be reduced. Then there are finitely many multiple points on ξ by 3.7.1. Since any satellite point is proximate to a multiple point (3.6.2) and the set of points on ξ proximate to a given point is finite (3.5.3), the claim follows. $\diamond$

Let ξ be now a reduced curve, let γ be a representative of a branch of ξ at O and let p be a point on γ infinitely near to O and simple on ξ. Denote by S_p the surface containing p as ordinary point, and by $\pi_p : S_p \longrightarrow S$ the corresponding composition of blowing-ups. By the hypothesis the strict transform of ξ with origin at p, $\tilde{\xi}_p$, is smooth and agrees with the strict transform of γ, $\tilde{\xi}_p = \tilde{\gamma}_p$: so one may choose a representative η of $\tilde{\xi}_p$ isomorphic to an open neighbourhood U of 0 in $\mathbb{C}$ and such that $\pi_p(\eta) \subset \gamma$. After identifying η and U we have:

Proposition 3.7.8 *The map $\pi_{p|\eta}$ is a uniformizing map of γ at O.*

PROOF: It is clear from its construction that $\pi_{p|\eta}$ is a holomorphic map into γ. Thus, according to the definition given in section 2.5, the next lemma will complete the proof. $\diamond$

Lemma 3.7.9 *If W is an open neighbourhood of $0 \in \mathbb{C}$ and $\psi : W \longrightarrow \gamma$ is a holomorphic map, there is a smaller neighbourhood of O, $W' \subset W$, and a holomorphic map $\psi' : W' \longrightarrow \eta$ such that $\psi_{|W'} = \pi_p \circ \psi'$. Furthermore, the germ of ψ' at 0 is uniquely determined by this property.*

PROOF: For the first part of the claim it is clearly enough to lift ψ to a holomorphic map $\tilde{\psi} : W_1 \longrightarrow \tilde{\gamma}$, where $W_1 \subset W$ and $\tilde{\gamma}$ is the strict transform of γ after blowing-up O, and then use induction.

Thus, choose local coordinates x, y at O such that γ is tangent to the x-axis, let q be the point on γ in the first neighbourhood of O, and take $\bar{x}, z = \bar{y}/\bar{x}$ as local coordinates at q. Assume that ψ is given by $\psi(t) = (x(t), y(t))$, $x(t), y(t) \in \mathbb{C}\{t\}$, in a suitable neighbourhood of $0 \in \mathbb{C}$. Since γ is tangent to the x-axis, the only side of its Newton polygon has slope bigger than -1, so that, after an easy computation from $g(x(t), y(t)) = 0$, g an equation of γ, we get $o_t y(t) > o_t x(t)$. Then it is enough to define $\tilde{\psi}$ by the rule $\tilde{\psi}(t) = (x(t), y(t)/x(t))$ using the coordinates $\bar{x}$ and z.

The second part of the claim is obvious from the fact that π_p restricts to a bijection between suitable neighbourhoods of O and p in γ and η, respectively (3.2.8). $\diamond$

Let ζ be the iterated strict transform of ξ by the blowing-ups π_i of 3.7.6 and $p_1, \ldots, p_s$ the points on ζ that are mapped to O, each p_j corresponding to the branch γ_j of ξ: the reader may easily see that the restriction of $\pi_1 \circ \cdots \circ \pi_r$ to the (disjoint) union of suitable neighbourhoods of the p_j in ζ is a uniformizing map of ξ.

Remark 3.7.10 Any non-reduced germ ξ has infinitely many multiple points, as all points on a multiple branch are multiple on ξ, by the additivity of multiplicity. However, since ξ_{red} has finitely many multiple points, it is clear that all but finitely many points on a r-fold branch of ξ have multiplicity r on ξ.

3.8 Equisingularity

In this section we introduce *equisingularity*, the main notion of equivalence of singularities of plane curves. Although classical geometricians that dealt with singularities seem to have had such a notion in mind (see [35], IV.I.9), the first precise definition was given by Zariski in [89]. In fact Zariski gave three different inductive definitions using pairings between the sets of branches of the germs and proved that they are equivalent. Our definition avoids induction by using infinitely near points, the reader should find no difficulty in proving that it is equivalent to Zariski's ones (see exercise 3.8).

Let ξ be a curve or a germ of curve. A (proper or infinitely near) point p on ξ will be called a *singular point* of ξ if and only if either

(a) p is multiple on ξ, or

(b) p is a satellite point, or

(c) p precedes a satellite point on ξ.

Notice that there may be simple and free points on a germ ξ preceding a satellite point still on ξ: these points are thus singular. The point in the first neighbourhood of the origin on $y^2 = x^3$ is the easiest example. Notice also that proper points are singular if and only if they are multiple (by 3.6.2) so that the notion of singular point introduced here extends the usual one (0.10) to infinitely near points. The easy proof of the next lemma is left to the reader.

Lemma 3.8.1 *Let p be an infinitely near point and E_p the germ at p of the exceptional divisor of the composition of the blowing-ups giving rise to p. The point p is a singular point of a germ ξ if and only if $[\bar{\xi}_p.E_p] > 1$.*

Assume that ξ is a non-empty reduced germ with origin at O: from 3.7.1 and 3.7.7 we know that there are finitely many singular points on ξ. Denote by $\gamma_1, \ldots, \gamma_s$ the branches of ξ and for each i denote by p_i the first point on γ_i which is non-singular for ξ. This implies that p_i and all points infinitely near to it on ξ are simple and free. Define $\mathcal{S}(\xi)$ as being the subset of $\mathcal{N}(\xi)$ containing the points p_i, $i = 1, \ldots, s$, and all points preceding some of them. In other words, $\mathcal{S}(\xi)$ contains all singular points on ξ and also the first non-singular point of ξ on each one of its branches. Obviously the points p_i are the maximal points of $\mathcal{S}(\xi)$ by the natural ordering. If ξ' is any representative of ξ we put $\mathcal{S}_O(\xi')$ for $\mathcal{S}(\xi)$. The points in $\mathcal{S}(\xi)$ together with their ordering and proximity relations encode all the information we want to keep, hence the next definition:

Definition: Two germs of curve ξ and ζ are said to be *equisingular* if and only if both are reduced and non-empty and there exists a bijection $\varphi : \mathcal{S}(\xi) \longrightarrow \mathcal{S}(\zeta)$ such that both φ and φ^{-1} preserve natural ordering and proximity of infinitely near points: for any $p, q \in \mathcal{S}(\xi)$, p is infinitely near (resp. proximate) to q if and only if $\varphi(p)$ is infinitely near (resp. proximate) to $\varphi(q)$.

The bijection φ will be referred to as an *equisingularity* or an *equisingularity map* between ξ and ζ. It is clear that to be equisingular is an equivalence

relation between non-empty reduced germs of curve: the corresponding classes
are called *equisingularity classes* or *equisingularity types* and so, if two germs
are equisingular, it is often said that they have the same equisingularity type.
We say that two curves ξ and ζ are *equisingular* at points O and O' if the
corresponding germs are so.

If needed, the notion of equisingularity may be easily extended to non-
reduced germs by taking account of the multiplicities of the branches. Nevertheless, we will not use such an extension here, and so, in the sequel, equisingular
germs will be implicitly assumed to be reduced. Also, non-reduced germs will
be taken as non-equisingular to any reduced germ.

The following examples may be easily checked. A germ ξ is equisingular to
a smooth germ if and only if it is smooth too. Smooth germs describe thus
an equisingularity class. Also nodes (0.11.1) describe an equisingularity class
and the same may be claimed of the ordinary singularities (0.11.4) of fixed
multiplicity. By contrast, two tacnodes (0.11.3) are equisingular if and only if
they have the same number of double points. The same condition holds for
the equisingularity of two cusps of the second order, in particular those with a
single double point (*ordinary cusps*, see 0.11.2) are all equisingular. Conditions
for equisingularity of other irreducible germs are not so easy, they are the main
subject of the forthcoming chapter 5.

An *equisingularity invariant* is any map defined on non-empty reduced germs
of curve that maps any two equisingular germs to the same image. One often
refers to these invariants by the image of the germ rather than by the map itself.
For instance theorem 3.8.6 below implies that the number of branches of a germ
(or rather the map that associates with each germ the number of its branches)
is an equisingularity invariant.

Ordering and proximity between infinitely near points being preserved by
analytic isomorphisms (3.5.1), it is clear that isomorphic germs are equisingular.
The converse is far from true, many examples of non-isomorphic equisingular
germs will be obtained in the forthcoming section 6.7 using polar germs.

The reader may have seen that there are no multiplicities involved in the
definition of equisingularity: this is because multiplicities are determined by
the proximity relations and so they are automatically preserved by an equisingularity, as we will see just after proving the next lemma:

Lemma 3.8.2 *If $p \in \mathcal{S}(\xi)$ is not maximal, then all points proximate to p on ξ
belong to $\mathcal{S}(\xi)$.*

PROOF: The points proximate to p on ξ and not in its first neighbourhood are
satellite points and therefore all belong to $\mathcal{S}(\xi)$. Regarding the points in the
first neighbourhood of p, they obviously belong to $\mathcal{S}(\xi)$ if $e_p(\xi) > 1$, as $\mathcal{S}(\xi)$
contains at least a simple point on each branch of ξ. Lastly, if $e_p(\xi) = 1$, then
there is a single point on ξ in the first neighbourhood of p and it needs to belong
to $\mathcal{S}(\xi)$ just because p is not maximal. ◇

Proposition 3.8.3 *If φ is an equisingularity between ξ and ζ, then $e_p(\xi) =
e_{\varphi(p)}(\zeta)$ for all $p \in \mathcal{S}(\xi)$.*

PROOF: We shall make a decreasing induction on the order of the neighbourhood the point p is belonging to. If p is maximal in $\mathcal{S}(\xi)$ by the natural order, then also $\varphi(p)$ is maximal in $\mathcal{S}(\zeta)$ and, since the maximal points in either $\mathcal{S}(\xi)$ or $\mathcal{S}(\zeta)$ are simple on the corresponding germs, the claim is true for any maximal point p.

Assume now that p is not maximal in $\mathcal{S}(\xi)$. By 3.8.2 all points on ξ proximate to p belong to $\mathcal{S}(\xi)$. Thus we have, by 3.5.3,

$$e_p(\xi) = \sum_{q \in T} e_q(\xi)$$

where T is the set of points $q \in \mathcal{S}(\xi)$ proximate to p. Since $\varphi(p)$ is not maximal, we have, for the same reason

$$e_{\varphi(p)}(\zeta) = \sum_{q' \in T'} e_{q'}(\zeta)$$

T' being now the set of points in $\mathcal{S}(\zeta)$ which are proximate to $\varphi(p)$. By definition φ restricts to a bijection between T and T' and then the claim follows from the induction hypothesis applied to the points in T. $\diamond$

Remark 3.8.4 It is clear that an equisingularity between the germs ξ and ζ may be uniquely extended to a bijection between $\mathcal{N}(\xi)$ and $\mathcal{N}(\zeta)$, both this bijection and its inverse preserving natural order, proximity and multiplicities.

Remark 3.8.5 The reader may easily see, using 3.5.3, that two irreducible germs are equisingular if and only if they have the same sequence of multiplicities. This result cannot be extended to non-irreducible germs (even if doing this seems to be a rather common mistake): there exist non-equisingular germs ξ and ζ and a pair of reciprocal bijections between $\mathcal{S}(\xi)$ and $\mathcal{S}(\zeta)$ preserving natural ordering and multiplicities. We shall give an example in the next section (example 3.9.2), as Enriques diagrams introduced there will allow an easier presentation.

After proving that multiplicities are determined by the equisingularity class of the germ, let us add some comments on equisingularity. The underlying idea of the definition of equisingularity is to classify germs of curve according to the way they intersect other germs, this being coherent with the viewpoint explained at the beginning of this chapter. Infinitely near points on a germ ξ are intended to provide, by means of the proximity relations and the Noether formula, a complete description of the intersection behaviour of ξ. Conversely, infinitely near points on a germ ξ, as well as their proximity relations, may be determined by intersecting ξ with other germs (see section 3.4 and [83]). Thus one may understand the definition of equisingularity as a precise formulation of the main but rather vague idea of taking as equivalent two singularities if they behave similarly when intersected with other germs.

From a quite different viewpoint, germs of plane curve may be classified as germs of topological subspaces of $\mathbb{C}^2 = \mathbb{R}^4$: reduced germs of curve ξ and ζ are said to be *topologically equivalent* if and only if there exist representatives ξ' of ξ and ζ' of ζ, defined in open subsets U and V respectively, so that the pairs (U, ξ') and (V, ζ') are homeomorphic. A very relevant fact is that, despite their apparently quite different meanings, equisingularity and topological equivalence of reduced germs of plane curve are equivalent. The proof was achieved, in one sense, by a very precise description of the topology of a pair (U, ξ), as a topological cone over a link in the three dimensional sphere (Brauner [10]). The converse came out after computing the fundamental groups and the Alexander polynomials of the components of these links (Burau [14], [15] and, independently, Zariski [86]). The reader may see also [69] and [33].

Back to the world of precise claims and proofs, we will show next that the equisingularity class of a given germ determines and is in turn determined by the equisingularity classes of its branches and their intersection multiplicities with each other.

Theorem 3.8.6 *Let ξ and ζ be non-empty and reduced germs of curve at points O and O' and assume that they have branches $\gamma_1, \ldots, \gamma_s$ and $\eta_1, \ldots, \eta_r$, respectively. If there is an equisingularity φ between ξ and ζ, then $s = r$ and, after a suitable reordering of the branches η_i, the restrictions φ_i of φ to $S(\gamma_i)$ are equisingularities between γ_i and η_i, for $i = 1, \ldots, s$. Furthermore we have $[\gamma_i . \gamma_j] = [\eta_i . \eta_j]$ for $i \neq j$, $i, j = 1, \ldots, s$.*

Conversely, assume that $s = r$, there are equisingularities φ_i between γ_i and η_i and furthermore $[\gamma_i . \gamma_j] = [\eta_i . \eta_j]$ for $i \neq j$, $i, j = 1, \ldots, s$. Then there exists an equisingularity φ between ξ and ζ whose restrictions are the φ_i, $i = 1, \ldots, s$.

PROOF: Notice first that the maximal points p_i in $S(\xi)$ being simple, they are in one to one correspondence with the branches of ξ, each p_i corresponding to the only branch, say γ_i, going through it. The same claim being true for ζ, call q_i the maximal point of $S(\zeta)$ that lies on η_i. Since φ restricts to a bijection between subsets of maximal points, we have $r = s$. The case $s = 1$ is obvious, hence we may assume $s > 1$ and, after reordering the branches η_i, also that $\varphi(p_i) = q_i$ for $i = 1, \ldots, s$.

It is clear that, for each i, the points in $S(\xi)$ (resp. $S(\zeta)$) lying on γ_i (resp. η_i) are just p_i (resp. q_i) and the points preceding it. Then it follows that φ restricts to a bijection $\bar{\varphi}_i$ from $S(\xi) \cap \mathcal{N}(\gamma_i)$ onto $S(\zeta) \cap \mathcal{N}(\eta_i)$. The same induction argument used in the proof of 3.8.3 applies here and shows that $\bar{\varphi}_i$ preserves the multiplicities on the branches. Thus, for any i and any $p \in S(\xi)$, $e_p(\gamma_i) = e_{\varphi(p)}(\eta_i)$. In particular, $\bar{\varphi}_i$, and hence also φ, restricts to a bijection between $S(\gamma_i)$ and $S(\eta_i)$ and clearly such a restriction is an equisingularity.

Fix now different indices i, j, with $i, j \leq s$. The points lying on both branches γ_i and γ_j are multiple on ξ, so they belong to $S(\xi)$ and one may write the Noether formula in the form

$$[\gamma_i . \gamma_j] = \sum_{p \in S(\xi)} e_p(\gamma_i) . e_p(\gamma_j).$$

Since a similar equality holds for the corresponding branches of ζ and we have seen above that φ preserves multiplicities on branches, it follows that $[\gamma_i.\gamma_j] = [\eta_i.\eta_j]$ as claimed.

For the converse we will use induction on s, the case $s = 1$ being obvious. We assume then $s > 1$ and put $\xi' = \gamma_1 + \cdots + \gamma_{s-1}$ and $\zeta' = \eta_1 + \cdots + \eta_{s-1}$. Assume as induction hypothesis that there is an equisingularity φ' between ξ' and ζ' whose restrictions are the φ_i, $i = 1, \ldots, s-1$. Let us still denote by φ' and φ_s the extensions of φ' and φ_s to bijections $\mathcal{N}(\xi') \longrightarrow \mathcal{N}(\zeta')$ and $\mathcal{N}(\gamma_s) \longrightarrow \mathcal{N}(\eta_s)$ as mentioned in 3.8.4. Note that the extended φ' still preserves the multiplicities of the points on the branches of ξ'.

Fix any $i < s$ and denote by p_j and q_j the points in the j-th neighbourhood of O which are on γ_s and γ_i respectively, $p_0 = q_0 = O$. Assume that h is such that $p_j = q_j$ if $j \leq h$ and $p_j \neq q_j$ for $j > h$: we have

$$[\eta_i.\eta_s] = [\gamma_i.\gamma_s] = \sum_{j=0}^{h} e_{p_j}(\gamma_i)e_{p_j}(\gamma_s) = \sum_{j=0}^{h} e_{\varphi'(q_j)}(\eta_i)e_{\varphi_s(p_j)}(\eta_s)$$

by using the hypothesis, then the Noether formula for $[\gamma_i.\gamma_s]$, and lastly the fact that φ' preserves multiplicities of points on γ_i.

On the other hand, using again that φ' restricts to an equisingularity between γ_i and η_i, it is clear that $\varphi_s(p_j)$ and $\varphi'(q_j)$, $j \geq 0$, are the points in the j-th neighbourhood of O' on η_s and η_j respectively. If h' is the highest index j for which $\varphi_s(p_j) = \varphi'(q_j)$, the Noether formula gives

$$[\eta_i.\eta_s] = \sum_{j=0}^{h'} e_{\varphi'(q_j)}(\eta_i)e_{\varphi_s(p_j)}(\eta_j).$$

By equating the right sides of the above displayed equalities we get $h = h'$ and therefore, using this fact for $i = 1, \ldots, s-1$, we see that $\varphi'(q) = \varphi_s(p)$ if and only if $q = p$, for any $q \in \mathcal{N}(\xi')$ and any $p \in \mathcal{N}(\gamma_s)$. This is enough to show that φ' and φ_s fit together to give a bijection from $\mathcal{N}(\xi)$ onto $\mathcal{N}(\zeta)$ which obviously gives an equisingularity. $\diamond$

Example 3.8.7 Consider the germs at $(0,0)$ of $\xi : x(y^3 - x^4) = 0$ and $\zeta :$ $(y^2 - x^3)(y^3 - x^2) = 0$. Since ξ has a smooth branch while ζ has not, ξ and ζ are not equisingular by 3.8.6. Nevertheless, both of them have a 4-fold point at the origin, followed by two simple points in its first neighbourhood. The reader may check that ξ contains two satellite points on the singular branch, in the second and third neighbourhoods of the origin, while ζ contains just one satellite point on each branch, both in the second neighbourhood of the origin. Notice that there is an obvious pair of reciprocal bijections between $\mathcal{N}(\xi)$ and $\mathcal{N}(\zeta)$ that preserve natural ordering and multiplicities, but not the proximity relation.

The easy proof of the next proposition is left to the reader.

Proposition 3.8.8 *Assume that ξ and ζ are equisingular germs and p and q are points on ξ and ζ that correspond by the equisingularity map. Then the strict transforms $\tilde{\xi}_p$ and $\tilde{\zeta}_q$ are equisingular too.*

3.9   Enriques diagrams

Let K be a finite and non-empty set of points equal or infinitely near to O, and assume that for each $p \in K$, K contains all points preceding p. Such a set K will be called a *cluster* of points infinitely near to O and O the *origin* of K. We have already dealt with the clusters $\mathcal{S}(\xi)$ for ξ non-empty and reduced.

There are many ways of representing K, the most obvious is like the one already described in section 3.3 for the set of points on a given germ: we draw a graph whose vertices are identified with the points of K and where two vertices p, q are joined by an edge if and only if one of them is in the first neighbourhood of the other. Such a graph is always a tree and it is useful to mark the point O (origin or root of the graph), or, equivalently, to put an orientation on the edges according to the natural ordering of infinitely near points.

A main flaw in the former representation is that it keeps no record of proximity. This may be avoided by similar but more expressive diagrams, introduced by Enriques and currently called Enriques diagrams ([35], IV.I). The *Enriques diagram of K* is a tree-graph which, as above, has a vertex for each point in K: the vertices are currently identified with (and hence named as) the points they represent and the vertex representing O is taken as the origin of the graph. Still the Enriques diagram of K has an edge joining each pair of vertices that represent points one of which is in the first neighbourhood of the other, but now the edges are of two different kinds, namely straight or curved. They are drawn according to the following rules:

- If q is in the first neighbourhood of p and it is proximate to no point preceding p (that is, q is free), then we join p and q by a smooth curved edge which, if $p \neq O$, has at p the same tangent as the edge ending at p.

- Assume that points p and q, q in the first neighbourhood of p have been represented. Then we represent all points proximate to p in the successive neighbourhoods of q, as well as the edges joining them, on a straight half-line which starts at q and is orthogonal to the edge pq at q. To avoid self-intersections of the diagram, such half-lines are alternately oriented to the right and to the left of the preceding edge.

Thus an Enriques diagram is a tree whose branches are composed of smooth curved parts and broken stairs-like parts: the free points are represented on the curved smooth part, while the satellite points are on the broken parts, the structure of the stairs showing the proximity relations. Figures 3.7 and 3.8 show some examples.

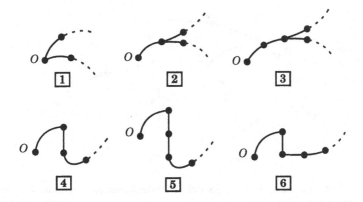

Figure 3.7: Enriques diagrams: a node (1), tacnodes (2,3), a cusp of order two (4) and two kinds of cusps of order three (5,6).

The singularity of a reduced germ ξ is represented by means of the Enriques diagram of $\mathcal{S}(\xi)$, which is called the *Enriques diagram of* ξ: it shows all points on ξ up to the first point on each branch which is simple and free and has no multiple or satellite points after it. Since for each branch γ of ξ there is represented at least one point on γ which is simple on ξ, it is clear that the branches of the Enriques diagram correspond to the branches of ξ.

The reader may easily see that an equisingularity between two germs induces a graph isomorphism between the corresponding Enriques diagrams that maps origin to origin and preserves the orientation, the type (straight or curved) of the edges and the angles at the vertices. Conversely, it is easy to see that any such graph isomorphism induces an equisingularity and hence the Enriques diagram of a germ ξ just encloses all the information about the equisingularity class of ξ.

As one can expect after 3.8.3, the multiplicities of the points on ξ need not be indicated on the Enriques diagram of ξ, as they may be easily obtained from the diagram itself as follows: by the definition of $\mathcal{S}(\xi)$, the maximal points of the Enriques diagram have all multiplicity one. If p is not a maximal point, assume, using reverse induction, that we have computed the multiplicities of all points infinitely near to p in the diagram. Then, by 3.8.2, the Enriques diagram contains all points on ξ proximate to p which, on the other hand, are just those represented on straight half-lines starting at the points in the first neighbourhood of p, such first neighbouring points included: by 3.5.3, we get $e_p(\xi)$ by adding up their multiplicities.

For example, figure 3.9 shows the multiplicities of the singular points on a germ whose Enriques diagram is the one already shown in figure 3.8.

Another remarkable feature of the Enriques diagrams comes from the different way in which free and satellite points are represented. One may imagine

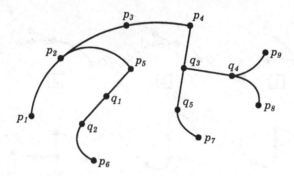

Figure 3.8: A more complex Enriques diagram: the points $p_1, \ldots, p_9$ are free, while the remaining ones are satellite. For instance, p_3, p_5, q_1 and q_2 are proximate to p_2. Notice that q_1 is the only satellite point in the first neighbourhood of p_5, while q_4 and q_5 are the two satellite points in the first neighbourhood of q_3, q_3 itself being a satellite point: q_4 is proximate to p_4 and q_5 is proximate to p_3.

that a free point p is allowed to move with respect to its preceding points by varying the curvature of the edge that ends at p. By contrast, if p is a satellite point, it is represented on a fixed position with respect to its preceding points, this position depending only on which points p is proximate to. This is just what the true infinitely near points do on the surfaces they are lying on: the satellite points in the exceptional divisor $E_{q'}$ of a blowing-up $S_q \longrightarrow S_{q'}$ are in a well defined position, as they are the double points of the exceptional divisor of $\pi_q : S_q \longrightarrow S$, while the free points describe a non-empty open set in $E_{q'}$.

Because of this, Enriques diagrams perform very well for representing, locally at a point, intersections of curves: once two curves share a free point p, they share also a certain number of satellite points of p their branches need to go through because of the proximity equalities (see 3.5.8). This is faithfully represented by the Enriques diagrams: once the diagrams representing the two curves are superposed up to the point p, an automatic superposition of satellite points infinitely near to p is produced, just as it happens to the true infinitely near points. Next is a precise example:

Example 3.9.1 Consider the irreducible germs at $O = (0,0)$ of $\xi : y^4 - x^5 = 0$ and $\zeta : (y + ax)^3 - bx^4 = 0$. The germ ξ has multiplicity 4 at O and a simple point p_1 in the first neighbourhood of O on the x-axis. Therefore, by 3.5.3, the points p_2, p_3 and p_4 in the second, third and fourth neighbourhoods of O on ξ are all simple and proximate to O, while the point p_5 in the fifth neighbourhood is simple and free. Similarly, ζ has multiplicity 3 at O, its first neighbouring point q_1 is simple and lies on $y + ax = 0$, the points q_2, q_3 in the second and third neighbourhoods are simple and proximate to O and the point q_4 in the first neighbourhood is simple and free.

If $a \neq 0$, ξ and ζ have different tangents so they share no point in the first

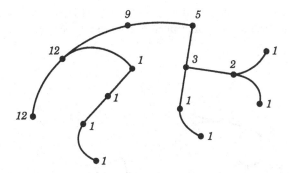

Figure 3.9: The multiplicities of a germ whose Enriques diagram is that in figure 3.8.

neighbourhood of O. If $a = 0$, both germs share the point in the first neighbourhood, $p_1 = q_1$ and then also the points in the second and third neighbourhood, $p_2 = q_2$, $p_3 = q_3$ because they are the only points in the first and second neighbourhood of $p_1 = q_1$ proximate to O. By contrast, since p_4 is also proximate to O and this is not the case for q_4, necessarily $p_4 \neq q_4$ no matter which value is taken by the coefficient b in the equation of ζ. One may confirm these results by computing the intersection multiplicities of ξ and ζ in both cases.

We see thus that, because of proximity, $p_1 = q_1$ implies $p_2 = q_2$ and $p_3 = q_3$ while p_4 cannot equal q_4. The reader may see in figure 3.10 how the Enriques diagram of $\xi + \zeta$ gives an accurate representation of these facts.

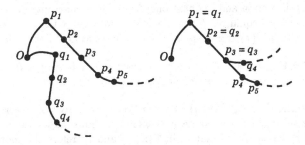

Figure 3.10: Enriques diagrams of the curve $\xi + \zeta$ of example 3.9.1 for $a \neq 0$ (left) and $a = 0$ (right).

The following example shows, as already claimed in section 3.8, that a pair of reciprocal bijections between $\mathcal{S}(\xi)$ and $\mathcal{S}(\zeta)$ preserving ordering and multiplicities does not make ξ and ζ equisingular.

Example 3.9.2 The Enriques diagrams of the germs at $(0,0)$, $\xi : x(y^2 - x^3 + x^2 y)(y^2 - x^3)(x - y^3) = 0$ and $\zeta : x(x - y^4)(y^2 - x^3)(y^2 + x^3) = 0$ are shown in figure 3.11: there is an obvious pair of reciprocal bijections between $\mathcal{S}(\xi)$ and $\mathcal{S}(\zeta)$ preserving ordering and multiplicities but, clearly enough, ξ and ζ are not equisingular.

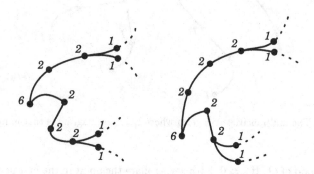

Figure 3.11: Enriques diagrams of the curves ξ and ζ in example 3.9.2.

In the forthcoming section 4.4, a quite different graphic representation of clusters, their dual graphs, will be introduced and related to the Enriques diagrams.

3.10   The ring in the first neighbourhood

Throughout this section and the next one, ξ will be a non-empty germ of curve at O defined by an equation f and we will denote by R the local ring of ξ: $R = \mathcal{O}_{\xi,O} = \mathbb{C}\{x,y\}/(f)$ if x, y are local coordinates at O. First of all we need:

Lemma 3.10.1 *If $f \in \mathbb{C}\{x\}[y]$ is the Weierstrass equation of ξ, the inclusion $\mathbb{C}\{x\}[y] \hookrightarrow \mathbb{C}\{x,y\}$ induces an isomorphism $\mathbb{C}\{x\}[y]/(f) \simeq R$.*

PROOF: Assume that $g \in \mathbb{C}\{x,y\}$. By the Weierstrass division theorem 1.8.8, one may write $g = hf + r$ where $r \in \mathbb{C}\{x\}[y]$. This shows that any class in $\mathbb{C}\{x,y\}/(f)$ has a representative in $\mathbb{C}\{x\}[y]$ and therefore the morphism is onto. Assume now that $g \in \mathbb{C}\{x\}[y]$ and that f divides g in $\mathbb{C}\{x,y\}$: $g = hf$, $h \in \mathbb{C}\{x,y\}$. Then, h is the quotient of the Weierstrass division (1.8.8) of g by f, by its uniqueness, and in particular $h \in \mathbb{C}\{x\}[y]$ again by 1.8.8. $\diamond$

Let $\pi : \bar{S} \longrightarrow S$ be the blowing-up of O, $p_1, \ldots, p_s$ the points in the first neighbourhood of O on ξ and R_i the local ring of the strict transform $\tilde{\xi}_{p_i}$ of ξ, $i = 1, \ldots, s$. We have the monomorphisms of local $\mathbb{C}$-algebras induced by π,

$$\pi_i^* : \mathcal{O}_{S,O} \longrightarrow \mathcal{O}_{\bar{S},p_i}$$

that map the germ at O of a function g to the germ at p_i of $g \circ \pi$.

Each π_i^* obviously induces a morphism

$$\kappa_i : R \longrightarrow R_i$$

and one can put all κ_i together to get a morphism

$$\kappa = \prod_i \kappa_i : R \longrightarrow \prod_{i=1}^{s} R_i.$$

Definition: The ring $R^1 = \prod_{i=1}^{s} R_i$, equipped with the morphism κ, is called the *the ring in the first neighbourhood* (or the *first neighbouring ring*) of O on ξ), and sometimes also the *ring in the first neighbourhood of R*.

We will prove, as part of theorem 3.10.2 below, that κ is injective: after this we will identify R and $\kappa(R)$ and the first neighbourhood ring R^1 will be considered as an extension of R. Theorem 3.10.2 gives a non intrinsic description of R^1 from which its main properties will easily follow. We assume, from now on in this section, that local coordinates x, y at O have been chosen in such a way that no branch of ξ is tangent to the y-axis. Then, using the same notations as in section 3.1, all points p_i belong to the open set $V \subset \bar{S}$ where $\bar{x}$ and $z = \bar{y}/\bar{x}$ may be taken as coordinates: assume that $p_i = (0, c_i)$ using such coordinates, so in particular we take $\bar{x}, z_i = z - c_i$ as local coordinates at p_i. For any $g \in \mathcal{O}_{S,O}$ we will denote by $\mathbf{g} \in R$ its class modulo f.

Theorem 3.10.2 *The morphism κ is injective. The ring R^1 is isomorphic, as extension of R, to $R[\mathbf{y}/\mathbf{x}]$.*

PROOF: Decompose ξ into its tangential components, $\xi = \xi_1 + \cdots + \xi_s$, the only principal tangent to ξ_i corresponding to the point p_i, $i = 1, \ldots, s$. Denote by f_i the Weierstrass equation of ξ_i, so that $f = f_1 \cdots f_s$ is the Weierstrass equation of ξ. Write $e_i = e_O(\xi_i)$, $e = e_O(\xi)$.

It is clear that all morphisms

$$\pi_i^* : \mathbb{C}\{x, y\} \longrightarrow \mathbb{C}\{\bar{x}, z - c_i\}$$

restrict to the same morphism

$$\pi^* : \mathbb{C}\{x\}[y] \longrightarrow \mathbb{C}\{\bar{x}\}[z]$$
$$g = g(x, y) \mapsto \bar{g} = g(\bar{x}, \bar{x}z).$$

Take $\tilde{f}_i = \bar{x}^{-e_i} \bar{f}_i$ and $\tilde{f} = \bar{x}^{-e} \bar{f}$, so that $\tilde{f} = \tilde{f}_1 \ldots \tilde{f}_s$. The strict transforms with origin at p_i of the germs ξ_j being empty for $j \neq i$, we have $\tilde{\xi}_{p_i} = (\tilde{\xi_i})_{p_i}$ and so $\tilde{f}_i$ may be taken as an equation of $\tilde{\xi}_{p_i}$. We will prove next that it is its Weierstrass equation relative to the local coordinates $\bar{x}, z_i$. Indeed, due to the fact that the y-axis is not tangent to ξ_i, its Newton polygon has no side with slope less than -1 and therefore its Weierstrass equation may be written

$$f_i = y^{e_i} + \sum_{j < e_i} a_j y^j,$$

where $a_j \in \mathbb{C}\{x\}$ and $o_x a_j \geq e_i - j$, $j = 0, \ldots, e_i - 1$. It follows that

$$\tilde{f}_i = z^{e_i} + \sum_{j < e_i} \bar{a}_j \bar{x}^{j - e_i} z^j,$$

all $\bar{a}_j \bar{x}^{j-e_i}$ belonging to $\mathbb{C}\{\bar{x}\}$. Since, clearly, $\mathbb{C}\{\bar{x}\}[z] = \mathbb{C}\{\bar{x}\}[z_i]$, the equation $\tilde{f}_i$ may be rewritten as

$$\tilde{f}_i = z_i^{e_i} + \sum_{j < e_i} b_j z_i^j,$$

where still all $b_j \in \mathbb{C}\{\bar{x}\}$. All we need to check now is that $b_j(0) = 0$ for $j = 0, \ldots, e_i - 1$. This follows from the fact that p_i is the only point on ξ_i in the first neighbourhood of O, as then, by 3.2.5,

$$e_i = e(\xi_i) = o_{z_i} \tilde{f}_i(0, z_i) = o_{z_i} (z_i^{e_i} + \sum_{j < e_i} b_j(0) z_i^j),$$

this forcing $b_j(0) = 0$, $j = 0, \ldots, e_i - 1$, as wanted.

Once we know that f and f_i are Weierstrass equations of, respectively, ξ and ξ_{p_i}, by 3.10.1 we may take $R = \mathbb{C}\{x\}[y]/(f)$, $R_i = \mathbb{C}\{\bar{x}\}[z]/(\tilde{f}_i)$ and κ_i to be the morphism induced by π^*. Since each $\tilde{f}_i$ divides $\tilde{f}$, it is clear that each morphism κ_i may be factored through the obvious morphism $\tau_i : \mathbb{C}\{\bar{x}\}[z]/(\tilde{f}) \longrightarrow \mathbb{C}\{\bar{x}\}[z]/\tilde{f}_i$ and hence we have $\kappa = \tau \circ \kappa'$,

$$\mathbb{C}\{x\}[y]/(f) \xrightarrow{\kappa'} \mathbb{C}\{\bar{x}\}[z]/(\tilde{f}) \xrightarrow{\tau} \prod_i \mathbb{C}\{\bar{x}\}[z]/(\tilde{f}_i)$$

where κ' is still induced by π^* and $\tau = \prod \tau_i$. Then we shall be done after proving the following claims:

(a) κ' is injective.

(b) τ is an isomorphism.

PROOF OF (a): Assume that $\bar{g} = g_1 \tilde{f}$, $g_1 \in \mathbb{C}\{\bar{x}\}[z]$: since the product of any element of $\mathbb{C}\{\bar{x}\}[z]$ by a suitable power of $\bar{x}$ is in the image of π^*, the former equality gives $\bar{x}^r \bar{g} = \bar{g}_2 \tilde{f}$ for a certain positive integer r and, thus, $x^r g = g_2 f$ in $\mathbb{C}\{x\}[y]$. Since x is not a factor of f by hypothesis (the y-axis is not a component of ξ), it follows that f divides g in $\mathbb{C}\{x\}[y]$ as wanted.

PROOF OF (b): By the Chinese remainder theorem ([48], II.2 for instance) it is enough to see that any two $\tilde{f}_i$ and $\tilde{f}_j$, $i \neq j$, are comaximal, i.e., that $(\tilde{f}_i, \tilde{f}_j) = \mathbb{C}\{\bar{x}\}[z]$. Assume that $\mathbf{m}$ is a maximal ideal of $\mathbb{C}\{\bar{x}\}[z]$ that contains both $\tilde{f}_i$ and $\tilde{f}_j$. Let $u = u(\bar{x}) \in \mathbb{C}\langle\langle\bar{x}\rangle\rangle$ be a root of $\tilde{f}_i$, as a polynomial in z. An easy computation shows that u has no terms of negative degree (because the monomial of the highest degree in $\tilde{f}_i$ has coefficient 1) and $u(0) = c_i$ (because $\tilde{f}_i(0, z) = (z - c_i)^{e_i}$). Since the same arguments apply to $\tilde{f}_j$, we see, by 1.5.11, that $\tilde{f}_i$ and $\tilde{f}_j$ have no common root as polynomials in z. Hence the Bezout identity (over the field of convergent Laurent series in $\bar{x}$, the quotient field of $\mathbb{C}\{\bar{x}\}$) gives rise to an equality

$$\bar{x}^r = g_i \tilde{f}_i + g_j \tilde{f}_j,$$

where $r \geq 0$ and $g_i, g_j \in \mathbb{C}\{\bar{x}\}[z]$. If $r = 0$, then $1 \in \mathbf{m}$ against the hypothesis of $\mathbf{m}$ to be maximal. Otherwise we have $\bar{x} \in \mathbf{m}$. As $\tilde{f}_i(0, z) \in (\tilde{f}_i, \bar{x})$ we have $\tilde{f}_i(0, z) \in \mathbf{m}$ and, similarly, $\tilde{f}_j(0, z) \in \mathbf{m}$. Since $f_i(0, z) = (z - c_i)^{e_i}$ and $f_j(0, z) = (z - c_j)^{e_j}$, using again the Bezout identity, this time in $\mathbb{C}[z]$, we get $1 \in \mathbf{m}$ as above. ◇

The first neighbourhood ring was introduced by Northcott ([64], [66]) as an algebraic (and intrinsic!) counterpart of the birational transformations used to blow up singularities. The original definition of Northcott is close to the characterization of 3.10.2. From now on, the first neighbourhood ring of R will be identified with $R[\mathbf{y}/\mathbf{x}]$. After this we have:

Corollary 3.10.3 *The first neighbourhood ring R^1 is an extension of R in its full quotient ring. Furthermore, R^1 is a finite R-module and may be generated over R by $1, \mathbf{y}/\mathbf{x}, \ldots, (\mathbf{y}/\mathbf{x})^{e-1}$, $e = e(\xi)$.*

PROOF: Recall that we are assuming that the y-axis is not tangent to ξ. In particular, the germ of the y-axis is not a branch of ξ, hence $\mathbf{x}$ does not divide zero in R and then the first part of the claim follows from 3.10.2. Let the Weierstrass equation of ξ be

$$f = y^e + \sum_{i=0}^{e-1} a_i y^i, \quad a_i \in \mathbb{C}\{x\},$$

where $e = e(\xi)$ because the y-axis is not tangent to ξ. Furthermore, by the same reason, the Newton polygon of f has no sides with slope less than -1, thus $o_x a_i \geq e - i$, $i = 1, \ldots, e-1$, and therefore the equation of ξ takes the form

$$f = y^e + \sum_{i=0}^{e-1} b_i x^{e-i} y^i, \quad b_i \in \mathbb{C}\{x\}.$$

After reducing modulo f, we have in R

$$0 = \mathbf{y}^e + \sum_{i=0}^{e-1} \mathbf{b}_i \mathbf{x}^{e-i} \mathbf{y}^i, \quad \mathbf{b}_i \in R$$

and hence

$$0 = (\mathbf{y}/\mathbf{x})^e + \sum_{i=0}^{e-1} \mathbf{b}_i (\mathbf{y}/\mathbf{x})^i$$

which is an integral dependence equation of $\mathbf{y}/\mathbf{x}$ over R. This equation shows that $R^1 = R[\mathbf{y}/\mathbf{x}]$ is generated by $1, \mathbf{y}/\mathbf{x}, \ldots, (\mathbf{y}/\mathbf{x})^{e-1}$ over R. ◇

Corollary 3.10.4 *The germ ξ is smooth if and only if $R = R^1$.*

PROOF: Obviously $R = R^1$ if and only if $\mathbf{y}/\mathbf{x} \in R$, that is, if and only if $\mathbf{x}$ generates the maximal ideal $(\mathbf{x}, \mathbf{y})$ of R. Then, since the y-axis is not tangent to ξ, the claim follows from 2.4.3. ◇

3.11   The rings in the successive neighbourhoods

Let us denote by $p_{j,i}$, $i = 1, \ldots, s_j$, $j \geq 1$ the points on ξ in the j-th neighbourhood of O. In particular, with respect to the notations of the preceding section we have $p_i = p_{1,i}$ and $s = s_1$. We denote by $R_{j,i}$ the local ring of $\tilde{\xi}_{p_{j,i}}$, the strict transform of ξ with origin at $p_{j,i}$.

Put $R^0 = R$ and $R^j = \prod_{i=1}^{s_j} R_{j,i}$ for $j > 0$. Notice that R^1 has been introduced above while, for $j > 1$, R^j is the product of the rings in the first neighbourhood of the rings $R_{j-1,i}$, $i = 1, \ldots, s_{j-1}$. Then, by 3.10.2, we have for each $j > 0$ a monomorphism

$$R^{j-1} \longrightarrow R^j,$$

namely the monomorphism τ from R into its first neighbouring ring for $j = 1$, and the product of the monomorphisms from the $R_{j-1,i}$ into their first neighbouring rings for $j > 1$. Thus each R^j will be considered as an extension of R^{j-1} (and therefore also of R) and called the *ring in the j-th neighbourhood of O on ξ*, and also the *ring in the j-th neighbourhood* or the *j-th neighbouring ring* of R. The neighbouring rings make thus an ascending chain of rings

$$R \subset R^1 \subset \cdots \subset R^{j-1} \subset R^j \subset \cdots$$

Remark 3.11.1 The reader may notice that after identifying the elements of R with its images in R^j, the class $\mathbf{g} \in R$ of a function $g \in \mathcal{O}_{S,O}$ is identified with the element of $R^j = \prod R_{j,i}$ whose components are the classes of $g \circ \pi_{p_{j,i}}$, $i = 1, \ldots, s_j$, $\pi_{p_{j,i}} : S_{p_{j,i}} \longrightarrow S$ being, as before, the composition of the blowing-ups giving rise to $p_{j,i}$.

The proof of the following algebraic lemma is easy and will be left to the reader.

Lemma 3.11.2 *If for $i = 1, \ldots, s$ the ring B_i is an extension of A_i within its full quotient ring, also $\prod_{i=1}^{s} B_i$ is an extension of $\prod_{i=1}^{s} A_i$ in its full quotient ring. Furthermore, if each B_i is a finitely generated A_i-module, then also $\prod_{i=1}^{s} B_i$ is finitely generated over $\prod_{i=1}^{s} A_i$.*

In our case, we have:

Corollary 3.11.3 *The rings R^j in the successive neighbourhoods of R are finite extensions of R contained in its full quotient ring.*

PROOF: It follows from 3.10.3 and 3.11.2 that R^j is finitely generated over R^{j-1} and contained in its full quotient ring. Then it is enough to use induction on j.
◇

Assume that the germ ξ is reduced: we know from 3.7.1 that all points in the k-th neighbourhood of O on ξ are simple if k is big enough. Therefore, for such a k the rings $R_{k,i}$ are (isomorphic to) one-variable power series rings, say $R_{k,i} = \mathbb{C}\{t_i\}$ and, because of 3.10.4, the sequence of neighbouring rings is stationary from the k-th term onwards: $R^k = R^j$ for all $j > k$. Moreover the rings $R_{k,i}$, and therefore also R^k, are integrally closed in its full quotient rings: this proves R^k to be the integral closure $\bar{R}$ of R in its full quotient ring. Conversely, if $R^k = \bar{R}$, it is integrally closed and then we have $R^k = R^j$ for $j > k$ by 3.11.3. Lastly, if $R^k = R^{k+1}$, then all points $p_{k,i}$ are simple on ξ by 3.10.4. Putting all together we have proved:

Theorem 3.11.4 *If ξ is a non-empty and reduced germ of curve at O and R^k denotes the ring in the k-th neighbourhood of O on ξ, then the following conditions are equivalent and all of them are satisfied if k is big enough:*

(i) All points in the k-th neighbourhood of O on ξ are simple on ξ.

(ii) $R^k = R^{k+1}$.

(iii) $R^k = R^j$ for $j > k$.

(iv) R^k is integrally closed in its full quotient ring.

(v) R^k is the integral closure $\bar{R}$ of R in its full quotient ring.

In particular the above construction of $\bar{R}$ makes evident a fact which is true for a large class of rings including all those which arise from algebraic or analytic varieties (see [58], p. 237 or [32], 13.3, for instance):

Corollary 3.11.5 *If R is the local ring of a reduced germ of curve, then the integral closure of R in its full quotient ring is a finite-generated R-module.*

Still assume that ξ is reduced and use the notations as above. The next corollary relates integral closure and uniformizing morphisms and gives a characterization of the elements of $\bar{R}$.

Corollary 3.11.6 *(a) The integral closure $\bar{R}$ may be identified with a product of power series rings, $\bar{R} = \prod_{j=1}^{r} \mathbb{C}\{t_j\}$, so that r is the number r of branches of ξ and the inclusion morphism $R \hookrightarrow \bar{R}$ is a uniformizing morphism of ξ.*

(b) If $g \in \mathcal{O}_{S,O}$ defines a germ ζ and its class $\mathbf{g}$ is identified with $(\mathbf{g}_1, \ldots, \mathbf{g}_r) \in \bar{R}$, then, after a suitable reordering of the branches $\gamma_1, \ldots, \gamma_r$ of ξ, $[\gamma_j \cdot \zeta] = o_{t_j} \mathbf{g}_j$ for $j = 1, \ldots, r$.

(c) Let $g, g' \in \mathcal{O}_{S,O}$ define germs ζ, ζ' and assume that ζ' has no common branch with ξ, so that $\mathbf{g}/\mathbf{g}'$ belongs to the total quotient ring of R. Then $\mathbf{g}/\mathbf{g}' \in \bar{R}$ if and only if $[\gamma.\zeta] \geq [\gamma.\zeta']$ for all branches γ of ξ.

PROOF: Part (a) easily follows from our construction of $\bar{R}$. Indeed, by the definition of R^k, $\bar{R} = R^k$ is the product of the local rings $R_{k,j}$ of the points $p_1, \ldots, p_r$ in the k-th neighbourhood of O on ξ. By 3.11.4, all these points are simple on ξ and therefore r is the number of branches of ξ and each $R_{k,j}$ may be identified with a ring of power series, $R_{k,j} = \mathbb{C}\{t_j\}$. Let π_j be the composition of the blowing-ups giving rise to p_j. By 3.7.8 the restriction of π_j to a suitable representative of the strict transform of γ_j is a uniformizing map of a representative of γ_j. Since, by 3.11.1, the composition of the inclusion and the j-th projection

$$\tau_j : R \hookrightarrow \bar{R} \to R_{k,j}$$

is induced by π_j, for $j = 1, \ldots, r$, $R \hookrightarrow \bar{R}$ is a uniformizing morphism of ξ as claimed.

For part (b), keep the notations as above and just notice that $\mathbf{g}_j = \tau_j(\mathbf{g})$: since we have just seen that τ_j is induced by a uniformizing morphism of γ_j, we have

$$[\gamma_j.\zeta] = o_{\gamma_j}(g) = o_{t_j}\mathbf{g}_j$$

by the definition of intersection multiplicity and 2.5.3, as wanted.

Claim (c) is a direct consequence of claim (b). ⋄

It is worth noticing that claim a) of 3.11.6 implies that any uniformizing morphism of ξ agrees, up to an isomorphism of R-algebras, with the inclusion morphism of the local ring R of ξ in its integral closure $\bar{R}$, as uniformization morphisms are essentially unique, see section 2.5.

Again assume that ξ is just a non-empty (non-necessarily reduced) germ of curve. Once we know all rings R^j to be finite extensions of R in its full quotient ring, it is clear that for any $j \geq 1$, the ideal of R,

$$(R : R^j) = \{\mathbf{g} \in R | \mathbf{g}R^j \subset R\}$$

contains at least a non-zero-divisor, namely the common denominator of a finite set generating R^j as R-module. The ideal $(R : R^j)$ is named the *conductor of R^j in R*. If ξ is reduced and $R^j = \bar{R}$, then we call it the *conductor of R* or the *conductor of ξ*. We have:

Proposition 3.11.7 *If $\mathbf{m}$ denotes the maximal ideal of R, for any $j > 0$, the conductor $(R : R^j)$ is either R or an $\mathbf{m}$-primary ideal.*

PROOF: We know that $(R : R^j)$ contains a non-zero-divisor. It is clear from 2.4.2 that the non-zero elements of any prime ideal of R different from $\mathbf{m}$ are zero-divisors. The maximal ideal $\mathbf{m}$ is then the only prime ideal of R that may contain $(R : R^j)$, from which the claim. ⋄

Remark 3.11.8 By its own definition, $(R : R^j)$ is also an ideal of R^j. Notice also that for r big enough, by 3.11.7, $(R : R^j)$ contains $\mathbf{m}^r$ and therefore also all powers $\mathbf{g}^r$, where $\mathbf{g}$ is any non-unit in R.

Since 3.10.3 gives a system of generators of R^1 over R, the conductor of R^1 in R may be explicitly determined:

Lemma 3.11.9 *If $e = e_O(\xi)$ and $n \geq e-1$, then $\mathbf{x}^n R^1 = \mathbf{m}^n$ and, in particular, $\mathbf{x}^n \in (R : R^1)$. The conductor of R^1 in R is $\mathbf{m}^{e-1}$.*

PROOF: The maximal ideal $\mathbf{m}$ of R being generated by $\mathbf{x}$ and $\mathbf{y}$, it is clear that $\mathbf{m}R^1 = \mathbf{x}R^1$ and therefore that $\mathbf{m}^n \subset \mathbf{x}^n R^1$. The opposite inclusion follows from 3.10.3, $\mathbf{x}^{e-1}$ being the common denominator of the generators of R^1 over R there shown. Hence we have $\mathbf{m}^n = \mathbf{x}^n R^1$ as claimed.

For the second half of the claim, the first one, already proved, easily gives $\mathbf{m}^{e-1} \subset (R : R^1)$ and so we just need to see that $\mathbf{m}^{e-1} \supset (R : R^1)$. Assume thus that there is $g \in \mathcal{O}_{S,O}$ such that $gR^1 \subset R$, and also that $g \notin \mathbf{m}^{e-1}$, from which we will get a contradiction. Since $f \in \mathcal{M}^e_{S,O}$, the last assumption is equivalent to $g \notin \mathcal{M}^{e-1}_{S,O}$. Let us denote by $[g]$ the initial form of g, $[g]$ being thus a homogeneous polynomial in x, y of degree $r < e - 1$. After a suitable (linear) change of coordinates, we may assume that still no branch of ξ is tangent to the y-axis and, furthermore, that x does not divide $[g]$. Since $\mathbf{y}/\mathbf{x} \in R^1$ (by 3.10.2) we have $\mathbf{y}g \in \mathbf{x}R$ and hence $yg = h_1 x + h_2 f$, $h_1, h_2 \in \mathcal{O}_{S,O}$. Then it is enough to equate the initial forms of both sides, using that $r + 1 < e$: we obtain that x divides $y[g]$, and hence also $[g]$ against our choice of coordinates. $\diamond$

The next theorem gives an interesting characterization of the intersection multiplicity of a couple of germs. We will prove it using properties of local rings of germs already obtained in this section. For other proofs, the reader may see [37], III.3 or [38], 1.2 and A.2.1.

Theorem 3.11.10 *If ξ and ζ are two germs of curve at O defined by equations $f, g \in \mathcal{O}_{S,O}$, then*

$$[\xi.\zeta] = \dim_{\mathbb{C}} \mathcal{O}_{S,O}/(f,g)$$

in the sense that if one of the sides is finite, then so is the other and they are equal.

PROOF: Both germs will be assumed to be non-empty, otherwise the claim is obvious. Put $\mathcal{O} = \mathcal{O}_{S,O}$. First, assume that f and g have a common irreducible factor, say h: then both germs have the common branch $h = 0$ and therefore $[\xi.\zeta] = \infty$. On the other hand $\dim_{\mathbb{C}} \mathcal{O}/(f,g) = \infty$ because $\mathcal{O}/(h)$ is a quotient of $\mathcal{O}/(f,g)$ and clearly $\dim_{\mathbb{C}} \mathcal{O}/(h) = \infty$.

So we assume from now on that f and g have no common factor and hence that $[\xi.\zeta]$ is finite. We will see first that $\dim_{\mathbb{C}} \mathcal{O}/(f,g)$ is additive in ζ (and hence also in ξ, by the obvious symmetry). Assume that $g = g_1 g_2$ and write $R = \mathcal{O}/(f)$ using the same notations as before. Since no factor of f divides g, the class of g_2 mod f does not divide zero in R. This easily gives the exactness of the sequence

$$0 \to R/g_1 \xrightarrow{\times g_2} R/g \to R/g_2 \to 0$$

from which it follows that

$$\dim_{\mathbb{C}} \mathcal{O}/(f,g) = \dim_{\mathbb{C}} \mathcal{O}/(f,g_1) + \dim_{\mathbb{C}} \mathcal{O}/(f,g_2),$$

as wanted.

Since the intersection multiplicity is additive too (2.6.1), it would be not restrictive to assume in the sequel that both germs ξ and ζ are irreducible. In fact we will assume in the sequel just that ξ is reduced, as this will be enough for the proof.

Take all notations as above. Let k be such that all points in the k-th neighbourhood of O on ξ are simple on ξ: each of them, $p_{k,i}$, $i = 1, \ldots, r$, is on a single branch γ_i of ξ and $R_{k,i} = \mathbb{C}\{t_i\}$. Assume that the class g of g in R is identified with $(g_1, \ldots, g_r) \in \bar{R} = R^k = \prod_{i=1}^r \mathbb{C}\{t_i\}$: it follows from 3.11.6 that

$$[\gamma_i.\zeta] = \dim_{\mathbb{C}} \mathbb{C}\{t_i\}/(g_i),$$

so that

$$[\xi.\zeta] = \dim_{\mathbb{C}} \bar{R}/g\bar{R}.$$

Take n big enough to have $g^n \in (R : \bar{R})$ (3.11.8): it is easy to see that the sequence

$$0 \to g^n \bar{R}/g^n R \to R/g^n R \to \bar{R}/g^n \bar{R} \to \bar{R}/R \to 0 \qquad (3.1)$$

is exact. On the other hand, since g does not divide zero in R, it cannot divide zero in $\bar{R}$ which is contained in the total quotient ring of R: multiplication by g gives thus an isomorphism $\bar{R}/R \simeq g^n \bar{R}/g^n R$ and the exact sequence 3.1 may be rewritten

$$0 \to \bar{R}/R \to R/g^n R \to \bar{R}/g^n \bar{R} \to \bar{R}/R \to 0. \qquad (3.2)$$

By using again that g does not divide zero, one may easily see by induction on n that

$$\dim \bar{R}/g^n \bar{R} = n \dim \bar{R}/g\bar{R} = n[\xi.\zeta].$$

In particular $\dim \bar{R}/g^n \bar{R}$ is finite, so from the exact sequence 3.2 we see that $\dim \bar{R}/R$ is finite too and hence

$$\dim_{\mathbb{C}} R/g^n R = \dim_{\mathbb{C}} \bar{R}/g^n \bar{R}.$$

Thus, since just as noticed above for $\bar{R}$,

$$\dim R/g^n R = n \dim R/gR,$$

we get

$$[\xi.\zeta] = \dim \bar{R}/g\bar{R} = \dim R/gR = \dim_{\mathbb{C}} \mathcal{O}/(f,g)$$

and hence the claim. ⋄

An important invariant of a reduced germ ξ may be introduced, in an algebraic way, as follows:

Definition: If ξ is a non-empty and reduced germ at O, then the integer $\delta = \delta(\xi) = \dim_{\mathbb{C}} \bar{R}/R$ is called the *order of singularity* of ξ, or the *order of singularity* at O of any representative ξ' of ξ. In such a case we write $\delta_O(\xi') = \delta(\xi)$.

It has been seen in the proof of 3.11.10 that $\delta < \infty$. Anyway this will follow from the computation of δ we shall make below.

It is clear from its own definition that δ measures how far R is from its integral closure $\bar{R}$ and, thus, in a certain sense, how deep the singularity of ξ is. In particular, $\delta = 0$ if and only if ξ is smooth (by 3.11.4). In fact δ may be used as a control function for resolution of singularities, instead of $d(\xi)$. A classical proof of the finiteness of the procedure of resolution of the singularities of a plane algebraic curve by ordinary quadratic transformations, due to Bertini, uses the sum of the orders of singularity at the points of the curve as a (global) control function. See [9] App. II.6, or [37], VII.4 , or also [35], IV.II where both the proofs of Bertini and Noether are presented.

If ξ is an algebraic projective and irreducible curve, the integer $\delta(\xi) = \sum_{p \in \xi} \delta_p(\xi)$ is closely related to the genus of ξ. For instance, if ξ is plane of degree n, $\delta(\xi)$ is just the difference between the genus of a generic (hence smooth) curve of degree n, and that of ξ itself (see exercise 6.8). In this form $\delta(\xi)$ was known since Noether [63], although the introduction of the local invariants $\delta(\xi_O)$ as above is due to Rosenlicht [70].

Now we start the computation of δ by computing what is called the *Hilbert-Samuel polynomial* of the ring R. Our curves being hypersurfaces, the computation is very easy.

Lemma 3.11.11 *If R is the local ring of the germ ξ and $\mathbf{m}$ its maximal ideal, then*

$$\dim_{\mathbb{C}} R/\mathbf{m}^n = en - \frac{e(e-1)}{2}$$

for all $n \geq e$, $e = e(\xi)$.

PROOF: Let f be an equation of ξ: by hypothesis we have $f \in \mathbf{m}^e - \mathbf{m}^{e+1}$ so that the reader should find no difficulty in proving that the sequence

$$0 \to \mathbb{C}\{x,y\}/(x,y)^{n-e} \xrightarrow{\times f} \mathbb{C}\{x,y\}/(x,y)^n \to R/\mathbf{m}^n \to 0$$

is exact for $n \geq e$. After this the claim is obvious. ⋄

The equality of 3.11.2 is still true for $n = e - 1$, as the reader may easily check, but no use of this fact will be made in the sequel.

Theorem 3.11.12 *If ξ is a non-empty and reduced germ of curve at O, then*

$$\delta(\xi) = \sum_p \frac{e_p(\xi)(e_p(\xi)-1)}{2},$$

the summation running on all points p equal or infinitely near to O and multiple on ξ (or, equivalently, on all points equal or infinitely near to O). In particular δ is an equisingularity invariant.

PROOF: Using the same notations as before, if $\bar{R} = R^k$ we have

$$\delta(\xi) = \dim_{\mathbb{C}} \bar{R}/R = \sum_{j=1}^{k} \dim_{\mathbb{C}} R^j/R^{j-1}.$$

If we denote by $R^1_{j,i}$ the first neighbouring ring of $R_{j,i}$, we have

$$R^j = \prod_{i=1}^{s_{j-1}} R^1_{j-1,i},$$

so that,

$$\dim_{\mathbb{C}} R^j/R^{j-1} = \sum_{i=1}^{s_{j-1}} \dim_{\mathbb{C}} R^1_{j-1,i}/R_{j-1,i}.$$

Then, it is enough to see that

$$\dim_{\mathbb{C}} R^1/R = \frac{e(e-1)}{2}$$

if R is the local ring of any reduced germ of curve, R^1 the ring in its first neighbourhood and e the multiplicity of the germ. In order to prove this, let us take the notations as in section 3.10: by 3.11.9, if n is big enough, the sequence

$$0 \to R/\mathbf{m}^n \to R^1/\mathbf{x}^n R^1 \to R^1/R \to 0$$

is exact, and by the definition of R^1 and 3.11.10,

$$\dim_{\mathbb{C}} R^1/\mathbf{x}^n R^1 = \sum_{i=1}^{s} \dim_{\mathbb{C}} R_i/\mathbf{x}^n R_i = \sum_{i=1}^{s} [\tilde{\xi}_{p_i}.nE_{p_i}] = ne,$$

the last equality coming from 3.2.5. Lastly, from 3.11.11, if n is big enough,

$$\dim_{\mathbb{C}} R/\mathbf{m}^n = ne - \frac{e(e-1)}{2},$$

so that the proof is complete. $\diamond$

The reader may have noticed that $\delta(\xi)$ appears in theorem 3.11.12 as a sum of terms depending on the infinitely near points on ξ, each multiple point giving rise to a strictly positive term, hence the usefulness of $\delta(\xi)$ as a control function for resolution of singularities.

An additivity formula for δ may easily be obtained from 3.11.12, specifically:

Corollary 3.11.13 *For any pair of non-empty and reduced germs of curve at O, ξ,ζ, sharing no branch,*

$$\delta(\xi + \zeta) = \delta(\xi) + \delta(\zeta) + [\xi.\zeta].$$

PROOF: Since, for any point p, $e_p(\xi + \zeta) = e_p(\xi) + e_p(\zeta)$,

$$e_p(\xi + \zeta)(e_p(\xi + \zeta) - 1) = e_p(\xi)(e_p(\xi) - 1) + e_p(\zeta)(e_p(\zeta) - 1) + 2e_p(\xi)e_p(\zeta).$$

The claim follows after adding up these equalities, for p multiple on $\xi + \zeta$, and using the Noether formula 3.3.1. $\diamond$

As is clear from 3.11.12, $\delta(\xi) = 1$ if the germ ξ is a either a node or an ordinary cusp. If ζ is a reduced algebraic curve, the sum $\delta(\zeta) = \sum_{p \in \zeta} \delta_p(\zeta)$ is often called, in old texts, the *virtual* or *apparent number of double points* of ζ, each proper singular point p of ζ being considered, in a certain sense, as equivalent to $\delta_p(\zeta)$ double points, either nodes or ordinary cusps.

3.12 Artin theorem for plane curves

A general theorem of M. Artin ([7], see also [85]) asserts that sufficiently good approximate formal solutions of a system of algebraic or analytic equations actually approximate true convergent solutions. We will present here a very particular version of it, namely for a single equation $f(x, y) = 0$ in two variables: in such a case the theorem easily follows from the theory already developed in this chapter, the involved bounds can be made quite explicit and the obstructions for an approximate solution to be close to a true one may be clearly seen from the singularity of the germ $f = 0$. The easy lemma that follows is the key to the theorem:

Lemma 3.12.1 *An irreducible germ γ is smooth if and only if it contains no satellite points.*

PROOF: That smooth germs do not contain satellite points has been seen in 3.6.2. Conversely, if $e_O(\gamma) > 1$, since γ contains finitely many multiple points (3.7.1), necessarily there are on γ points p and p' with p' in the first neighbourhood of p and $e_p(\gamma) > e_{p'}(\gamma)$: then, by 3.5.3, the point in the first neighbourhood of p' is proximate to p and hence a satellite point. $\diamond$

Fix $f \in \mathbb{C}\{x, y\}$. In chapter 2, by means of the Newton–Puiseux algorithm, we have solved for y, locally at the origin, the equation $f(x, y) = 0$. Assume we are now interested in analytic solutions only, that is, in series $s = s(x) \in \mathbb{C}\{x\}$, $s(0) = 0$, such that $f(x, s(x)) = 0$ identically in x. Clearly such solutions are just the Puiseux series of the smooth branches of the germ $\xi : f = 0$ at $O = (0, 0)$ that are not tangent to the y-axis. Let us say that a non-invertible series $\bar{s} \in \mathbb{C}[[x]]$ is an *approximate solution* of $f(x, y) = 0$ up to the order β if and only if $o_x f(x, \bar{s}(x)) > \beta$. Then we have:

Theorem 3.12.2 (Artin) *Given $f \in \mathbb{C}\{x, y\}$, for any $\alpha \in \mathbb{N}$ there is a $\beta \in \mathbb{N}$ such that for any $\bar{s} \in \mathbb{C}[[x]]$, approximate solution of $f(x, y) = 0$ up to the order β, there is a true analytic solution $s \in \mathbb{C}\{x\}$ such that $o_x(s - \bar{s}) > \alpha$.*

PROOF: The claim is trivially satisfied if f is invertible, as invertible series have no approximate solutions up to the order 0. Thus assume f non-invertible and denote by ξ the germ $f = 0$ at $O = (0,0)$. A path of free points on ξ will be any finite set of points $p_0, \ldots, p_r$ on ξ, from which $p_0 = O$ and each p_i is a free point in the first neighbourhood of p_{i-1} for $i > 0$. It follows from 3.12.1 above that a path of free points on ξ can be indefinitely enlarged by adding successive free points on ξ if and only if all its points lie on a smooth branch of ξ. Therefore, if a path of free points on ξ does not lie on a smooth branch of ξ, then it is contained in some maximal path. Furthermore, it is clear that there are finitely many maximal paths of free points on ξ, at most one per branch.

Fix now $\alpha \in \mathbb{N}$. Define β_1 to be the maximal value of the sums $e_{p_0}(\xi) + \cdots + e_{p_{\alpha-1}}(\xi)$ for all paths of α free points $p_0, \ldots, p_{\alpha-1}$ lying on a smooth branch of ξ ($\beta_1 = 0$ if ξ has no smooth branches). Furthermore, define β_0 as being the maximal value of the sums $e_{p_0}(\xi) + \cdots + e_{p_r}(\xi)$ for all maximal paths of free points on of ξ ($\beta_0 = 0$ if there are no maximal paths) and take $\beta = \max(\beta_0, \beta_1)$.

Let now $\bar{s} \in \mathbb{C}[[x]]$, $\bar{s}(0) = 0$ and assume that $o_x(f(x, \bar{s}(x))) > \beta$. Notice first that this hypothesis and the claim $o_x(s - \bar{s}) > \alpha$ remain unmodified if we take a suitable partial sum of $\bar{s}$ instead of $\bar{s}$ itself, so it is not restrictive to assume in the sequel that $\bar{s} \in \mathbb{C}\{x\}$. Let $\bar{\zeta}$ be the germ of curve at O with equation $y - \bar{s}(x) = 0$, which clearly is smooth and non-tangent to the y-axis. We have $o_x(f(x, \bar{s}(x))) = [\xi.\bar{\zeta}]$: we assume in the sequel this intersection multiplicity to be finite, as otherwise $s = \bar{s}$ clearly satisfies the claim. Let $O = q_0, \ldots, q_n$ be the points on both ξ and $\bar{\zeta}$: since $\bar{\zeta}$ is smooth, by 3.12.1, $q_0, \ldots, q_n$ is a path of free points on ξ. Since by the Noether formula (3.3.1)

$$e_{q_0}(\xi) + \cdots + e_{q_n}(\xi) = [\xi.\bar{\zeta}] > \beta \geq \beta_0,$$

by the definition of β_0, $q_0, \ldots, q_n$ cannot be contained in a maximal path and therefore they lie on a smooth branch ζ of ξ.

Since also

$$e_{q_0}(\xi) + \cdots + e_{q_n}(\xi) = [\xi.\bar{\zeta}] > \beta \geq \beta_1,$$

it is clear from the definition of β_1 that $q_0, \ldots, q_n$ cannot be contained in the only path of α free points on ζ, so necessarily $n + 1 > \alpha$. In particular $n > 0$, so that ζ is tangent to $\bar{\zeta}$ and therefore it is not tangent to the y-axis. The Puiseux series of ζ belongs to $\mathbb{C}\{x\}$, by 2.2.9, and is thus a true analytic solution of $f(x, y) = 0$. Furthermore, using again the Noether formula,

$$o_x(s - \bar{s}) = [\zeta.\bar{\zeta}] \geq n + 1 > \alpha$$

as wanted. ◇

Remark 3.12.3 The reader may easily see that the proof still works if the paths of free points on ξ whose second point is on the y-axis are excluded from the definitions of β_0 and β_1, which in some cases gives a smaller value of β.

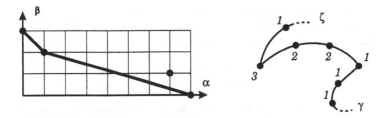

Figure 3.12: The Newton polygon and the Enriques diagram of the germ ξ in example 3.12.5.

Remark 3.12.4 Assume that f has no multiple factor and therefore that ξ is reduced. It is clear that β, as defined in the proof of 3.12.2, may be easily computed from the Enriques diagram of ξ. One may easily give other bounds: for instance 3.12.2 still holds if one takes $\beta = 2\delta(\xi) + \alpha$ as this value of β always is bigger (and often far bigger) than the one used in the proof.

Example 3.12.5 Take $f = xy^2 + y^3 - x^7y - x^8$. Then $\bar{s} = 0$ is an approximate solution up to the order 7 which does not approximate any true solution. In fact it is clear from its Newton polygon that the germ $\xi : f = 0$ has in this case two branches. One of them, say ζ, is smooth and has slope -1, while the other, call it γ, is not smooth. It is not difficult to see that γ has double points in the first and second neighbourhoods and a simple point in the third one, all of them on the x-axis, which forces the point in the fourth neighbourhood to be a satellite one. Then we have $\beta \geq \beta_0 = 8$ for any α, and the theorem may not be applied to $\bar{s}$: in fact $\bar{s} = 0$ corresponds to the x axis which approximates the non-smooth branch γ but not the smooth one. By contrast, since there are no smooth branches through the satellite point on γ in the fourth neighbourhood (3.12.1) any approximate solution up to the order 8 or higher corresponds to a smooth germ close to ζ and hence approximates the only true solution in this case, which incidentally is $y = -x$.

3.13 Exercises

3.1 *Ordinary quadratic transformations.* Let $\mathbb{P}, \bar{\mathbb{P}}$ be projective planes with homogeneous coordinates x_0, x_1, x_2 and y_0, y_1, y_2, respectively. The correspondence Q between $\mathbb{P}$ and $\bar{\mathbb{P}}$ defined by the relations

$$x_0 y_0 = x_1 y_1 = x_2 y_2$$

is called an *ordinary quadratic transformation.* The vertices of the triangles of reference are its *fundamental points* and the triangles themselves its *fundamental triangles.*

(a) Check that the graph S of Q is a smooth surface in $\mathbb{P} \times \bar{\mathbb{P}}$ and that the projections restrict to analytic morphisms $\varphi : S \longrightarrow \mathbb{P}$ and $\bar{\varphi} : S \longrightarrow \bar{\mathbb{P}}$.

(b) For each fundamental point $p_i \in \mathbb{P}$, $\bar{p}_i \in \bar{\mathbb{P}}$, $i = 1, 2, 3$, put $E_i = \varphi^{-1}(p_i)$, $\bar{E}_i = \bar{\varphi}^{-1}(\bar{p}_i)$ and describe all intersections $E_i \cap E_j$, $E_i \cap \bar{E}_j$, $\bar{E}_i \cap \bar{E}_j$.

(c) Prove that φ and $\bar{\varphi}$ restrict to isomorphisms

$$S - E_0 \cup E_1 \cup E_2 \longrightarrow \mathbb{P} - \{p_0, p_1, p_2\}$$
$$S - \bar{E}_0 \cup \bar{E}_1 \cup \bar{E}_2 \longrightarrow \bar{\mathbb{P}} - \{\bar{p}_0, \bar{p}_1, \bar{p}_2\}.$$

(d) The claim being similar for the other fundamental points, fix $p = p_0$. Prove that there is an open neighbourhood U of p in $\mathbb{P}$ so that $\varphi^{-1}(U)$ and the surface obtained from U by blowing up p are isomorphic over U. By contrast, show that no restriction of Q to a pointed neighbourhood $U - \{p\}$ of p is injective and so, even locally at p, Q does not behave like blowing up p.

(e) Let C be an algebraic curve on $\mathbb{P}$ with no side of the fundamental triangle as a principal tangent at $p = p_0$. Compare $\bar{\varphi}(\varphi^{-1}(C))$ (the total transform of C by Q) and the total transform of C by blowing up p.

The reader may find more information on ordinary quadratic transformations in [35], IV.II.13, [37], VII.4 and [75], IV.3, among many others.

3.2 Let p be a point equal or infinitely near to O.

(a) Prove that the rule

$$\gamma \sim \gamma' \Leftrightarrow [\gamma \cdot \gamma'] > \sum_{q \leq p} e_q(\gamma) e_q(\gamma')$$

defines an equivalence relation in the set T of all irreducible germs through p.

(b) Prove that there is a one to one map from $T/\sim$ onto the first neighbourhood of p, the equivalence class of a germ γ being mapped to the point in the first neighbourhood of p that γ is going through.

(c) If p' is the point on the germ γ in the first neighbourhood of p, prove that

$$e_{p'}(\gamma) = \min_{\gamma' \sim \gamma} \{[\gamma \cdot \gamma'] - \sum_{q \leq p} e_q(\gamma) e_q(\gamma')\}.$$

3.3 Prove that a germ of curve ξ contains no satellite points if and only if for each point p on ξ

$$e_p(\xi) = \sum e_q(\xi),$$

the summation ranging on the points q in the first neighbourhood of p.

3.4 Let $\varphi : S(\xi) \longrightarrow S(\zeta)$ be an equisingularity between reduced germs ξ and ζ. Prove that for each $p \in S(\xi)$ there is a one to one correspondence preserving multiplicities between the set of principal tangents to $\tilde{\xi}_p$ and the set of principal tangents to $\tilde{\zeta}_{\varphi(p)}$.

3.5 Consider the germs of curve $\gamma : y^2 - x^5 = 0$ and $\zeta : y^3 - x^5 = 0$. Prove that there is a bijection $\varphi : S(\gamma) \longrightarrow S(\zeta)$ so that for any $p, q \in S(\gamma)$, $\varphi(p)$ is infinitely near (resp. proximate) to $\varphi(q)$ if p is infinitely near (resp. proximate) to q, while γ and ζ are not equisingular.

3.6 Let $\mathcal{P} = \{p_i\}_{i \in \mathbb{N}}$ be a sequence of points with $p_0 = O$ and p_i in the first neighbourhood of p_{i-1} for $i > 0$. Prove that there is at most one irreducible germ γ with origin at O containing all points in $\mathcal{P}$. Show by an example that there exist sequences $\mathcal{P}$ from which no germ contains all points. (Sequences all whose points are lying on an irreducible germ will be characterized in 5.7.6.)

3.7 If ξ is a reduced germ of curve, the *maximal contact* of ξ is defined as the maximum (maybe ∞) of the intersection multiplicities of ξ with smooth germs. Prove that ξ has finite maximal contact if and only if it has no smooth branch. Prove also that the maximal contact is an equisingularity invariant.

3.8 Prove the equivalence between definition 3.8.1 and each one of the three definitions of equisingularity given by Zariski in [89].

3.9 Let ξ be a germ of curve whose Enriques diagram is that shown in figure 3.9. Draw the Enriques diagrams of its branches and compute their multiplicity sequences.

3.10 Describe all equisingularity classes of irreducible germs up to multiplicity three.

3.11 Prove that if a germ of curve ξ has an ordinary singularity (0.11.4) then it satisfies the equivalent conditions of 3.11.4 for $k = 1$.

3.12 *Sufficient condition for Noether's $Af + B\varphi$* (see [37], 7.5, prop. 3). Assume that the germ of curve $\xi : f = 0$, with origin at O, has an ordinary singularity, and that $g, h \in \mathcal{O}_O$ define germs $\zeta_1 : g = 0$, $\zeta_2 : h = 0$ so that

$$[\zeta_2 \cdot \gamma] \geq [\zeta_1 \cdot \gamma] + e_O(\xi) - 1$$

for any branch γ of ξ. Prove that h belongs to the ideal generated by f and g in $\mathcal{O}_O$. (*Hint:* notations being as in section 3.11, prove that g divides h in R^1 and that h/g belongs to the conductor $(R : R^1)$).

3.13 Prove that reduced germs ξ, ζ with $\delta(\xi) = \delta(\zeta)$ need not be equisingular, even if they belong to the same pencil, using the germs

$$x^2 - \alpha y^2 + y^3 = 0, \quad \alpha \in \mathbb{C}.$$

Chapter 4

Virtual multiplicities

Let O be a point on a smooth surface S. As in the preceding chapter, denote by $\mathcal{O} = \mathcal{O}_{S,O}$ the local ring of S at O and by $\mathcal{M}$ its maximal ideal. We say that a family of conditions imposed on germs of curve at O is *linear* if and only if the germs satisfying these conditions describe a linear system, that is (section 2.7) if and only if their equations describe the set of non-zero elements of an ideal of $\mathcal{O}$. Given any positive integer ν, it is clear that the condition of having multiplicity at least ν at O is a linear one, the corresponding ideal of $\mathcal{O}$ being just $\mathcal{M}^\nu$. One may ask the germs to go through given infinitely near points with prescribed multiplicities: if K is a cluster with origin at O and for each $p \in K$ there is given an integer ν_p, we may consider the family of conditions on ξ,

$$e_p(\xi) \geq \nu_p \text{ for all } p \in K.$$

In general such a family is not linear. Indeed, for an easy example, take O and the point p in its first neighbourhood on the x-axis as the points of K; then it is clear that all germs $\xi_\lambda : x^2 - y^2 + \lambda y = 0$, $\lambda \neq 0$, have at both O and p multiplicity non-less than (in fact equal to) one. Nevertheless, the germ ξ_0 does not go through p. In this chapter we introduce families of linear conditions that, although weaker, are to a certain extent similar to the conditions of going through the points of a cluster with prescribed multiplicities.

4.1   Curves through a weighted cluster

Definition: A *system of virtual multiplicities* for (the points of) a cluster K is a map $\nu : K \longrightarrow \mathbb{Z}$. We will usually write $\nu_p = \nu(p)$ and call ν_p the *virtual multiplicity* of the point p. A pair $\mathcal{K} = (K, \nu)$, where K is a cluster and ν a system of virtual multiplicities for it, will be called a *weighted cluster*.

If $\mathcal{K} = (K, \nu)$ is a weighted cluster, the points in K will be called the points of $\mathcal{K}$ and we will sometimes abuse the notations by writing $p \in \mathcal{K}$ for $p \in K$. We will also say that p is a ν_p-fold point of $\mathcal{K}$. To avoid confusions, the multiplicity

119

of a curve or germ of curve ξ at point p, $e_p(\xi)$, as defined in section 3.3, will be sometimes called the *effective* multiplicity of ξ at p. The *depth* of a cluster K with origin at O is defined as the maximum of the orders of the neighbourhoods of O which contain some point of K. The depth of a weighted cluster $\mathcal{K} = (K, \nu)$ is the depth of its underlying cluster K.

The main definitions in this chapter are as follows:

Definition: We say that a curve ξ *goes through the point O with the virtual multiplicity* ν_O (or through the depth-zero weighted cluster (O, ν_O)) if and only if $e_O(\xi) \geq \nu_O$.

Definition: Assume that the curve ξ goes through O with virtual multiplicity ν_O. Then, the *virtual transform of* ξ relative to the virtual multiplicity ν_O is

$$\check{\xi} = \tilde{\xi} + (e_O(\xi) - \nu_O)E$$

where $\tilde{\xi}$ and E denote, respectively, the strict transform of ξ and the exceptional divisor of blowing up O.

Notice that also

$$\check{\xi} = \bar{\xi} - \nu_O E$$

so that one may understand the virtual transform $\check{\xi}$ as being obtained from ξ like the strict transform, but formally considering O as a ν_O-fold point of ξ. In particular we have $\check{\xi} = \tilde{\xi}$ if and only if $e_O(\xi) = \nu_O$.

Assume now that $\mathcal{K} = (K, \nu)$ is a weighted cluster which has origin at O and positive depth. Denote by p_i, $i = 1, \ldots, s$ the points of $\mathcal{K}$ in the first neighbourhood of O.

For each $i = 1, \ldots, s$, denote by K_i the cluster with origin at p_i that contains p_i and all points infinitely near to it in K. For $i = 1, \ldots, s$, the restriction of ν to K_i is a system ν_i of virtual multiplicities for K_i. It gives rise to a weighted cluster $\mathcal{K}_i = (K_i, \nu_i)$ whose depth is strictly less than the depth of $\mathcal{K}$.

Then we complete the first definition above using induction on the depth of K:

Definition: If $\mathcal{K} = (K, \nu)$ has positive depth, we say that a curve ξ *goes through $\mathcal{K}$* if and only if

(a) ξ goes through O with virtual multiplicity ν_O and,

(b) the virtual transform of ξ relative to the virtual multiplicity ν_O goes through $\mathcal{K}_i$, for $i = 1, \ldots, s$.

We will equivalently say that ξ goes through the cluster K, or through the points of K, with the virtual multiplicities ν.

The second definition above needs to be completed too. Keep the notations as before and assume that ξ goes through $\mathcal{K}$:

Definition: The *virtual transform* $\check{\xi}_p$ of ξ with origin at p, relative to the virtual multiplicities ν, is defined as being the germ of curve $\check{\xi}_p = \xi_O$ if $p = O$.

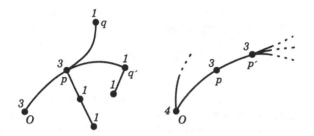

Figure 4.1: The Enriques diagram of a weighted cluster (on the left) and that of a germ of curve through it (on the right). The point p' is not assumed to agree with either q or q'.

Otherwise and using induction on the order of the neighbourhood p belongs to, we assume that p lies in the first neighbourhood of q and define $\check{\xi}_p$ as the germ at p of the virtual transform relative to the virtual multiplicity ν_q of any representative of $\check{\xi}_q$.

The above definitions may be used on germs of curve as well: if ξ is a germ of curve, we say that ξ *goes through* $\mathcal{K} = (K, \nu)$ if and only if one of its representatives ξ' does so. In such a case the virtual transform of ξ' with origin at $p \in K$ will be denoted by $\check{\xi}_p$ and called the *virtual transform* of ξ with origin at p, relative to ν. Obviously neither the condition of going through $\mathcal{K}$ nor the virtual transforms $\check{\xi}_p$ depend on the representative ξ' of ξ we are using.

Back to the example given at the beginning of this chapter, the reader may see that all germs ξ_λ go through the cluster $\{O, p\}$ with the virtual multiplicities $\{1, 1\}$. In particular ξ_0 shows that a germ ξ need not have $e_q(\xi) \geq \nu_q$ for all $q \in K$ in order to go through (K, ν).

Now, after the definitions, our first job is to show that the family of conditions of going through a weighted cluster is a linear one. Denote by $\mathcal{O}_p = \mathcal{O}_{S_p,p}$ the local ring at p of the surface S_p on which p lies as a proper point.

Proposition 4.1.1 *The equations of the germs going through a weighted cluster $\mathcal{K} = (K, \nu)$ describe the set of non-zero elements of an ideal $H_\mathcal{K}$ of $\mathcal{O}$ which is either $\mathcal{M}$-primary or the whole of $\mathcal{O}$. Furthermore, for each $p \in K$ there is a morphism of $\mathcal{O}$-modules $\psi_p : H_\mathcal{K} \longrightarrow \mathcal{O}_p$ which is continuous for the Zariski topology and maps each $f \in H_\mathcal{K}$ to an equation $\psi_p(f)$ of the virtual transform $\check{\xi}_p$ of $\xi : f = 0$.*

PROOF: Let us set $\nu' = \max\{\nu_O, 0\}$ and use the notations as above. If K has depth zero, then $H_\mathcal{K} = \mathcal{M}^{\nu'}$ and the claim is obvious.

If K has positive depth, for each point $p_i \in K$ in the first neighbourhood of O we fix a local equation at p_i of the exceptional divisor and call it x_i. Let

$$\varphi_i : \mathcal{O} \longrightarrow \mathcal{O}_{p_i}$$

be the morphisms induced by blowing up. They are continuous by 2.7.6. Since $\varphi_i(\mathcal{M}^{\nu'}) \subset x_i^{\nu'}\mathcal{O}_{p_i}$ (3.2.1) we can define

$$\psi_i : \mathcal{M}^{\nu'} \longrightarrow \mathcal{O}_{p_i}$$

by the rule $\psi_i(f) = x_i^{-\nu_O}\varphi_i(f)$. If $f \in \mathcal{M}^{\nu'}$ is the equation of a curve ξ, then $\psi_i(f)$ is an equation of the germ at p_i of the virtual transform of ξ relative to the multiplicity ν_O. Furthermore, since x_i may be taken as a local coordinate at p_i, dividing by $x_i^{\nu_O}$ obviously gives a continuous map (for the Zariski topologies) $x_i^{\nu_O}\mathcal{O}_{p_i} \longrightarrow \mathcal{O}_{p_i}$, and hence ψ_i is continuous too.

Thus, assuming by induction the claim to be true for the clusters $\mathcal{K}_i$, we obviously have

$$H_\mathcal{K} = \bigcap_i \psi_i^{-1}(H_{\mathcal{K}_i})$$

and hence the first half of the claim. Then, for the morphisms ψ_p, either take the corresponding ψ_i, if $p = p_i$ is in the first neighbourhood of O, or compose the suitable ψ_i with one of the morphisms provided by the induction hypothesis.
◇

Notice also that 4.1.1 implies that there exists s_0 such that $\mathcal{M}^s \subset H_\mathcal{K}$ for $s \geq s_0$, so that all germs of curve ξ with $e_O(\xi) \geq s_0$ go through $\mathcal{K}$. The ideal $H_\mathcal{K}$ is thus either the whole ring $\mathcal{O}$ or a $\mathcal{M}$-primary ideal. Hence, by 2.7.5, $H_\mathcal{K}$ is a Zariski-closed set and a germ of curve goes through $\mathcal{K}$ if and only if the coefficients of any of its equations satisfy a well determined finite system of homogeneous linear equations. Of course such equations involve only finitely many coefficients of the series.

The first half of 4.1.1 may be equivalently stated by saying that the germs through a weighted cluster describe a linear system without fixed part (see section 2.7). It follows also from 4.1.1 that if germs generating a linear system go through $\mathcal{K}$, then all germs in the linear system actually go through $\mathcal{K}$.

The next proposition is similar to 3.3.4, but for the fact that virtual multiplicity and virtual transform are used for one of the curves instead of the effective and strict ones. We shall often call it the *virtual Noether formula*.

Proposition 4.1.2 *Assume that ξ and ζ are curves defined in a neighbourhood of O so that ζ goes through O with the virtual multiplicity ν_O. Then we have*

$$[\xi.\zeta]_O = e_O(\xi)\nu_O + \sum_{p \in E}[\tilde{\xi}.\check{\zeta}]_p,$$

where E is the exceptional divisor of blowing up O, and $\tilde{\xi}$ and $\check{\zeta}$ are the strict and virtual transforms, respectively, of ξ and ζ, O taken with the virtual multiplicity ν_O.

PROOF: By the definition of virtual transform,

$$\check{\zeta} = \tilde{\zeta} + (e_O(\zeta) - \nu_O)E,$$

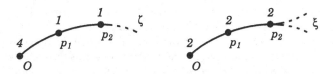

Figure 4.2: The example in 4.1.3: the germ ζ and the points of K with their virtual multiplicities on the left, and the germ ξ on the right.

so we have

$$e_O(\xi)\nu_O + \sum_{p \in E}[\tilde{\xi}.\tilde{\zeta}]_p = e_O(\xi)\nu_O + \sum_{p \in E}[\tilde{\xi}.\tilde{\zeta}]_p + (e_O(\zeta) - \nu_O)\sum_{p \in E}[\tilde{\xi}.E]_p.$$

Then, since $\sum_{p \in E}[\tilde{\xi}.E]_p = e_O(\xi)$ by 3.2.10, the claim follows from 3.3.4. ◇

Remark 4.1.3 Assume that a weighted cluster $\mathcal{K} = (K, \nu)$ and a germ of curve ζ are given. It easily follows from 4.1.2 that if a germ ξ goes through $\mathcal{K}$, then

$$[\zeta.\xi] \geq \sum_{p \in K} e_p(\zeta)\nu_p.$$

Nevertheless, the converse fails to be true, even if ζ is irreducible and all points of K belong to ζ. To see an easy example, take $K = \{O, p_1, p_2\}$, where p_1 and p_2 are on the x-axis, in the first and the second neighbourhood of O, respectively, and $\nu_O = 4$, $\nu_{p_1} = \nu_{p_2} = 1$. Let ζ be the x-axis and ξ have equation $(y - x^3)(y + x^3) = 0$, so that $e_O(\xi) = e_{p_1}(\xi) = e_{p_2}(\xi) = 2$: clearly ξ does not go through (K, ν) but satisfies the above inequality.

4.2 When virtual multiplicities are effective

Let $\mathcal{K} = (K, \nu)$ be a weighted cluster. We will say that a curve or germ ξ *goes through $\mathcal{K}$ with effective multiplicities equal to the virtual ones* if and only if $e_p(\xi) = \nu_p$ for all points $p \in K$. We will equivalently say that ξ has effective multiplicities equal to the virtual ones at the points of $\mathcal{K}$. It is clear that such curves or germs are in particular going through $\mathcal{K}$.

We have already seen that there may be germs ξ going through (K, ν) and such that $e_p(\xi) \neq \nu_p$, and even $e_p(\xi) < \nu_p$, for some $p \in K$. In fact, since the effective multiplicities $e_p(\xi)$ of a germ ξ need to satisfy the proximity equalities (3.5.3) and no restrictions have been imposed on the virtual multiplicities ν, it is easy to give examples of weighted clusters through which no germ goes with effective multiplicities equal to the virtual ones: indeed, it is enough to have in the cluster a point whose virtual multiplicity is less than the sum of the virtual multiplicities of its proximate points in the cluster, or, in particular, points p and q with q in the first neighbourhood of p and $\nu_q > \nu_p$.

We will show in this section that the proximity equalities provide a set of necessary and sufficient conditions for the existence of germs through a weighted cluster with effective multiplicities equal to the virtual ones. First of all we will deal with the germs whose effective multiplicities differ from the virtual ones: assume as before that $\mathcal{K} = (K, \nu)$ is a weighted cluster.

Proposition 4.2.1 *The equations of the germs ξ that go through $\mathcal{K}$ and have $e_p(\xi) \neq \nu_p$ for some $p \in K$ are the non-zero elements of the union of a finite set of ideals $I_1, \ldots, I_r \subset \mathcal{O}$. Furthermore, each I_j is either $\mathcal{M}$-primary or the whole of $\mathcal{O}$.*

PROOF: We use the same notations as in the proof of 4.1.1 and, once again, induction on the depth of $\mathcal{K}$.

If $\nu_O < 0$, the claim is obviously satisfied by taking $r = 1$ and $I_1 = \mathcal{O}$. Thus we assume $\nu_O \geq 0$ in the sequel.

If $\mathcal{K}$ has depth zero, the claim is also obvious: $r = 1$ and $I_1 = \mathcal{M}^{\nu_O + 1}$.

Assume thus that $\mathcal{K}$ has positive depth and, by induction, that the clusters $\mathcal{K}_i$ and the restrictions of ν satisfy the claim: let $I_{i,1}, \ldots, I_{i,r_i}$ be the ideals of the claim for $\mathcal{K}_i$. Since the curves ξ we are looking for have either $e_O > \nu_O$ or $e_p(\xi) \neq \nu_p$ for some p in one of the $\mathcal{K}_i$, it is clear that the claim is true if one takes $I_1 = \mathcal{M}^{\nu_O + 1}$ and the other I_j to be $\psi_i^{-1}(I_{i,j})$, $i = 1, \ldots, s$, $j = 1, \ldots, r_i$. ◇

After 4.2.1 we see that the germs that do not have effective multiplicities equal to the virtual ones at all points of $\mathcal{K}$ are distributed in finitely many linear systems. Nevertheless, as noticed before, such linear systems may cover the whole of the germs through $\mathcal{K}$. The next condition will be used for characterizing the weighted clusters $\mathcal{K}$ for which there is a germ of curve having effective multiplicities equal to the virtual ones at all points of $\mathcal{K}$.

Definition: A *consistent cluster* is a weighted cluster $\mathcal{K} = (K, \nu)$ such that, for all $p \in K$,

$$\nu_p \geq \sum_q \nu_q,$$

the summation running on all points $q \in K$ which are proximate to p.

In particular we will refer to the inequality above as the *proximity inequality at p* and so, if $\mathcal{K}$ is consistent we will equivalently say that $\mathcal{K}$ satisfies the proximity inequalities.

Any empty sum being equal to zero by definition, the reader may notice that if $\mathcal{K} = (K, \nu)$ is consistent, then $\nu_p \geq 0$ for all $p \in \mathcal{K}$.

Theorem 4.2.2 *If there is a germ of curve going through $\mathcal{K} = (K, \nu)$ with effective multiplicities equal to the virtual ones, then $\mathcal{K}$ is consistent. Conversely, assume that a finite set T of points infinitely near to O and not in K is fixed:*

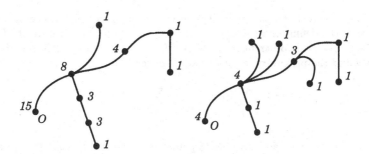

Figure 4.3: Examples of a consistent (left) and a non-consistent (right) weighted cluster.

if $\mathcal{K}$ is consistent, there exists a germ of curve going through $\mathcal{K}$ with effective multiplicities equal to the virtual ones and missing all points in T.

PROOF: The first half of the claim is obvious from 3.5.3. For the converse we will use induction on the depth of $\mathcal{K}$. After substituting for each point in T the first point preceding it that does not belong to K, we may assume without loss of generality that any point in T belongs to the first neighbourhood of some point in K. Let $q_1, \ldots, q_r$ be the points in the first neighbourhood of O that belong to T.

Assume first that the depth of $\mathcal{K}$ is zero. Then the only proximity inequality implies that $\nu_O \geq 0$, so that any non-zero element of $\mathcal{M}^{\nu_O} - \mathcal{M}^{\nu_O+1}$ defines a germ of curve ξ with $e_O(\xi) = \nu_O$. Since it is clear that ξ may be chosen missing all points q_j, $j = 1, \ldots, r$, (by 3.2.2), the claim is satisfied.

If $\mathcal{K}$ has positive depth, let p_i, $i = 1, \ldots, s$ be the points of $\mathcal{K}$ in the first neighbourhood of O and take the clusters $\mathcal{K}_i$, $i = 1, \ldots, s$, as in the proofs of 4.1.1 and 4.2.1. If E is the exceptional divisor of blowing up O, for each $i = 1, \ldots, s$, let q'_i be the first point on E infinitely near to p_i which does not belong to K. Then we define the set T_i as containing all points in T that are infinitely near to p_i and the point q'_i. Assume by induction that, for each i, there is a germ ξ_i with origin at p_i, going through the points of $\mathcal{K}_i$ with effective multiplicities equal to the virtual ones and missing all points in T_i. Then, ξ_i does not go through q'_i and we have, by the Noether formula,

$$[\xi_i . E_{p_i}] = \sum_{\substack{p \in \mathcal{K}_i \\ p \text{ prox. to } O}} e_p(\xi_i) = \sum_{\substack{p \in \mathcal{K}_i \\ p \text{ prox. to } O}} \nu_p.$$

In particular it is clear that no one of the ξ_i has the germ of E as a component. By 3.2.11, there is, for each $i = 1, \ldots, s$, a germ ζ_i, with origin at O and a single principal tangent, whose strict transform at p_i is ξ_i. Take $\xi' = \zeta_1 + \cdots + \zeta_s$: since each ζ_i has a single principal tangent, by 3.2.2 and 3.2.11, we have $\tilde{\xi}'_{p_i} = \xi_i$ and $\tilde{\xi}'_p = \emptyset$ if $p \neq p_i$, $i = 1, \ldots, s$. By its own construction it is clear that $e_p(\xi') = \nu_p$

for all $p \in K - \{O\}$. Furthermore, it is also clear that ξ' contains no point in T, because of the definition of the ξ_i and the fact that ξ' has no point in the first neighbourhood of O different from the p_i , $i = 1, \ldots, s$. From 3.2.5, the above equality and the proximity inequality at O we get

$$e_O(\xi') = \sum_i [\xi_i . E_{p_i}] = \sum_{\substack{p \in \mathcal{K} \\ p \text{ prox. to } O}} \nu_p \leq \nu_O.$$

Thus it is enough to add $\nu_O - e_O(\xi')$ different smooth branches to ξ', each missing all points $p_1, \ldots, p_s, q_1, \ldots, q_r$, to get a germ ξ satisfying the claim. ◇

Remark 4.2.3 It follows from 4.2.2 that the germs going through a consistent cluster $\mathcal{K}$ with effective multiplicities equal to the virtual ones share no point other than those in $\mathcal{K}$.

Corollary 4.2.4 *If $\mathcal{K}$ is consistent, then the equations of the germs going through $\mathcal{K}$ with effective multiplicities equal to the virtual ones describe a non-empty Zariski-open set in $H_\mathcal{K}$.*

PROOF: Follows from 4.2.1, 2.7.5 and 4.2.2. ◇

In corollary 4.2.7 we will see that, in addition, generic germs through $\mathcal{K}$ have all their singular points in K. The next two propositions will be useful to this end.

Proposition 4.2.5 *Assume that a germ ξ goes through $\mathcal{K} = (K, \nu)$ with effective multiplicities equal to the virtual ones. For each $p \in K$ put $\rho_p = \nu_p - \sum \nu_q$, the summation being on the points q in K which are proximate to p. Then ξ has no singular points outside of K if and only if, for each $p \in K$, ξ has just ρ_p different points in the first neighbourhood of p not in K and furthermore all these points are free.*

PROOF: Since ξ goes through $\mathcal{K}$ with effective multiplicities equal to the virtual ones, for any $p \in K$, the proximity equality 3.5.3 gives

$$\rho_p = \sum_q e_q(\xi), \tag{4.1}$$

the summation being on the points q on ξ that do not belong to K and are proximate to p.

Assume that ξ has no singular points outside of K. Then all points q above are free points and therefore need to belong to the first neighbourhood of p. Thus they are just the points on ξ in the first neighbourhood of p that do not belong to K. Furthermore, such points are simple on ξ, as they are non-singular. Hence the equality above shows the number of these points to be just ρ_p as claimed.

For the converse, assume that p' is on ξ and $p' \notin K$: denote by p the last point in K the point p' is infinitely near to. By hypothesis there are $\rho = \rho_p$ different points $q_1, \ldots, q_\rho$ in the first neighbourhood of p and not in K and all these points are free. Then the equality 4.1 shows that the points $q_1, \ldots, q_\rho$ are the only points on ξ that are proximate to p and do not belong to K, and also that all of them are simple on ξ. The point p' being equal or infinitely near to one of the q_i, say to q_1, it is also simple on ξ. Assume that p' is satellite: then $p' \neq q_1$ and, by 3.6.2, p' needs to be proximate to some multiple point, let us call it p_0. Since q_1 is simple, p_0 precedes q_1, and so q_1 itself (like all points between p_0 and p') is proximate to p_0. Since q_1 is free by hypothesis, necessarily $p_0 = p$ and p' is then proximate to p. We have already seen that the points q_i are the only points on ξ and not in K that are proximate to p: this gives $p' = q_1$ against the hypothesis. $\diamond$

If a curve or germ of curve ξ goes through $\mathcal{K}$ with effective multiplicities equal to the virtual ones and has no singular points outside of $\mathcal{K}$, we will say for short that ξ *goes sharply* through $\mathcal{K}$.

In the sequel the integer

$$\rho_p = \nu_p - \sum_{\substack{q \in K \\ q \text{ prox. to } p}} \nu_q$$

will be called the *excess* of $\mathcal{K}$ at p. Note that $\rho_p \geq 0$ for all $p \in K$ if and only if $\mathcal{K}$ is consistent. In such a case, for each $p \in K$, the germs going sharply through $\mathcal{K}$ have just ρ_p points in the first neighbourhood of p not belonging to $\mathcal{K}$ and these points are all non-singular (4.2.5).

Proposition 4.2.6 *The following claims hold true if $\mathcal{K} = (K, \nu)$ is consistent:*

(a) *All germs going sharply through $\mathcal{K}$ are reduced.*

(b) *If ξ goes sharply through $\mathcal{K}$, then, for each $p \in K$, ξ has just ρ_p branches through p missing all points after p in K. Hence ξ has a total of $\sum_{p \in K} \rho_p$ branches.*

(c) *Any two germs going sharply through $\mathcal{K}$ are equisingular.*

PROOF: If ξ goes sharply through $\mathcal{K}$, then all points on ξ not in K are simple points. Thus ξ is reduced by 3.7.10. Furthermore, since each simple point belongs to a single branch of ξ, claim (b) follows from 4.2.5.

Assume that ξ_1, ξ_2 are germs going sharply through $\mathcal{K}$. For $i = 1, 2$, let K_i be the cluster containing all points in K and all points on ξ_i in the first neighbourhood of some point in K. Since ξ_i has no singular points outside of K, clearly $S(\xi_i) \subset K_i$. For any $p \in K$ both germs have the same number of points in the first neighbourhood of p and not in K, namely, as noticed above, ρ_p different points. Thus the identity map of K may be easily extended to a bijection between K_1 and K_2 which in turn restricts to an equisingularity between ξ_1 and ξ_2. $\diamond$

We show next that if $\mathcal{K}$ is consistent, then generic germs through $\mathcal{K}$ go sharply through it.

Corollary 4.2.7 *If the weighted cluster $\mathcal{K} = (K, \nu)$ is consistent, then the equations of the germs going sharply through it describe a non-empty Zariski-open set in $H_\mathcal{K}$.*

PROOF: Let us see first that there exists a germ going sharply through $\mathcal{K}$. For this define a second weighted cluster $\mathcal{K}'$ as follows: take all points of K with their own virtual multiplicities and furthermore, for each $p \in K$, pick ρ_p free points in the first neighbourhood of p not already in K, all taken with virtual multiplicity one. It is clear that $\mathcal{K}'$ is consistent too and hence, by 4.2.2, there exists a germ through $\mathcal{K}'$ with effective multiplicities equal to the virtual ones. Then, by 4.2.5, such a germ goes sharply through $\mathcal{K}$.

It remains to see that the equations of the germs going sharply through $\mathcal{K}$ describe a Zariski-open set in $H_\mathcal{K}$. We already know, from 4.2.4, that the equations of the curves through $\mathcal{K}$ with effective multiplicities equal to the virtual ones describe an open set $U \subset H_K$. For each $p \in K$ denote by π_p the composition of the blowing-ups giving rise to p. Call $\mathcal{K}_p$ the weighted cluster consisting of p and all points infinitely near to p in K, all taken with the same virtual multiplicities as in $\mathcal{K}$. The morphism of 4.1.1 is in fact a morphism

$$\psi_p : H_\mathcal{K} \longrightarrow H_{\mathcal{K}_p}$$

and maps the equation of any germ through $\mathcal{K}$ to an equation of its virtual transform with origin at p. Let $U_p \subset H_{\mathcal{K}_p}$ be the set of all equations of all germs ζ, with origin at p and going through $\mathcal{K}_p$, such that

(a) $e_p(\zeta) = \nu_p$,

(b) ζ has ρ_p different points not belonging to $\mathcal{K}_p$ in the first neighbourhood of p, and

(c) all these points are free.

It is clear that after seeing that U_p is an open subset of $H_{\mathcal{K}_p}$, for all $p \in K$, we will have the claim proved. Indeed, the morphisms ψ_p are continuous for the Zariski topology (4.1.1) and the set we claim to be open is, by 4.2.5,

$$U \cap \left(\bigcap_{p \in K} \psi_p^{-1}(U_p) \right).$$

Now fix $p \in K$. We shall prove that U_p is open in $H_{\mathcal{K}_p}$. Let x, y be local coordinates at p. Assume that if $p \neq O$ they are chosen so that the exceptional divisor of π_p has local equation at p either x or xy, according as p is a free or a satellite point (section 3.6). We will use homogeneous coordinates z_0, z_1 on the exceptional divisor E of blowing up p as in section 3.2. In particular, if

p is free, the satellite point in its first neighbourhood is $[0, 1]$, while if p is a satellite point, then the satellite points in its first neighbourhood are $[0, 1]$ and $[1, 0]$. Assume that $p_{1,j} = [\alpha_j, \beta_j]$, $j = 1, \ldots, s$, are the points of $\mathcal{K}$ in the first neighbourhood of p. If ζ is a germ of curve through $\mathcal{K}_p$ and f is any equation of ζ, we write

$$f = f_{\nu_p} + \cdots + f_i + \cdots$$

where each f_i is a homogeneous polynomial of degree i in x, y. The condition $e_p(\zeta) = \nu_p$ is obviously equivalent to $f_{\nu_p} \neq 0$ which clearly gives rise to an open subset in $H_{\mathcal{K}_p}$.

Denote by $p_{i,j}$, $i = 2, \ldots, r_j$ the points of $\mathcal{K}_p$ proximate to p in the successive neighbourhoods of $p_{1,j}$, and by δ_j the sum of its virtual multiplicities. One may inductively use 4.1.2 in order to obtain $[\tilde{\zeta}_{p_{1,j}}.E_{p_{1,j}}] \geq \delta_j$ where $\tilde{\zeta}_{p_{1,j}}$ is the strict transform of ζ at $p_{1,j}$. Then it follows from 3.2.2 that, for each j, f_{ν_p} has the factor $\beta_j x - \alpha_j y$ with multiplicity at least δ_j. Write

$$g = \prod_{j=1}^{s} (\beta_j x - \alpha_j y)^{\delta_j}$$

and

$$f_{\nu_p} = f' g.$$

Since $f'(z, 1)$ is the quotient of the Euclidean division of $f_{\nu_p}(z, 1)$ by $g(z, 1)$, it is clear that the coefficients of f' are polynomial functions in the coefficients of f_{ν_p}. The degree of f' is just $\nu_p - \sum_j \delta_j = \rho_p$ so that, using 3.2.2 again, ζ has ρ_p different points in the first neighbourhood of p and not in $\mathcal{K}_p$ if and only if f' has no multiple factor and $f'(\alpha_j, \beta_j) \neq 0$ for $j = 1, \ldots, s$. Furthermore, conditions for all such points to be free are $f'(0, 1) \neq 0$ if p is free, or $f'(0, 1) \neq 0$, $f'(1, 0) \neq 0$ if p is satellite. All these conditions giving rise to Zariski-open subsets, the proof is complete. $\diamond$

One may even add to 4.2.7 the condition of the germs to miss finitely many already fixed points not in K, i.e.:

Corollary 4.2.8 *Let $\mathcal{K} = (K, \nu)$ be a consistent cluster and assume that there is given a finite set of points T so that $K \cap T = \emptyset$. Then there is a a non-empty Zariski-open set $U \subset H_{\mathcal{K}}$ so that for any $f \in U$, the germ $\xi : f = 0$ goes sharply through $\mathcal{K}$ and no point in T belongs to ξ.*

PROOF: It is clearly non-restrictive to assume that all points $q \in T$ lie in the first neighbourhood of some point in K (for each $q \in T$, take the first point preceding q and not in K instead of q itself, if needed). For each $q \in T$ let $\mathcal{K}_q$ be the weighted cluster obtained from $\mathcal{K}$ by adding the point q with virtual multiplicity one. We know from 4.2.2 that the Zariski-open set

$$V = H_{\mathcal{K}} - \bigcup_{q \in T} H_{\mathcal{K}_q}$$

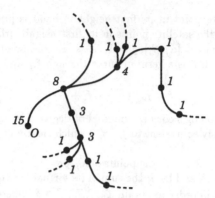

Figure 4.4: The Enriques diagram and the effective multiplicities of the germs going sharply through the consistent cluster of figure 4.3.

is non-empty. If V' is the open set of the equations of all germs going sharply through $\mathcal{K}$ (4.2.7), then $U = V \cap V'$ satisfies the claim, as if the point q belongs to a germ going sharply through $\mathcal{K}$ then this germ necessarily goes through $\mathcal{K}_q$ too. ◇

We close this section by proving an existence result that will be useful later on:

Lemma 4.2.9 *Assume that there are given a point p infinitely near to O, a finite set T of points in the first neighbourhood of p and a positive integer n. Then there exists an irreducible germ γ such that $e_p(\gamma) = n$ and the point on γ in the first neighbourhood of p is free and does not belong to T. In particular, if $n = 1$ all points after p on γ are simple and free.*

PROOF: Choose any free point $p_1 \notin T$ in the first neighbourhood of p and, if $n > 1$, take $p_2, \ldots, p_n$ to be the points proximate to p in the first, $\ldots$, $(n-1)$-th neighbourhoods of p_1. Let K be the cluster consisting of the point p, all points preceding it, and the points $p_1, \ldots, p_n$. Define a system of virtual multiplicities for K in the following way: $\nu_{p_1} = \cdots = \nu_{p_n} = 1$, $\nu_p = n$ and, if q precedes p, by a decreasing recurrent procedure we take

$$\nu_q = \sum_{\substack{p' \in K \\ p' \text{ prox. to } q}} \nu_{p'}.$$

The reader may easily check that the resulting weighted cluster $\mathcal{K} = (K, \nu)$ has all excesses zero, but for its last point which has excess one: $\rho_q = 0$ if $q \in K$, $q \neq p_n$ and $\rho_{p_n} = 1$. Then $\mathcal{K}$ is in particular consistent and it is enough to take γ as being any germ going sharply through it. Clearly the condition $e_p(\gamma) = n$ is fulfilled. Furthermore, γ is unibranched by 4.2.6 and p_1, which has been chosen free and not in T, is the point on γ in the first neighbourhood of p because $\nu_{p_1} = 1$. $\diamond$

4.3 Blowing up all points in a cluster

Let K be a cluster. In this section we define an inductive procedure that is essentially unique and that may be understood as the successive blowing up of all points in K, provided that after each blowing-up $S_j \longrightarrow S_{j-1}$ we identify the points not yet blown up with their corresponding ones on the surface S_j.

In fact, for using induction, it is better to deal with many clusters together, so assume that $K_1, \ldots, K_s$ are clusters with origins $O_1, \ldots, O_s \in S$ and put $K = K_1 \cup \cdots \cup K_s$. We call a total ordering $\prec$ on K *admissible* if and only if for any $p, q \in K$, q infinitely near to p, we have $p \prec q$. The existence of an admissible ordering on K may be easily proved by the reader using induction. Assume that an admissible ordering $\prec$ on K has been fixed and let p_1 be the first point: p_1 needs to be a proper point on S, i.e. the origin of one of the clusters K_i. After reordering the clusters if needed, let us assume that $i = s$, and thus, $p_1 = O_s$. Denote by $q_1, \ldots, q_r$ the points of K_s in the first neighbourhood of O_s ($r = 0$ if $K_s = \{O_s\}$). Let $\pi_1 : S^{(1)} \longrightarrow S$ be the blowing-up of O_s: if $K = \{O_s\}$ we stop here by taking $\pi_K = \pi_1$ and $S_K = S^{(1)}$. If $K \neq \{O_s\}$, then let φ be the inverse of the isomorphism induced by π_1 in the complementary set of the exceptional divisor. Identify each cluster K_i, $i = 1, \ldots, s-1$ with $K_i^{(1)} = \varphi(K_i)$, which is a cluster with origin $\varphi(O_i) \in S^{(1)}$. Furthermore, for $i = 1, \ldots, r$, let $K_{s+i-1}^{(1)}$ be the cluster with origin q_i that contains q_i itself and all points infinitely near to q_i in K_s. We take $K^{(1)} = \bigcup_{i=1}^{s+r-1} K_i^{(1)}$ which clearly may be identified with $K - \{O_1\}$ and we restart from $S^{(1)}$ and $K^{(1)}$ with the admissible ordering coming from $\prec$ through the identification. If ℓ is the cardinal of K, after ℓ steps there are no further points to blow up and we have obtained a sequence of blowing-ups

$$S^{(\ell)} \xrightarrow{\pi_\ell} S^{(\ell-1)} \to \ldots \to S^{(1)} \xrightarrow{\pi_1} S.$$

Then we take $S_K = S^{(\ell)}$ and $\pi_K = \pi_1 \circ \cdots \circ \pi_\ell$: we call π_K the *blowing-up of (all points in)* K and, thus, we say that S_K has been obtained by blowing up K.

Obviously the exceptional divisor of π_K (as introduced in section 3.5) is

$$F_K = \pi_K^{-1}(\{O_1, \ldots, O_s\})$$

and 3.5.5 applies to it. We know from 3.5.6 that each component of F_K is the strict transform of (and hence isomorphic to) the exceptional divisor of a π_j. Since in the above procedure the centre p'_j of each π_j has been identified with the j-th point p_j in K, the components of F_K are in one to one correspondence with the points of K. We write F_p for the component of F_K corresponding to $p \in K$: F_p is the strict transform of the exceptional divisor of the blowing-up whose centre has been identified with p.

If ξ is a curve we call $\bar{\xi}_K = \pi_K^*(\xi)$ the *total transform* of ξ after blowing up K. We define the *strict transform* $\tilde{\xi}_K$ of ξ by π_K, or by blowing up K, as being the curve obtained from $\bar{\xi}_K$ by removing all its components contained in F_K. Clearly both transforms may be also obtained as the iterated total and strict transform of ξ by the blowing-ups composing π_K, and both the germs of $\bar{\xi}_K$ and $\tilde{\xi}_K$ at F_K depend only on the germ of ξ at $\{O_1, \ldots, O_s\}$ and not on the curve ξ itself: they will be called the *total* and *strict transform* of the germ of ξ at $\{O_1, \ldots, O_s\}$ by π_K or by blowing up K.

Our goal is to prove that all the above notions are independent of the admissible ordering $\prec$ we have used for defining them. To this end we first state the next lemma.

Lemma 4.3.1 *Let p_1, p_2 be distinct points on S and $\pi_i : S_i \longrightarrow S$, $i = 1, 2$ the corresponding blowing-ups. Put $q_2 = \pi_1^{-1}(p_2)$ and $q_1 = \pi_2^{-1}(p_1)$. Let $\pi_{1,2} : S_{1,2} \longrightarrow S_1$ and $\pi_{2,1} : S_{2,1} \longrightarrow S_2$ be the blowing-ups of q_2 and q_1, respectively. Then there is a unique isomorphism $\varphi : S_{1,2} \longrightarrow S_{2,1}$ over S, i.e., one that makes commutative the diagram*

$$
\begin{array}{ccc}
S_{1,2} \overset{\varphi}{\cong} S_{2,1} & \xrightarrow{\ \pi_{2,1}\ } & S_2 \\
\big\downarrow{\scriptstyle \pi_{1,2}} & & \big\downarrow{\scriptstyle \pi_2} \\
S_1 & \xrightarrow{\ \pi_1\ } & S.
\end{array}
$$

PROOF: Put $E_1 = \pi_1^{-1}(p_1)$, $F_1 = \pi_{1,2}^{-1}(E_1)$, $F_2 = \pi_{1,2}^{-1}(q_2)$, $E_2 = \pi_2^{-1}(p_2)$, $G_2 = \pi_{2,1}^{-1}(E_2)$, and $G_1 = \pi_{2,1}^{-1}(q_1)$.

The uniqueness of φ is obvious: $\bar{\pi}_1 = \pi_1 \circ \pi_{1,2}$ induces an isomorphism from $S_{1,2} - F_1 \cup F_2$ onto $S - \{p_1, p_2\}$, and similarly for $\bar{\pi}_2 = \pi_{2,1} \circ \pi_2$; then φ is determined by the commutativity condition on a dense subset of $S_{1,2}$, and hence on the whole of it.

The restriction of π_1 is an isomorphism

$$\pi_{1|S_1 - E_1} : S_1 - E_1 \longrightarrow S - \{p_1\}$$

that maps q_2 to p_2. Thus, it lifts to an S-isomorphism

$$S_{1,2} - F_1 \longrightarrow S_2 - \{q_1\}$$

that composed with $\pi_{2,1}^{-1}$ gives an S-isomorphism

$$\varphi_1 : S_{1,2} - F_1 \longrightarrow S_{2,1} - G_1.$$

On the other hand, we have another isomorphism

$$\pi_{2|S - p_2}^{-1} : S - p_2 \longrightarrow S_2 - E_2$$

mapping p_1 to q_1. Therefore it induces an S-isomorphism

$$S_1 - \{q_2\} \longrightarrow S_{2,1} - G_2,$$

which, after composition with $\pi_{1,2}|_{S_{1,2}-F_2}$, gives

$$\varphi_2 : S_{1,2} - F_2 \longrightarrow S_{2,1} - G_2,$$

still an S-isomorphism.

It is clear that $F_1 \cap F_2 = \emptyset$ so that $S_{1,2} = (S_{1,2} - F_1) \cup (S_{1,2} - F_2)$. Moreover φ_1 and φ_2 patch together just because they are S-isomorphisms and $\bar{\pi}_{1,2}$ and $\bar{\pi}_{2,1}$ give isomorphisms outside of $F_1 \cup F_2$ and $G_1 \cup G_2$ respectively. So the lemma is proved. $\diamond$

Now we will prove that π_K, F_k and the decomposition $F_K = \sum_{p \in K} F_p$ are essentially independent of the admissible ordering.

Proposition 4.3.2 *If $\pi_K : S_K \longrightarrow S$ and $\pi'_K : S'_K \longrightarrow S$ are blowing-ups of K obtained from admissible orderings $\prec$ and $\prec'$ on K, there is an unique S-isomorphism $\varphi : S_K \longrightarrow S'_K$. Furthermore, for any $p \in K$, $\varphi(F_p) = F'_p$, where F_p and F'_p are the components of the exceptional divisors F_K and F'_K of π_K and π'_K corresponding to p.*

PROOF: As in preceding cases, the uniqueness of φ follows from the condition of S-morphism, which determines it on a dense subset.

For the existence of φ, we use induction on the cardinal $\sharp(K)$ of K, the claim being obvious if $\sharp(K) = 1$.

If both orderings have the same first element, say p, the claim easily follows from the induction hypothesis applied after blowing up p.

Assume that the ordering $\prec$ is

$$\prec \ p_1, p_2, \ldots, p_i, q, \ldots, p_\ell$$

where q is the first element by the ordering $\prec'$. Since q is a proper point on S, the ordering

$$\prec_1 \ p_1, q, p_2, \ldots, p_i, \ldots p_\ell$$

obtained from $\prec$ by moving q to the second place is also admissible. Thus, we know that the claim is true for the orderings $\prec$ and $\prec_1$, because they have the same first element, and hence after taking $\prec_1$ instead of $\prec$, it is enough to make the proof assuming that the first element by $\prec'$ is the second one by $\prec$.

The symmetric reasoning allows us to assume that also the first element by $\prec$ is the second one by $\prec'$, and so that we have

$$\prec : p_1, p_2, p_3, \ldots, p_i, \ldots p_\ell$$
$$\prec' : p_2, p_1, p'_3, \ldots, p'_i, \ldots p_\ell.$$

In such a case the reader may see that the claim is an easy consequence of 4.3.1 and the induction hypothesis applied after blowing up p_1 and p_2. $\diamond$

Remark 4.3.3 Obviously the S-isomorphism φ of 4.3.2 maps the total (resp. strict) transform by π_K of any curve on S to its total (resp. strict) transform by π'_K.

The next lemma will be useful in the sequel.

Lemma 4.3.4 *If ξ is a connected curve on S, $\pi_K^*(\xi)$ is connected also.*

PROOF: The reader may easily prove the claim in the case of blowing up a single point, after the results of section 3.2. Then, the general case follows using induction. ◇

Assume that ξ is a reduced curve. The reader may have noticed that the resolution of singularities described in section 3.7 is just blowing up the union of all clusters of multiple points on ξ. It is now clear, after 4.3.2, that such a procedure does not depend on the ordering according to which the points are blown up. In the same way, resolution of singularities of a reduced germ of curve ξ_O (or local resolution of ξ at O) is achieved by blowing up the cluster of all multiple points on ξ_O, the strict transform of ξ_O after such blowing-up being composed of finitely many smooth germs.

The examples of the germs $x^3y - y^3 = 0$ and $x^3 - y^2 = 0$ at $O = (0,0)$ show that after blowing up all multiple points on a reduced curve or germ of curve ξ, its strict transform may meet the exceptional divisor at a double point or may be tangent to some of its components. The reader may easily verify that both phenomena are due to the existence of simple satellite points on ξ, so that after blowing up all singular points on ξ, its strict transform still is smooth and furthermore it meets transversally the exceptional divisor at all of their intersection points.

The blowing-up π_K of a finite union of clusters K is called a *resolution* of the singularities of a curve or germ of curve ξ if and only if the strict transform $\tilde{\xi}_K$ of ξ is smooth. If in addition $\tilde{\xi}_K$ and the exceptional divisor F_K meet transversally at all their common points, then π_K is called an *embedded resolution* of ξ. We have seen thus that blowing up all multiple points on a reduced curve or germ of curve ξ is a resolution of ξ, while blowing up all its singular points is an embedded resolution.

4.4   Exceptional divisors and dual graphs

Keep all notations and conventions as in the preceding section. In the present one, once it has been seen that blowing up K gives rise to an essentially unique morphism $\pi_K : S_K \longrightarrow S$, we study the exceptional divisor F_K of π_K and use it to introduce the dual graph of K, a new graphical representation of the cluster K.

We already know that two distinct irreducible components F_q, $F_{q'}$ of F_K either do not meet or meet transversally at a unique point which then belongs to no other component of F_K (3.5.5). It is clear that $\pi_K^{-1}(O_i) \cap \pi_K^{-1}(O_j) = \emptyset$ if $i \neq j$. Furthermore we have:

Proposition 4.4.1 *For each $i = 1, \ldots, s$, $\pi_K^{-1}(O_i)$ is connected.*

PROOF: By 4.3.2, it is not restrictive to assume that O_i is the first point we blow up. After the first blowing-up, the inverse image of O_i is a connected curve (the exceptional divisor), so it is enough to apply 4.3.4 to it. ⋄

Next we find out which components of the exceptional divisor actually meet, which is closely related to proximity between their corresponding points, i.e.:

Proposition 4.4.2 *If F_q and $F_{q'}$, $q, q' \in K$, $q \neq q'$, are irreducible components of F_K, then $F_q \cap F_{q'} \neq \emptyset$ if and only if one of the following conditions is satisfied:*

(a) *q' is proximate to q and the point proximate to q in the first neighbourhood of q' does not belong to K.*

(b) *q is proximate to q' and the point proximate to q' in the first neighbourhood of q does not belong to K.*

In other words, distinct components F_q and $F_{q'}$ meet if and only if one of the points q, q' is maximal among the points in K that are proximate to the other. In fact, as we will see in a short while, from $F_q \cap F_{q'} \neq \emptyset$ one cannot decide which point is proximate to the other.

PROOF OF 4.4.2: Again we use induction on $\sharp(K)$, the case $\sharp(K) = 1$ being obvious.

Let us assume that the admissible ordering we have chosen for blowing up K is

$$\prec p_1, p_2, \ldots, p_\ell$$

and take $K^{(1)}$ as in the definition of π_K in section 4.3. If both q and q' are different from p_1, the claim follows from the induction hypothesis because the identification of $K^{(1)}$ with $K - \{p_1\}$ clearly preserves proximity and, on the other hand, $F_{K^{(1)}} + F_{p_1} = F_K$.

Thus, assume that one of our points, say q, is p_1. If $q' = p_i$, then, by 4.3.2, it is non-restrictive to assume that $p_2, \ldots, p_{i-1}$ are just the points infinitely near to p_1 preceding q'. Let us denote, as before, by π_j, $j = 1, \ldots, i$, the blowing-up of centre p_j. Put E_1 for the iterated strict transform under $\pi_2, \ldots, \pi_{i-1}$ of the exceptional divisor of blowing up p_1. Write $\tilde{E}_1$ for the strict transform of E_1 after blowing-up p_i and E for the exceptional divisor of blowing up p_i. By definition, $q' = p_i$ is proximate to $q = p_1$ if and only if $q' \in E_1$, which is in turn equivalent to $\tilde{E}_1 \cap E \neq \emptyset$. Notice that in such a case $\tilde{E}_1 \cap E = \{q''\}$ where q'' is the point proximate to q in the first neighbourhood of q' and the intersection is transverse.

Since F_q and $F_{q'}$ are the iterated strict transforms of $\tilde{E}_1$ and E by the remaining blowing-ups, we have the claim: if q' is not proximate to q, then $\tilde{E}_1 \cap E = \emptyset$ and so $F_q \cap F_{q'} = \emptyset$. If q' is proximate to q and $q'' \notin K$, then the composition of the remaining blowing-ups is an isomorphism near q'' and hence $F_q \cap F_{q'} \neq \emptyset$. Lastly, if $q'' \in K$, it is blown up (one may assume without restriction $q'' = p_{i+1}$), after which the strict transforms of $\tilde{E}_1$ and E do not meet and so again $F_q \cap F_{q'} = \emptyset$. ⋄

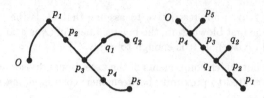

Figure 4.5: The Enriques diagram of a cluster and its corresponding dual graph.

Remark 4.4.3 It is clear from 4.4.2 that the proximity relation in K determines the incidence between components of the exceptional divisor. Conversely, proximity may be recovered from the natural ordering on K and the incidence between components of F_K: given any $q \in K$, we determine first the points $q_j \in K$ that are infinitely near to q and such that F_{q_j} meets F_q; then, according to 4.4.2, the points proximate to q in K are the points infinitely near to q that precede or equal one of the q_j.

For simplicity, we assume from now on that K is just a cluster with origin O. Incidence between irreducible components of F_K is currently represented by means of a graph, which is called the *dual graph* of K. It is defined by taking a vertex for each irreducible component of F_K and joining two vertices by an edge if and only if their corresponding components of F_K meet. Since the irreducible components of F_K are in one to one correspondence with the points of K, using 4.4.2 one may equivalently say that the dual graph is built by taking one vertex $\dot{p}$ for each point $p \in K$, and joining the vertices $\dot{p}$ and $\dot{q}$ by an edge if and only if one of the points p, q is maximal among the points in K that are proximate to the other. The point corresponding to the origin O of K is distinguished and called the root of the graph.

It should be noticed that though the vertices of the dual graph are in one to one correspondence with the points of K, its edges are not representing the relation *to be in the first neighbourhood of*, hence the dual graph looks very different from the Enriques diagram of K, see figure 4.5.

The next proposition describes how a dual graph grows up by the addition of a point to the cluster.

Proposition 4.4.4 *Let K be a cluster, p a maximal point of K (by the natural ordering) and K' the cluster $K' = K - \{p\}$. The dual graph of K is obtained from that of K' by the following modifications:*

(a) *If p is free, say proximate to a single point $q \in K'$, then we add a new vertex p corresponding to p, and join it to $\dot{q}$ by a new edge.*

(b) *If p is satellite, and therefore it is proximate to points $q_1, q_2 \in K$, the dual graph of K' has an edge joining q_1 to q_2: then we drop this edge, add a new vertex p corresponding to p, and join it to $\dot{q}_1$ and $\dot{q}_2$ by a couple of new edges.*

PROOF: That the set of vertices of the dual graph of K comes from that of K' by adding a new vertex corresponding to p, is clear. In case (a), p is proximate to q and to no other point in K'. Hence $\dot{p}$ and $\dot{q}$ need to be joined by a new edge and no further modification is needed, because for any $q', q'' \in K'$, q' is maximal among the points proximate to q'' in K' if and only if it has the same property in K. In case (b) one of the points q_1, q_2 is infinitely near to the other, as p is proximate to both. Assume, for instance, that $q_1 < q_2$. Then p lies in the first neighbourhood of q_2 and q_2 is also proximate to q_1. The point p is thus maximal among the points proximate to q_1 in K, while q_2 has this property in K'. Hence the reason for dropping the edge $\dot{q}_1 \dot{q}_2$ (which certainly exists) and adding the new one $\dot{q}_1 \dot{p}$. The point p being also proximate to q_2, the new edge $\dot{q}_2 \dot{p}$ has to be added also. To close, as in case (a), no further modification is needed, as p is proximate to no point other than q_1, q_2 and therefore for any $q' \neq q_2$ its condition of being maximal or not among the points proximate to, say, q'' is not modified by the addition of p. ⋄

Corollary 4.4.5 *The dual graph of a cluster K is simply connected.*

PROOF: The dual graph is connected by 4.4.1. That it contains no non-trivial closed path is easily seen using induction on the number of points in K and 4.4.4 above. ⋄

Figure 4.6: The Enriques diagram of a cluster giving the same dual graph as that in figure 4.5.

The dual graph of K may be easily obtained from its Enriques diagram after 4.4.2. Nevertheless, natural ordering and proximity on K cannot be fully recovered from its dual graph alone: both the clusters of figures 4.5 and 4.6 give rise to the same dual graph, with the same root. For this reason, further information is usually added to the dual graph in a number of different ways (some quite fancy, see for instance [13], 8.4). An obvious way is to keep the ordering on the vertices induced by the natural ordering on K, as then 4.4.3 directly applies. A better way is to label each vertex $\dot{p}$ with the order of the neighbourhood of O that contains p. Figure 4.7 shows an example. We describe next an inductive procedure to recover natural ordering and proximity (and hence the Enriques diagram) from a dual graph labeled in this way.

We make induction on the number of points in K. The case $K = \{O\}$ being obvious, we assume $K \neq \{O\}$ and pick $p \in K$ belonging to the neighbourhood

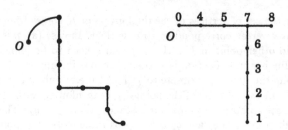

Figure 4.7: The Enriques diagram and the dual graph of a cluster, the orders of the neighbourhoods are indicated on the latter.

of O of the highest order. In particular, $p \neq O$, p is a maximal point in K and therefore no point in K is proximate to it. A vertex $\dot{q}$ is then joined to p by an edge if and only if the point p is proximate to q. This allows just two possibilities, namely, p is proximate to one or two points and so there are either one or two edges with $\dot{p}$ as a vertex. In both cases we may reverse 4.4.4 in order to get the dual graph of $K' = K - \{p\}$ from the dual graph of K: if there is a single edge ending at $\dot{p}$, just drop this edge and $\dot{p}$ itself, while if $\dot{p}$ is joined by edges to different vertices $\dot{q}_1$ and $\dot{q}_2$, then drop $\dot{p}$ and these edges and add a new edge joining $\dot{q}_1$ to $\dot{q}_2$. Now, once we have the dual graph of K', we may assume by induction that natural ordering and proximity among the points in K' have been recovered from it. Thus, we just need to determine the point whose first neighbourhood contains p and the points p is proximate to. But this is clear in both cases. Indeed, if $\dot{p}$ lies on a single edge, say with ends $\dot{p}, \dot{q}$, then p lies in the first neighbourhood of q and is proximate to no other point. If there are two edges with end $\dot{p}$, say joining $\dot{p}$ to vertices $\dot{q}_1, \dot{q}_2$, then p is proximate to both q_1 and q_2, these points lie in different neighbourhoods of O, as both precede p, and so the one lying in the highest neighbourhood of O contains p in its first neighbourhood.

Another labeling for dual graphs will be described at the end of the forthcoming section 4.5.

The dual graph of $\mathcal{S}(\xi)$ (section 3.8) may be used to represent the singularity of a reduced germ of curve ξ. It is then called the *dual graph* or the *resolution graph* of ξ. The reader may see that the vertices corresponding to the maximal points of $\mathcal{S}(\xi)$ are on a single edge of the dual graph, just because they are free and no point in $\mathcal{S}(\xi)$ is infinitely near to one of them. In order to distinguish such vertices, they are often pictured as an arrowhead or a star, as shown in figure 4.8.

4.5   The total transform of a curve

Many of the facts presented in section 4.2 will be considered again in the present one, using this time total transforms of curves instead of the virtual or strict

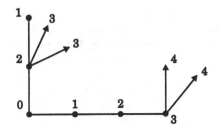

Figure 4.8: The dual graph of the germ of $\zeta : x(x - y^4)(y^2 - x^3)(y^2 + x^3) = 0$ at the origin with indication of the orders of the neighbourhoods. The corresponding Enriques diagram has been shown in figure 3.11.

ones. Once a cluster K with origin O has been given, we assume that an admissible total ordering in K has been fixed, after which we take K as a set of indices. We will consider vectors $\alpha = (\alpha_p)_{p \in K} \in \mathbb{Z}^K$ as single-column matrices and also square matrices $(\alpha_q^p)_{(p,q) \in K \times K}$. In particular, given two such vectors α and α', we will write $\alpha \leq \alpha'$ if and only if $\alpha_p \leq \alpha_p'$ for all $p \in K$, and $\alpha < \alpha'$ if and only if $\alpha \leq \alpha'$ and $\alpha \neq \alpha'$.

The proximity relation in K may be coded in a $K \times K$ square matrix M defined by taking the entry m_q^p in the p-th row and q-th column as

$$m_q^p = \begin{cases} 1 & \text{if } p \text{ is proximate to } q \\ 0 & \text{otherwise.} \end{cases}$$

It is clear that the only non-zero entries of M are below the diagonal, so in particular M is a nilpotent matrix. To be precise, we have $M^{\ell-1} = 0$ if ℓ is the cardinal of K. In fact it is better to consider a second matrix obtained from M:

Definition (Du Val [31]): The *proximity matrix* of K is

$$P_K = I - M,$$

where I denotes the $K \times K$ unit matrix.

The matrix M being nilpotent, it is clear that P_K is invertible over $\mathbb{Z}$. In the sequel we write P instead of P_K if no reference to the cluster K is needed.

If ξ is a curve, still we denote by $\bar{\xi}_K$ and $\tilde{\xi}_K$ the total and strict transforms of ξ after blowing up K. Let us write

$$\bar{\xi}_K = \tilde{\xi}_K + \sum_{p \in K} v_p(\xi) F_p$$

where the F_p are, as in the preceding section, the components of the exceptional divisor F_K of π_K. Each $v_p(\xi)$ is a non-negative integer which will be called the *value of ξ at F_p* or the *p-value* of ξ. As it will be easy to see after the proof of the next proposition 4.5.1, for a fixed ξ the integer $v_p(\xi)$ depends only on the point p (and the points preceding it) and not on the whole of K.

Since the values $v_p(\xi)$ are determined by the germ of ξ at O, we also call $v_p(\xi)$ the p-value of the germ of ξ at O and write $v_p(\xi_O) = v_p(\xi)$. If f is any equation of ξ_O, we write $v_p(f)$ for $v_p(\xi)$ and call it the p-value of f. In particular $v_p(f) = 0$ if f is invertible (i. e., if ξ_O is empty) and it will be useful to extend the definition to the whole of $\mathcal{O} = \mathcal{O}_{S,O}$ by taking $v_p(0) = \infty$. The reader may easily see that v_p induces a discrete valuation of the ring $\mathcal{O}$. (For the definition of discrete valuation, see for instance [48], XII.6. The notion of valuation is also recalled at the beginning of the forthcoming section 8.1. Sections 8.1 and 8.2 are devoted to the study and classification of all valuations of the ring $\mathcal{O}$. Valuations such as v_p above will be classified as type 1 in section 8.2.)

If ξ is a curve or a germ of curve, we use vectorial notation and put $v_K(\xi) = (v_p(\xi))_{p \in K}$ for the vector of the p-values of ξ and also $e_K(\xi) = (e_p(\xi))_{p \in K}$ for the vector of the effective multiplicities of ξ at the points of K. The proximity matrix of K relates multiplicities and values:

Proposition 4.5.1 *For any curve or germ of curve ξ at O,*

$$e_K(\xi) = P_K v_K(\xi).$$

PROOF: Of course it is enough to prove the claim for ξ a curve defined in a neighbourhood of O. Let p be any point in K and denote by K' the cluster of the points in K preceding p by the admissible ordering. Consider on K' the restriction of the admissible ordering we have in K. Since values remain unmodified by further blowing-ups, we have

$$\bar{\xi}_{K'} = \tilde{\xi}_{K'} + \sum_{q \in K'} v_q(\xi) F'_q$$

where the F'_q are the components of the exceptional divisor of blowing up K' and the $v_q(\xi)$ are the values of ξ at the components of the exceptional divisor of blowing up K corresponding to points in K'.

Let p' be the point in $S_{K'}$ which is identified with p. After blowing up all points in K', p' is the next point to blow up in order to continue blowing up K. The total transform of ξ after blowing up $K' \cup \{p\}$ is the total transform of $\bar{\xi}_{K'}$ after blowing up p', that is,

$$\widetilde{\bar{\xi}_{K'}} + e_{p'}(\bar{\xi}_{K'}) E^{p'}$$

where $\widetilde{}$ denotes the strict transform and $E^{p'}$ the exceptional divisor after blowing up p'. The multiplicity of $E^{p'}$ in such a total transform being unaffected by further blowing-ups, we see that $e_{p'}(\bar{\xi}_{K'}) = v_p(\xi)$, the value of ξ at F_p.

Thus, we will compute $e_{p'}(\bar{\xi}_{K'})$: since $p' \in F'_q$ if and only if p is proximate to q, we have two possibilities, namely

(a) p is a free point, and hence it is proximate to a single point q: then we have

$$v_p(\xi) = e_{p'}(\bar{\xi}_{K'}) = e_p(\xi) + v_q(\xi).$$

(b) p is a satellite point, and hence it is proximate to just two points q_1 and q_2: in such a case

$$v_p(\xi) = e_{p'}(\bar{\xi}_{K'}) = e_p(\xi) + v_{q_1}(\xi) + v_{q_2}(\xi).$$

Using the definition of the matrix M, these equalities give

$$v_K(\xi) = e_K(\xi) + M v_K(\xi)$$

and hence the claim. ◇

Assume now that a system ν of virtual multiplicities for K has been given, we will take it as a column vector indexed by K. Then, one may consider the multiplicities of the components of F_K in the iterated virtual transform of a curve ξ, relative to the virtual multiplicities ν. If ξ goes through $\mathcal{K} = (K, \nu)$, we denote by $\check{\xi}_{\mathcal{K}}$ such an iterated virtual transform. Let us write

$$\check{\xi}_{\mathcal{K}} = \tilde{\xi}_K + \sum_{p \in K} u_p(\xi) F_p$$

and $u_{\mathcal{K}}(\xi) = (u_p(\xi))$. As in former cases, $u_p(\xi)$ and $u_{\mathcal{K}}(\xi)$ are determined by the germ ξ_O and we will use them on germs as well: $u_p(\xi_O) = u_p(\xi)$ and $u_{\mathcal{K}}(\xi_O) = u_{\mathcal{K}}(\xi)$.

Let p be a point in K and take the same conventions as in the proof of 4.5.1 above. Let ν' be the restriction of ν to K' and $\mathcal{K}' = (K', \nu')$. Inductively, we assume that ξ goes through $\mathcal{K}'$ and write

$$\check{\xi}_{\mathcal{K}'} = \tilde{\xi}_{K'} + \sum_{q \in K'} u_q(\xi) F_q'.$$

As in the proof of 4.5.1, one may compute

$$e_{p'}(\check{\xi}_{\mathcal{K}'}) = e_p(\xi) + u_q(\xi)$$

if p is a free point, proximate to a single point q, while

$$e_{p'}(\check{\xi}_{\mathcal{K}'}) = e_p(\xi) + u_{q_1}(\xi) + u_{q_2}(\xi)$$

if p is a satellite point, proximate to points q_1 and q_2.

Define

$$u_p(\xi) = e_{p'}(\check{\xi}_{\mathcal{K}'}) - \nu_p = e_p(\xi) + u_q(\xi) - \nu_p$$

for p free, and

$$u_p(\xi) = e_{p'}(\check{\xi}_{\mathcal{K}'}) - \nu_p = e_p(\xi) + u_{q_1}(\xi) + u_{q_2}(\xi) - \nu_p$$

for p satellite. It is clear that the further condition of going through p with the assigned virtual multiplicity ν_p is equivalent to

$$u_p(\xi) \geq 0.$$

If such condition is satisfied, it is also clear that the exceptional divisor of blowing up p' must be taken $u_p(\xi)$ times in the iterated virtual transform, while the multiplicities of the other components of the whole exceptional divisor are unmodified. Thus we have given an inductive proof of the next result:

Theorem 4.5.2 *Let* $\mathcal{K} = (K, \nu)$ *be a weighted cluster. A germ of curve* ξ *goes through* $\mathcal{K}$ *if and only if*

$$P_K^{-1}(e_K(\xi) - \nu) \geq 0,$$

and in such a case the components of the vector

$$u_{\mathcal{K}}(\xi) = P_K^{-1}(e_K(\xi) - \nu)$$

are the multiplicities of the components of the exceptional divisor in the virtual transform $\tilde{\xi}_{\mathcal{K}}$.

We shall call $\bar{\nu} = P_K^{-1}\nu$ the *virtual values* of the weighted cluster $\mathcal{K} = (K, \nu)$. It is clear that a system of virtual multiplicities determines a system of virtual values for the same cluster and conversely. Thus a weighted cluster may be defined by giving its points and then, equivalently, either a system of virtual multiplicities ν, or the corresponding system $P_K^{-1}\nu$ of virtual values.

As for multiplicities, the values of a curve or germ will be often called *effective values* in order to avoid confusion with the virtual ones. The conditions for going through $\mathcal{K} = (K, \nu)$ may be easily given using values instead of multiplicities:

Corollary 4.5.3 *A curve or germ of curve* ξ *goes through* $\mathcal{K} = (K, \nu)$ *if and only if*

$$v_K(\xi) \geq \bar{\nu}.$$

PROOF: Follows from 4.5.1, 4.5.2 and the definition of the virtual values. ⋄

The elements of $H_{\mathcal{K}}$ are thus easily characterized in terms of values:

Corollary 4.5.4 *We have* $H_{\mathcal{K}} = \{f \in \mathcal{O} | v_p(f) \geq \bar{\nu}_p \text{ for any } p \in K\}$.

Since the v_p are valuations of $\mathcal{O}$, it easily follows from 4.5.4 that $H_{\mathcal{K}}$ is an ideal, as already seen in 4.1.1. Other algebraic properties of $H_{\mathcal{K}}$, in particular that of being an integrally closed ideal, follow from 4.5.4 too and will be dealt with in the forthcoming chapter 8.

In the sequel we will say that the system of virtual values $\bar{\nu}$ of a weighted cluster $\mathcal{K}$ is consistent (or satisfies proximity) if and only if $\mathcal{K}$ itself is consistent. All results of section 4.2 may be easily translated in terms of values. It is easy to see that the proximity inequalities that define consistent clusters may be written in the form

$$\nu^t P_K \geq 0$$

where ^{t} indicates transposition of matrices. Using values instead of multiplicities, the former inequality reads

$$\bar{\nu}^t P_K^t P_K \geq 0$$

and is thus the condition for the virtual values $\bar{\nu}$ to be consistent.

The above vectorial inequality is often written in the form

$$\bar{\nu}^t N_K \leq 0$$

where the matrix N_K is defined as $N_K = -P_K^t P_K$. It is clear from its own definition that N_K is a symmetric matrix that defines a negative-definite symmetric bilinear form.

An easy computation using 4.4.2 gives the following description of the matrix N_K:

Proposition 4.5.5 *The entries $n_{p,q}$ of the matrix N_K are:*

$$n_{p,q} = 1 \ \text{if } p \neq q \ \text{and } F_p \cap F_q \neq \emptyset$$
$$n_{p,q} = 0 \ \text{if } p \neq q \ \text{and } F_p \cap F_q = \emptyset$$
$$n_{p,p} = -r - 1$$

if r is the number of points proximate to p in K.

The reader familiar with intersections of divisors on a surface, may see that, for all $p, q \in K$, $n_{p,q}$ is the intersection number of F_p and F_q on S_K (see [62], for instance). Because of this the matrix N_K is often called the *intersection matrix* of F_K.

We leave as an exercise to check that the following translations of 4.2.2, 4.2.4 and 4.2.7 are right.

Theorem 4.5.6 *Given a weighted cluster K let $\bar{\nu}$ be its system of virtual values. Then there exist germs with effective values equal to the virtual ones at the components of F_K if and only if $\bar{\nu}$ is consistent.*

Corollary 4.5.7 *Notations being as above, if the system of virtual values $\bar{\nu}$ is consistent, then the germs of the equations of the curves ξ that have effective values equal to the virtual ones at the components of F_K describe a non-empty Zariski-open set in H_K. The same claim is true if the curves ξ are in addition asked to have their strict transforms $\tilde{\xi}_K$ intersecting F_K transversally at all intersection points.*

Some facts about the inverse of the proximity matrix are presented in the next lemma, they will be useful later on. Denote by 1_p the column vector whose q-th entry is 1 if $q = p$ and is 0 otherwise.

Lemma 4.5.8 *The matrix P_K^{-1} has all its entries above the diagonal equal to zero, those on the diagonal are all equal to one and those below the diagonal are all non-negative. Furthermore all entries in the first column are positive.*

PROOF: Let A_p denote the p-th row of P_K^{-1}. The equality $P_K P_K^{-1} = I$ and the definition of P_K easily give the inductive relations

$$A_p = 1_p^t + \sum_{q < p} \delta_{p,q} A_q,$$

where $\delta_{p,q} = 1$ if p is proximate to q, and $\delta_{p,q} = 0$ otherwise. From them the claim easily follows. $\diamond$

The diagonal entries $n_{p,p}$ of the intersection matrix N_K of F_K are often taken as labels for the vertices p of the dual graph of K. Note that the other entries $n_{p,q}$, $p \neq q$, can be read directly from the dual graph, by 4.5.5. The vertices corresponding to maximal points of K have label -1, again by 4.5.5. Thus the reader may easily modify the procedure already described in section 4.3 in order to recover natural ordering and proximity in K from a dual graph labeled in this way (labels need to be recalculated after each step).

4.6   Unloading

Assume that there is given a non-consistent weighted cluster $\mathcal{K}$. We know from 4.2.2 and 4.5.6 that no germ of curve goes through $\mathcal{K}$ with effective multiplicities (or values) equal to the virtual ones. In this section we will describe a procedure due to Enriques ([35], IV.II.17) that allow us to explicitly compute the effective multiplicities (or values) of generic germs through $\mathcal{K}$. Following Enriques, we call it *unloading*, because at each step some amount of virtual multiplicity is unloaded onto a point from those proximate to it.

Let us begin with a definition:

Definition: Weighted clusters $\mathcal{K}$ and $\mathcal{K}'$, are said to be *equivalent* if and only if $H_{\mathcal{K}} = H_{\mathcal{K}'}$.

The properties of an equivalence relation being obviously satisfied, we write $\mathcal{K} \sim \mathcal{K}'$ for the equivalence of weighted clusters, Notice that one may equivalently say that $\mathcal{K} \sim \mathcal{K}'$ if and only if any germ through $\mathcal{K}$ goes through $\mathcal{K}'$ and conversely.

If two weighted clusters $\mathcal{K} = (K, \nu)$ and $\mathcal{K}' = (K, \nu')$, with the same underlying cluster K, are equivalent, then we will equivalently say that their systems of virtual multiplicities or values are equivalent.

Next we will describe what we call an *unloading step*: it applies to any non-consistent weighted cluster $\mathcal{K}$ to give an equivalent weighted cluster. Later on, we will show that any sequence of successive unloading steps is necessarily finite, and hence that performing successive unloading steps from $\mathcal{K}$ eventually produces a consistent cluster $\mathcal{K}^i$ which has the same points as $\mathcal{K}$ and is equivalent to it. Then, by 4.2.2, the virtual multiplicities (values) of $\mathcal{K}^i$ are the effective multiplicities (values) of generic germs through $\mathcal{K}^i$ and, by the equivalence of $\mathcal{K}$ and $\mathcal{K}^i$, also the effective multiplicities (values) of generic germs through $\mathcal{K}$.

An unloading step may be equivalently made using either multiplicities or values. Since both ways are useful, we shall give a double description.

Unloading multiplicities: Let $\mathcal{K} = (K, \nu)$ be a weighted cluster and assume that the proximity inequality at a point $p \in K$ is not satisfied, that is, that

$$\rho_p = \nu_p - \sum_{q \text{ prox. to } p} \nu_q < 0.$$

Denote by r_p the number of points proximate to p in K and define n as being the least integer non-less than $-\rho_p/(r_p+1)$. *Unloading multiplicities on* p is to define a new cluster $\mathcal{K}' = (K, \nu')$ by the rules:

$$\nu'_p = \nu_p + n$$
$$\nu'_q = \nu_q - n \text{ if } q \text{ is proximate to } p$$
$$\nu'_q = \nu_q \text{ otherwise.}$$

As above, denote by 1_p the vector all of whose components equal zero except for the one corresponding to p which equals one. The cluster K being fixed throughout this section, we just write P for its proximity matrix and $N = -P^t P$. Recall from 4.5.5 that $1_p^t N 1_p = -r_p - 1$, r_p still being the number of points proximate to p in K.

Unloading values: Let $\mathcal{K}$ be as above and denote by $\bar{\nu}$ its system of virtual values. Still assume that the proximity inequality at p is not satisfied, that is, that the p-th entry of $\bar{\nu}^t N$ is positive:

$$\bar{\nu}^t N 1_p > 0.$$

Define n as being the least integer such that

$$(\bar{\nu}^t + n 1_p^t) N 1_p = -n(r_p + 1) + \bar{\nu}^t N 1_p \leq 0.$$

Unloading values on p is to take

$$\bar{\nu}' = \bar{\nu} + n 1_p$$

as a new system of virtual values for K, thus defining a new weighted cluster $\mathcal{K}'$.

First of all let us check that both unloadings give rise to the same weighted cluster $\mathcal{K}'$, whether multiplicities or values are unloaded. Indeed, from $\nu = P\bar{\nu}$ it follows that $\rho_p = \nu^t P 1_p = -\bar{\nu}^t N 1_p$, so that in both cases the same integer n has been taken. Then an easy computation shows that $\nu' = P\bar{\nu}'$, that is, that $\bar{\nu}'$ is the system of virtual values of (K, ν'), as wanted. In the sequel we will often just say that $\mathcal{K}'$ is obtained from $\mathcal{K}$ by *unloading on* p. Notice that unloading on p can be made only if the proximity inequality at p is not satisfied. Enriques called the next theorem the *unloading principle* (*principio di scaricamento*).

Theorem 4.6.1 *Let $\mathcal{K}$ be a weighted cluster and assume that the proximity inequality at one of its points p is not satisfied. Then $\mathcal{K}$ and the weighted cluster $\mathcal{K}'$ it gives rise to by unloading on the point p, are equivalent.*

PROOF: Using values will make an easier proof. Thus assume that $\mathcal{K} = (K, \nu)$ and $\mathcal{K}' = (K, \nu')$ have virtual values $\bar{\nu}$ and $\bar{\nu}'$, respectively. We have

$$H_{\mathcal{K}} = \{f \in \mathcal{O} | v(f) \geq \bar{\nu}\}$$
$$H_{\mathcal{K}'} = \{f \in \mathcal{O} | v(f) \geq \bar{\nu}'\}$$

and $\bar{\nu}' = \bar{\nu} + n1_p \geq \bar{\nu}$, so that it is clear that $H_K \supset H_{K'}$. Assume thus that $f \in H_K$, $f \neq 0$. We have $v(f) \geq \bar{\nu}$ and, after the definition of $\bar{\nu}'$, we need only show that $v_p(f) \geq \bar{\nu}'_p = \bar{\nu}_p + n$.

Write $N = (n_p^q)$. Since for $q \neq p$ all n_p^q are either 0 or 1 (4.5.5), we have

$$\sum_{q \neq p} v_q(f) n_p^q \geq \sum_{q \neq p} \bar{\nu}_q n_p^q.$$

On the other hand, $v(f)$ is obviously a consistent system of virtual values for K, by 4.5.6, and so

$$v(f)^t N 1_p = \sum_{q \in K} v_q(f) n_p^q \leq 0.$$

From these inequalities

$$-v_p(f) n_p^p \geq \sum_{q \neq p} \bar{\nu}_q n_p^q,$$

hence,

$$(\bar{\nu}_p - v_p(f)) n_p^p \geq \sum_{q \in K} \bar{\nu}_q n_p^q = \bar{\nu}^t N 1_p,$$

which we may write

$$\left(\bar{\nu}^t + (v_p(f) - \bar{\nu}_p) 1_p^t \right) N 1_p \leq 0.$$

Then, the definition of n gives

$$v_p(f) - \bar{\nu}_p \geq n$$

so that

$$v_p(f) \geq \bar{\nu}_p + n = \bar{\nu}'_p$$

as wanted. ◇

Enriques used multiplicities in his original statement of the unloading principle ([35], IV.II.17) and gave no proof of it. For a proof using multiplicities, see [20].

Now, as already explained, we will see that finitely many successive unloading steps allow us to reach a consistent cluster. It is worth noting that this is managed no matter which point p is chosen to unload on at each step, and so the procedure does not need a special strategy.

Theorem 4.6.2 *Assume that K is a non-consistent weighted cluster. Put $K = K^0$ and, inductively, as far as K^{i-1} is not consistent, define K^i from K^{i-1} by unloading on a suitable point. Then we have:*

(a) There is an m so that K^m is consistent, has the same points as K and is equivalent to it.

(b) There is a non-empty Zariski-open set $U \subset H_{\mathcal{K}}$ such that for any $f \in U$, the germ defined by f goes sharply through $\mathcal{K}^m$.

(c) $\mathcal{K}^m$ is the only consistent cluster which is equivalent to $\mathcal{K}$ and has the same points. In particular it does not depend on the choice of the points on which the unloadings are performed.

In the sequel we will say that $\mathcal{K}^m$ above is the consistent cluster obtained from $\mathcal{K}$ by an *unloading procedure* or just by *unloading*. The underlying clusters being the same, we will also say that the system of virtual multiplicities (values) of $\mathcal{K}^m$ is the consistent one obtained from the system of virtual multiplicities (values) of $\mathcal{K}$ by unloading.

The reader may notice that the behaviour of generic germs through the non-consistent weighted cluster $\mathcal{K} = (K, \nu)$ is now well determined by claim (b) above. In particular, by 4.2.6, all of them have the same equisingularity type which in turn is explicitly determined after performing the unloading procedure.

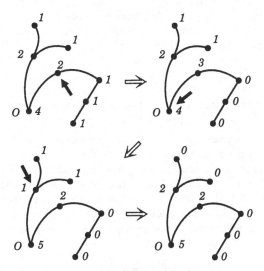

Figure 4.9: A sequence of unloading steps from a non-consistent system of virtual multiplicities (top left) to the corresponding consistent one (bottom right). The black arrows indicate the points onto which multiplicities are unloaded.

PROOF OF 4.6.2: To prove (a), notice first that, by construction, all weighted clusters $\mathcal{K}^i$ are equivalent to $\mathcal{K}$ and have the same underlying cluster. If we write $\bar{\nu}^i$ for the system of virtual values of $\mathcal{K}^i$, from the definition of unloading values it follows that

$$\bar{\nu}^{i-1} < \bar{\nu}^i.$$

On the other hand, let f be any non-zero element of $H_{\mathcal{K}}$ and write $v(f)$ for the vector of its p-values, $p \in K$. The weighted clusters $\mathcal{K}$ and $\mathcal{K}^i$ being equivalent,

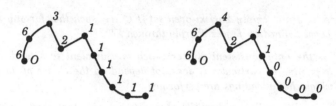

Figure 4.10: A system of virtual multiplicities (left) and the consistent one obtained by unloading (right).

we have $H_{\mathcal{K}} = H_{\mathcal{K}^i}$ for all i, in particular $f \in H_{\mathcal{K}^i}$ which means

$$\bar{\nu}^i \leq v(f)$$

for all i. This shows that the strictly increasing sequence of systems of values $\bar{\nu}^i$ needs to be finite. Since unloading may be applied as far as the system of values is non-consistent, a consistent system of values must be obtained after finitely many unloading steps, which proves part (a).

Because of the equivalence of $\mathcal{K}$ and $\mathcal{K}^m$, $H_{\mathcal{K}} = H_{\mathcal{K}^m}$ and so part (b) is a direct consequence of 4.2.7 applied to $\mathcal{K}^m$. Lastly, since the same argument applies to any consistent cluster (K, μ) equivalent to $\mathcal{K}$, part (c) obviously follows as μ needs to be the system of effective multiplicities of all germs $f = 0$ for f in a non empty Zariski-open subset of $H_{\mathcal{K}}$. $\diamond$

In practice, the easiest way of performing an unloading procedure is to unload multiplicities using an Enriques diagram. Figures 4.9 and 4.10 show two examples. Values may be unloaded using a dual graph which has been labeled using the diagonal entries of the intersection matrix, as described at the end of section 4.5, as then the whole intersection matrix N is easily read from the labeled cluster, by 4.5.5. The reader may perform the unloading procedure of figure 4.9 using values instead of multiplicities. The systems of values the procedure gives rise to are

$$\bar{\nu} = (4, 6, 7, 14, 2, 6, 7, 7)$$
$$\bar{\nu}' = (4, 7, 7, 14, 2, 6, 7, 7)$$
$$\bar{\nu}'' = (5, 7, 7, 14, 2, 6, 7, 7)$$
$$\bar{\nu}''' = (5, 7, 7, 14, 2, 7, 7, 7),$$

where the admissible ordering has been taken so that the points on the longest branch of the Enriques diagram precede the remaining ones.

It has been seen above that unloading gives an effective construction of the only consistent weighted cluster (K, μ) equivalent to a given non-consistent (K, ν). Just proving the existence and uniqueness of such a weighted cluster is far easier. Indeed, for the existence it is enough to take $\bar{\mu}_p = \min\{v_p(f) | f \in H_{\mathcal{K}}\}$ and then the uniqueness follows as in the proof of 4.6.2. We leave the details to the reader.

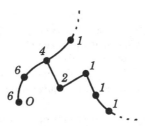

Figure 4.11: The equisingularity type of generic germs through the weighted clusters of figure 4.10.

Enriques and Chisini's book, [35], IV.II.17, describes two further principles to be used, together with unloading, for determining the effective behaviour of generic germs through a non-consistent cluster $\mathcal{K}$. One of them, called *principio di scorrimento* (*moving principle*), is a particular case of unloading. The other, named *principio della massima separazione dei rami* (*highest splitting principle*), claims that generic germs through $\mathcal{K}$ split into the maximum possible number of branches, which is equivalent to having no singular points outside of $\mathcal{K}$ and so is guaranteed by 4.6.2.

Remark 4.6.3 When performing unloading, the reader may find it easier to drop, before each unloading step, all 0-fold points that are either maximal or followed by 0-fold points only. This modified procedure also gives rise to an equivalent consistent cluster after finitely many steps, as dropping points as indicated clearly gives rise to an equivalent cluster, and the argument used in the proof of 4.6.2 still applies with obvious modifications.

We close this section by showing that 0-fold points make the only difference between equivalent consistent clusters.

Lemma 4.6.4 *If $\mathcal{K} = (K, \nu)$ and $\mathcal{Q} = (Q, \mu)$ are consistent equivalent clusters, then $\nu_p = \mu_p$ for all $p \in K \cap Q$, $\nu_p = 0$ if $p \in K - Q$ and $\mu_p = 0$ if $p \in Q - K$.*

PROOF: Since $H_{\mathcal{K}} = H_{\mathcal{Q}}$, by 4.2.7, there is a germ ξ that goes sharply through both clusters and so, for $p \in K \cap Q$, $\nu_p = e_p(\xi) = \mu_p$. If $p \in K - Q$, the equality $H_{\mathcal{K}} = H_{\mathcal{Q}}$ and 4.2.8 show that there is a germ ξ going sharply through both clusters and missing p, which clearly implies $\nu_p = e_p(\xi) = 0$. The same argument applies if $p \in Q - K$. $\diamond$

4.7 The number of conditions

Let $\mathcal{K}$ be a weighted cluster. We already know from 4.1.1 and 1.8.10 that $\dim_{\mathbb{C}}(\mathcal{O}_{S,O}/H_{\mathcal{K}}) < \infty$. In the sequel we will call the integer $\dim_{\mathbb{C}}(\mathcal{O}_{S,O}/H_{\mathcal{K}})$

the *codimension* of $\mathcal{K}$. Once local coordinates x, y at O are chosen, $H_{\mathcal{K}}$ may be defined as a linear subspace of $\mathcal{O}_{S,O} = \mathbb{C}\{x, y\}$ by $\dim_{\mathbb{C}}(\mathcal{O}_{S,O}/H_{\mathcal{K}})$ independent linear equations in the coefficients of the series. Such equations may be interpreted as the conditions to be satisfied by (the equation of) a germ for it to go through $\mathcal{K}$. Because of this, the codimension of $\mathcal{K}$ is also called, in an old fashioned but still expressive phrase, the *number of conditions* imposed by $\mathcal{K}$ to the germs of curve.

If $\mathcal{K} = (K, \nu)$ is a weighted cluster, we define the *virtual codimension* or *virtual number of conditions* $c(\mathcal{K})$ of $\mathcal{K}$ by the equality

$$c(\mathcal{K}) = \sum_{p \in K} \frac{\nu_p(\nu_p + 1)}{2}.$$

Our main goal in this section is to show that codimension and virtual codimension agree for consistent clusters, thus giving an easy way for computing the former. We will prove thus:

Proposition 4.7.1 *If* $\mathcal{K} = (K, \nu)$ *is a consistent cluster, then*

$$c(\mathcal{K}) = \sum_{p \in K} \frac{\nu_p(\nu_p + 1)}{2} = \dim_{\mathbb{C}}(\mathcal{O}_{S,O}/H_{\mathcal{K}}).$$

PROOF: We will prove the equality using induction on $c(\mathcal{K})$. The claim being obvious for $c(\mathcal{K}) = 0$, we assume $c(\mathcal{K}) > 0$ and choose a point $p \in K$ such that $\nu_p > 0$ and no point $q \in K$ infinitely near to p has $\nu_q > 0$. Let us define from $\mathcal{K}$ a second weighted cluster $\mathcal{Q} = (Q, \mu)$: take all points in $\mathcal{K}$ with the same virtual multiplicities, but for the point p, which is taken with virtual multiplicity $\mu_p = \nu_p - 1$; furthermore, take $\nu_p - 1$ different free points not already in K, all in the first neighbourhood of p and virtually counted once. It is clear that $\mathcal{Q}$ is consistent too and an elementary computation shows that $c(\mathcal{Q}) = c(\mathcal{K}) - 1$. Thus, by the induction hypothesis,

$$\dim_{\mathbb{C}}(\mathcal{O}/H_{\mathcal{Q}}) = c(\mathcal{Q}) = c(\mathcal{K}) - 1.$$

Define a third weighted cluster $\mathcal{Q}' = (Q', \mu')$ by adding to $\mathcal{Q}$ a further free point q in the first neighbourhood of p and not already in Q, still taken with virtual multiplicity one. Let S' be the surface the point q is lying on: there is a linear monomorphism (4.1.1)

$$\varphi : H_{\mathcal{Q}} \longrightarrow \mathcal{O}_{S',q}$$

mapping each $f \in H_{\mathcal{Q}}$ to an equation of the virtual transform with origin at q (relative to μ or μ', there is no difference) of the germ defined by f. Then we have $H_{\mathcal{Q}'} = \varphi^{-1}(\mathcal{M}_{S',q})$ and, since $\dim_{\mathbb{C}}(\mathcal{O}_{S',q}/\mathcal{M}_{S',q}) = 1$, $\dim_{\mathbb{C}}(H_{\mathcal{Q}}/H_{\mathcal{Q}'}) \leq 1$, i.e., the new simple point adds at most a single condition.

On the other hand, since $\mathcal{Q}$ is consistent, it follows from 4.2.2 that there exists a germ going through $\mathcal{Q}$ with effective multiplicities equal to the virtual ones and missing q. Such a germ does not go through $\mathcal{Q}'$, hence $H_{\mathcal{Q}} \neq H_{\mathcal{Q}'}$ and thus $\dim_{\mathbb{C}}(H_{\mathcal{Q}}/H_{\mathcal{Q}'}) = 1$. It follows that

$$\dim_{\mathbb{C}}(\mathcal{O}_{S,O}/H_{\mathcal{Q}'}) = c(\mathcal{Q}) + 1 = c(\mathcal{K})$$

which gives the claim, because $\mathcal{K}$ comes from $\mathcal{Q}'$ by a single unloading on p and hence $H_{\mathcal{Q}'} = H_{\mathcal{K}}$. $\diamond$

Proposition 4.7.1 does not hold true for arbitrary weighted clusters. For an easy counterexample it is enough to take the point O and three different points in its first neighbourhood, all with virtual multiplicity one. However, the equality of codimension and virtual codimension still holds true for some non-consistent clusters, as for the one in figure 4.10, and thus does not characterize the consistent ones. In the remaining part of this section we will have a further look at the relationship between codimension and virtual codimension. The next proposition shows how virtual codimension behaves by unloading.

Proposition 4.7.2 *Assume that the weighted cluster $\mathcal{K}'$ comes from $\mathcal{K} = (K, \nu)$ by unloading on p. Then $c(\mathcal{K}') \leq c(\mathcal{K})$ and the equality holds if and only if the excess ρ_p of $\mathcal{K}$ at p is -1.*

PROOF: Take the notations as in 4.6.1. Direct computation gives

$$c(\mathcal{K}') - c(\mathcal{K}) = \frac{n}{2}(2\rho_p + 2 + (n-1)(r_p + 1)).$$

Since, by the definition of n, $\rho_p = -1$ implies $n = 1$, it is clear that in such a case $c(\mathcal{K}') - c(\mathcal{K}) = 0$.

Again from the definition of n, $n - 1 < -\rho_p/(r_p + 1)$ which forces $2\rho_p + 2 + (n-1)(r_p + 1) \leq \rho_p + 1 \leq 0$, as necessarily $\rho_p \leq -1$, and so $c(\mathcal{K}') \leq c(\mathcal{K})$.

To close, if $c(\mathcal{K}') - c(\mathcal{K}) = 0$, then the above inequalities obviously give $\rho_p = -1$, as claimed. $\diamond$

In the sequel, unloading on a point p with $\rho_p = -1$ will be called a *tame unloading*.

Corollary 4.7.3 *Let $\mathcal{K}$ be a non-consistent cluster and $\mathcal{K}^1, \ldots, \mathcal{K}^m$ weighted clusters obtained from $\mathcal{K}$ by successive unloading steps so that $\mathcal{K}^m$ is consistent. Then $c(\mathcal{K}) \geq \dim_{\mathbb{C}}(\mathcal{O}/H_{\mathcal{K}})$ and the equality holds if and only if all unloadings leading from $\mathcal{K}$ to $\mathcal{K}^m$ are tame.*

PROOF: By the first half of 4.7.2 and 4.7.1,

$$c(\mathcal{K}) \geq c(\mathcal{K}^1) \geq \cdots \geq c(\mathcal{K}^m) = \dim_{\mathbb{C}}(\mathcal{O}/H_{\mathcal{K}^m}) = \dim_{\mathbb{C}}(\mathcal{O}/H_{\mathcal{K}}).$$

Then it is enough to apply the second half of 4.7.2. $\diamond$

4.8   Adjoint germs and curves

Assume for a while that X is an irreducible algebraic plane curve. The adjoint curves of X are a certain kind of algebraic plane curve used since Riemann and Noether in the intrinsic geometry of linear series on X, and in particular for proving the Riemann–Roch theorem: the main property of the adjoint curves is that they may be used for describing all complete linear series on X, and in particular the canonical series and the differentials of the first kind on X (see for instance [37], ch. 8).

If the curve X has no singularities other than nodes, then the adjoint curves of X may be easily defined as the algebraic curves going through the nodes. More generally, if all singular points of X are ordinary (i.e., with the number of their different tangents equal to their multiplicity, 0.11.4), a curve Y is an adjoint of X if and only if it has a point of multiplicity at least $(e - 1)$ at each (proper) e-fold point of X. These are the cases in which adjoint curves are mostly considered, as the theory of linear series may be developed on a birational model of the curve having ordinary singular points, or even nodes, as its only singularities (see again [37]).

In the general case the adjoint curves of X are defined by means of local conditions at each (proper) singular point of X. In this section we take these local conditions to define local adjoints at a point, so that the global adjoints of any reduced algebraic curve X on a smooth surface may be then defined as the curves that are adjoints at each (proper) multiple point of X.

Let ξ be a non-empty and reduced germ of curve at a point O on a surface S. Define a weighted cluster $\Delta(\xi)$ by taking O and all multiple points on ξ infinitely near to O, each point p being taken with virtual multiplicity $e_p(\xi) - 1$.

Definition: If ξ is a reduced and non-empty germ of curve at O, a germ of curve ζ is an *adjoint germ* (or just an adjoint) of ξ if and only if it goes through $\Delta(\xi)$.

The definition applies to curves as well: if ξ and ζ are curves and ξ is reduced at O we put $\Delta_O(\xi) = \Delta(\xi_O)$ and say that ζ is an *adjoint curve of ξ at O* if and only if ζ_O is an adjoint germ of ξ_O, which is obviously the same as saying that ζ goes through $\Delta_O(\xi)$. Hence, ζ is an adjoint of ξ at O if and only if it goes through the multiple points p of ξ equal or infinitely near to O with the virtual multiplicities $e_p(\xi) - 1$. This is just the classical definition as it appears in [35], V.II.16 or [74], VIII.4, for instance. Inspired by this definition, some modern authors have misunderstood the term *virtual multiplicity* and given naive incorrect definitions of adjoint curves wrongly labeled as 'classical'.

The geometric notion of adjoint germs of a germ ξ and the more algebraic one of conductor (section 3.11) are closely related, such a relationship being in fact the underlying reason for all properties of adjoint curves useful in the geometry of linear series. We have:

Theorem 4.8.1 *Let $\xi : f = 0$ be a non-empty and reduced germ of curve at O. A germ of curve $\zeta : g = 0$, $g \in \mathcal{O}_{S,O}$, is an adjoint of ξ if and only if the class $\mathbf{g}$ of g mod f belongs to the conductor of ξ.*

PROOF: Put $\mathcal{O} = \mathcal{O}_{S,O}$ and $\mathcal{M} = \mathcal{M}_{S,O}$. Take the notations as in sections 3.10 and 3.11, in particular ξ has equation f, $R = \mathcal{O}_{S,O}/(f)$ is the local ring of ξ, and for any $g \in \mathcal{O}_{S,O}$, **g** denotes the class of g modulo f. Let R^i be the rings in the successive neighbourhoods of R, and in particular $R^k = \bar{R}$, the integral closure of R in its total quotient ring (section 3.11). Denote by p_i, $i = 1, \ldots, s$ the points on ξ in the first neighbourhood of O and write R_i for the local ring of $\tilde{\xi}_{p_i}$, $i = 1, \ldots, s$. By its definition $R^1 = \prod_{i=1}^s R_i$. Write $\bar{R}_i$ for the integral closure of R_i in its total quotient ring: as we know from section 3.11, $\bar{R}$ may be identified with the product of the $\bar{R}_i$, $i = 1, \ldots, s$ so that for each i we have a commutative diagram of $\mathcal{O}$-algebras

$$
\begin{array}{ccccc}
R & \longrightarrow & R^1 & \longrightarrow & \bar{R} \\
& & \downarrow & & \downarrow \\
& & R_i & \longrightarrow & \bar{R}_i
\end{array}
$$

where the horizontal arrows are inclusions of rings and the vertical ones projections onto the i-th factor.

We will use induction on the number of multiple points of ξ equal or infinitely near to O, the claim being obvious if ξ is smooth.

Assume first that $\mathbf{g} \in (R : \bar{R})$. Then, in particular $\mathbf{g} \in (R : R^1)$ so that, by 3.11.9, $\mathbf{g} \in \mathbf{m}^{e-1}$ (**m** being the maximal ideal of R) and therefore $g \in \mathcal{M}^{e-1}$, where $e = e_O(\xi)$. This shows that ζ goes through O with the virtual multiplicity $e - 1$, as required by the definition of adjoint curve.

Consider any $\mathbf{h} \in \bar{R}$: since $\mathbf{gh}$ belongs to the conductor of ξ, the same argument used above for **g** shows that $\mathbf{gh} \in \mathbf{m}^{e-1}$ and hence we see that $\mathbf{g}\bar{R} \subset \mathbf{m}^{e-1}$.

Since $g \in \mathcal{M}^{e-1}$, we may write $g = x^{e-1}g'$, where $g' \in \mathcal{O}[y/x]$ and may be taken as a local equation of the virtual transform of ζ at each p_i. Thus in R^1 we have $\mathbf{g} = \mathbf{x}^{e-1}\mathbf{g}'$, $\mathbf{g}'$ the class of g'. On the other hand, from $\mathbf{g}\bar{R} \subset \mathbf{m}^{e-1}$, since $\mathbf{m}^{e-1}R^1 = \mathbf{x}^{e-1}R^1$ (3.11.9), we obtain $\mathbf{g}\bar{R} \subset \mathbf{x}^{e-1}R^1$ and thus $\mathbf{g}'\bar{R} \subset R^1$. This gives $\mathbf{g}'\bar{R}_i \subset R_i$ for $i = 1, \ldots, s$ and hence, by the induction hypothesis, g' defines an adjoint of $\tilde{\xi}$ at each p_i, which proves that ζ is an adjoint of ξ.

For the converse, assume that ζ is an adjoint of ξ at O. Then in particular $g \in \mathcal{M}^{e-1}$ and $g' = gx^{-e+1}$ defines an adjoint of $\tilde{\xi}_{p_i}$ for all $i = 1, \ldots, s$. Thus, by induction, the class $\mathbf{g}_i'$ of g' in R_i lies in $(R_i : \bar{R}_i)$ and then $\mathbf{g}' = (\mathbf{g}_i')_{i=1,\ldots,s} \in (R^1 : \bar{R})$. Since we know from 3.11.9 that $\mathbf{x}^{e-1}R^1 \subset R$, it follows that $\mathbf{g}\bar{R} = \mathbf{x}^{e-1}\mathbf{g}'\bar{R} \subset R$ as wanted. $\diamond$

In particular we get the codimension of the conductor:

Corollary 4.8.2 *Notations and hypothesis being as above,*

$$\dim_{\mathbb{C}} R/(R : \bar{R}) = \delta(\xi).$$

PROOF: It is enough to use 4.7.1 and 3.11.12. $\diamond$

Corollary 4.8.3 *We have*

$$\dim_{\mathbb{C}} \bar{R}/(R:\bar{R}) = 2\delta(\xi).$$

Furthermore, if g is any equation of an adjoint germ ζ that goes through $\Delta(\xi)$ with effective multiplicities equal to the virtual ones and contains no other point on ξ, then $g\bar{R} = (R:\bar{R})$.

PROOF: The first part of the claim follows from the obviously exact sequence

$$0 \longrightarrow R/(R:\bar{R}) \longrightarrow \bar{R}/(R:\bar{R}) \longrightarrow \bar{R}/R \longrightarrow 0$$

after using 4.8.2 and the definition of $\delta(\xi)$. For the second part, it is clear that $g\bar{R} \subset (R:\bar{R})$. The intersection multiplicity $[\xi.\zeta]$ may be easily computed from the hypothesis using the Noether formula (3.3.1), which gives

$$[\xi.\zeta] = \sum e_p(\xi)(e_p(\xi) - 1) = 2\delta(\xi).$$

Since we have seen in the proof of 3.11.10 that $[\xi.\zeta] = \dim \bar{R}/g\bar{R}$, using the first part of the claim we get

$$\dim \bar{R}/(R:\bar{R}) = 2\delta(\xi) = \dim \bar{R}/g\bar{R},$$

which gives the wanted equality. ⋄

The reader may notice that adjoint germs in the conditions of 4.8.3 actually exist, as $\Delta(\xi)$ is consistent and therefore 4.2.2 applies to it.

As in section 3.11, identify the local rings in the k-th neighbourhood to power series rings so that $\bar{R} = \prod_{j=1}^{r} \mathbb{C}\{t_j\}$ and take, for each branch γ_j of ξ,

$$c_j = \sum_p (e_p(\xi) - 1)e_p(\gamma_j),$$

the summation running on the multiple points of ξ lying on γ_j.

Corollary 4.8.4 *The conductor $(R:\bar{R})$ is generated, as ideal of $\bar{R}$, by the elements $t_1^{c_1}, \ldots, t_r^{c_r}$.*

PROOF: If $\zeta : g = 0$ is taken as in 4.8.3, by the Noether formula 3.3.1, $[\gamma_j.\zeta] = c_j$. Thus, by 3.11.6, $\mathbf{g} = (t_1^{c_1}u_1, \ldots, t_r^{c_r}u_r) = (u_1, \ldots, u_r)(t_1^{c_1}, \ldots, t_r^{c_r})$ where all u_i are invertible. Then $(u_1, \ldots, u_r)$ is invertible too and the claim follows from 4.8.3. ⋄

4.9 Noether's $Af + B\varphi$ theorem

As in preceding section, let O be a point on a smooth surface S and write $\mathcal{O} = \mathcal{O}_{S,O}$. Assume that f, g are elements of $\mathcal{O}$ that have no common factor.

Let h be any element in $\mathcal{O}$. The $Af + B\varphi$ theorem provides a sufficient condition for h to belong to the ideal (f, g) generated by f, g in $\mathcal{O}$, or, which is the same, for the existence of $A, B \in \mathcal{O}$ such that $h = Af + Bg$. The theorem is named after such an equality, which was originally written by Noether using φ instead of g. In fact classical claims are global, for $S = \mathbb{P}_2$, the complex projective plane, and deal with homogeneous polynomials instead of analytic functions. Anyway the core of the theorem is local and the global claim is easily obtained from the local one (see for instance [37], V.5).

Consider the germs $\xi : f = 0$ and $\zeta : g = 0$. Sufficient conditions for $h \in (f, g)$ depend on the way the germ $\eta : h = 0$ goes through the points lying on both ξ and ζ. For most of the applications it is enough to deal with particular cases of the theorem in which one has a supplementary hypothesis on the singularities or the relative position of ξ and ζ: for instance one may assume, as a supplementary hypothesis, that one of the germs ξ, ζ has an ordinary singularity (see exercise 3.12), or else that ξ and ζ share no principal tangent. The latter case was called by the Italians the *caso semplice* (the simple case): then O is the only point lying on both germs ξ and ζ. The usual classical proofs dealt with the *caso semplice* only, the general case being vaguely justified afterwards by continuity arguments (see for instance [35] Vol. III, p. 131 or [74], chap. VIII, sect. 3 and the historical note therein). In fact a precise claim for the general case needs to use virtual multiplicities and the first proof of it seems to be due to Northcott [65].

Theorem 4.9.1 (Noether's $Af + B\varphi$) *Let $f, g \in \mathcal{O}$ be non-invertible germs of analytic function sharing no factor and call ξ and ζ the germs $f = 0$ and $g = 0$ respectively. Let T be the set of points (effectively) lying on both ξ and ζ, and put $\mu(p) = e_p(\xi_1) + e_p(\xi_2) - 1$ for $p \in T$. $\mathcal{T} = (T, \mu)$ is a weighted cluster and if $h \in \mathcal{O}$ defines a germ η going through $\mathcal{T}$, then there exist $A, B \in \mathcal{O}$ so that $h = Af + Bg$.*

PROOF: Since by hypothesis ξ and ζ are non-empty and have no common branch, T is a non-empty finite set (3.3.2) and $\mathcal{T}$ is a weighted cluster. We will use induction on the depth $d(\mathcal{T})$ of $\mathcal{T}$.

Write $e = e_O(\xi)$ and $e' = e_O(\zeta)$. As in the preceding section we call R the local ring of ξ at O, $R = \mathcal{O}/(f)$ and for any $a \in \mathcal{O}$ we write a for its class modulo f. Let $p_1, \ldots, p_r$ be the points on ξ in the first neighbourhood of O. Denote by $\mathcal{O}_i$ the local ring of p_i, all $\mathcal{O}_i$ being identified with extensions of $\mathcal{O}$. Write $\tilde{\xi}_i$ and $\tilde{\zeta}_i$ for the strict transforms with origin at p_i of ξ and ζ, respectively, and R_i for the local ring of $\tilde{\xi}_i$.

Choose local coordinates x, y at O so that no branch of ξ is tangent to $x = 0$. Then x may be taken as a local equation of the exceptional divisor at any of the points p_i and so we may take $\tilde{f} = x^{-e} f$ and $\tilde{g} = x^{-e'} g$ as local equations of $\tilde{\xi}_i$ and $\tilde{\zeta}_i$ for any $i = 1, \ldots, r$. In the same way, $\tilde{h} = x^{-e-e'+1} h$ is a local equation of the virtual transform $\tilde{\eta}_i$ of η with origin at p_i.

We are going to prove that for each i there is an equality $\tilde{h}_i = A_i \tilde{f}_i + B_i \tilde{g}_i$, $A_i, B_i \in \mathcal{O}_i$. Indeed, this is obvious if p_i does not belong to ζ, as then $\tilde{g}_i$ is invertible in $\mathcal{O}_i$. This is in particular the case for all i if $d(T) = 0$. Hence, if p_i belongs to ζ then $d(T) > 0$ and we are allowed to use induction on $d(T)$. Write T_i for the (now non-empty) set of points in T that are equal or infinitely near to p_i. Notice that T_i is the set of points lying on both $\tilde{\xi}_i$ and $\tilde{\zeta}_i$. It is clear from the hypothesis that still $\check{\eta}_i$ goes through the points $q \in T_i$ with the virtual multiplicities $\mu_q = e_q(\tilde{\xi}_i) + e_q(\tilde{\zeta}_i) - 1$ and so $\tilde{f}, \tilde{g}$ and $\check{h}$ are elements of $\mathcal{O}_i$ satisfying the hypothesis of the claim. Furthermore, $d(T_i) \leq d(T) - 1$ and hence, by induction, we have the wanted equalities $\check{h}_i = A_i \tilde{f}_i + B_i \tilde{g}_i$, $A_i, B_i \in \mathcal{O}_i$ for $i = 1, \ldots, r$.

After multiplication by $x^{e+e'-1}$ the above equalities give, still in $\mathcal{O}_i$, $h = x^{e'-1} A_i f + x^{e-1} B_i g$, $i = 1, \ldots, r$. Let R_i be the local ring of $\tilde{\xi}_i$, $R_i = \mathcal{O}_i/(\tilde{f}_i)$. If we use $[\ ]_i$ to denote classes mod $\tilde{f}_i$, from the latter equalities we get

$$[h]_i = [x^{e-1}]_i [B_i]_i [g]_i$$

in R_i, $i = 1, \ldots, r$. Now consider the first neighbourhood ring $R^1 = \prod_{i=1}^{s} R_i$ (see section 3.10). We know from 3.10.3 that R^1 is an extension of R and furthermore, using the notations already introduced, for any $a \in \mathcal{O}$, $\mathbf{a} = ([a]_1, \ldots, [a]_s)$. Thus the equalities displayed above give rise to a single one in R^1, namely

$$\mathbf{h} = \mathbf{x}^{e-1} \beta \mathbf{g},$$

where $\beta = ([B_1]_1, \ldots, [B_s]_s)$. Now, by 3.11.9, $\mathbf{x}^{e-1}\beta \in R$, hence we have $\mathbf{h} \in \mathbf{g}R$ and the claim is proved. $\diamond$

Remark 4.9.2 Under the hypothesis of 4.9.1 and using the same notations, the reader may easily verify that the germ of curve $B = 0$ goes through the points $p \in T$ with the virtual multiplicities $e_p(\xi) - 1$. Then, the roles of f and g being symmetric, also $A = 0$ goes through the points $p \in T$ with the virtual multiplicities $e_p(\zeta) - 1$.

4.10   Exercises

4.1 Describe the cluster K of all singular points on $\xi : y^4 - \lambda x^{11} = 0$ for a fixed and non-zero $\lambda \in \mathbb{C}$. Take the effective multiplicities of ξ, $e_p(\xi)$, as virtual multiplicities for the points in K. Check that the resulting weighted cluster $\mathcal{K}$ does not depend on λ provided $\lambda \neq 0$. Directly check that both the non-reduced germs $y^4 = 0$ and $x^{11} = 0$ go through $\mathcal{K}$. Give a second proof of the same fact using 4.1.1.

4.2 Let $\mathcal{K} = (K, \nu)$ be a consistent cluster and ξ a germ of curve. Put $\nu'_p = \nu_p + e_p(\xi)$, $p \in K$, thus defining a second weighted cluster $\mathcal{K}' = (K, \nu')$. Prove that a germ ζ goes through $\mathcal{K}$ if and only if $\xi + \zeta$ goes through $\mathcal{K}'$. Prove also that if $\mathcal{K}$ is consistent, so is $\mathcal{K}'$, and show by an example that the converse is false.

4.3 Prove that for any point equal or infinitely near to O there is an irreducible germ of curve that contains p as a simple point and no singular point infinitely near to p.

4.4 Let $\{n_i\}$, $i \in \mathbb{N}$, be a non-increasing sequence of positive integers, all but finitely many of them being equal to one and satisfying the conditions of 3.5.9. Fix r so that $n_i = 1$ for $i \geq r$ and define a totally ordered cluster $K = \{O = p_0, p_1, \ldots, p_r\}$ such that, after taking n_i as the virtual multiplicity of p_i, $i = 0, \ldots, r$, it becomes a consistent cluster with all excesses equal to zero but the last one which equals 1, Then deduce from 4.2.2 and 4.2.5 the existence of an irreducible germ of curve whose sequence of multiplicities is $\{n_i\}$.

4.5 Let K be a cluster and $p, q \in K$, $p \neq q$. Notations being as in section 4.3, assume that the components F_p and F_q of F_K meet at a point t. Also assume that q is proximate to p (otherwise p is proximate to q, see 4.3.6). Prove that through the identifications of 3.5.6, the point t is identified with the point in the first neighbourhood of p that precedes or equals q, and with the point in the first neighbourhood of q that is proximate to p.

4.6 Give an example of a pair of non-equisingular unibranched germs having the same unlabeled resolution graph.

4.7 Let ξ and ζ be germs of curve and K a cluster, all with origin at a point O. Using the notations of section 4.5, prove that

$$\sum_{p \in F_q} [\tilde{\zeta}_K . F_q]_p = -v_K(\zeta)^t N_K 1_q$$

for any $q \in K$, and also that

$$[\xi.\zeta] = \sum_{p \in F_K} [\tilde{\xi}_K . \tilde{\zeta}_K]_p - v_K(\xi)^t N_K v_K(\zeta).$$

4.8 Let K be a cluster and $\bar{S}$ the surface obtained after blowing up all points in K. For each $p \in K$, denote by F_p (resp. F_p^*) the strict (resp. total) transform on $\bar{S}$ of the exceptional divisor of blowing up p. Prove that $\{F_p^*\}_{p \in K}$ is a basis of the free $\mathbb{Z}$-module generated by the $\{F_p\}_{p \in K}$ and also that the components of the F_p in the basis $\{F_q^*\}$ are the columns of the proximity matrix P_K.

4.9 Let $\mathcal{K} = (K, \nu)$ be a consistent cluster and take ℓ to be the number of points $p \in K$ where $\mathcal{K}$ has excess $\rho_p > 0$. Prove that the number of branches $r(\zeta)$ of a germ ζ going through $\mathcal{K}$ with effective multiplicities equal to the virtual ones satisfies

$$\ell \leq r(\zeta) \leq \sum_{p \in K} \rho_p,$$

and also that both extremal values may be attained.

4.10 Notations being as in exercise 4.9, give an example of a consistent cluster $\mathcal{K}$ and germs ξ, ζ through $\mathcal{K}$ so that

$$r(\xi) > \sum_{p \in K} \rho_p, \quad r(\zeta) < \sum_{p \in K} \rho_p.$$

Compare with 4.2.6 (b).

4.11 Take points O, p in the first neighbourhood of O and q proximate to O in the first neighbourhood of p, with respective virtual multiplicities 1, 6 and 1, to make a weighted cluster $\mathcal{K}$.

(a) Compute the proximity matrix and the virtual values of $\mathcal{K}$.

(b) Get a consistent cluster equivalent to $\mathcal{K}$, by unloading multiplicities, and notice that negative virtual multiplicities do appear in intermediate steps of the unloading procedure.

4.12 Let P be the proximity matrix of a cluster K with origin O. Prove that the entries in the first column of P^{-1} are the values $v_p(\gamma)$, $p \in K$, of a suitable smooth germ γ at O. Use this fact to prove that all of them are positive, which has been already proved in 4.5.8.

4.13 As above, let P be the proximity matrix of a cluster K. Prove that no entry of P^{-1} below the diagonal equals zero if and only if K is totally ordered by the natural ordering.

4.14 Assume that ξ and ζ are germs of curve going through consistent clusters $\mathcal{K} = (K, \nu)$ and $\mathcal{Q} = (Q, \tau)$, respectively. Prove that

$$[\xi.\zeta] \geq \sum_{p \in K \cap Q} \nu_p \tau_p.$$

(*Hint:* Choose ξ' going sharply through $\mathcal{K}$ and missing all points on ζ not in K and use 4.1.3 twice to prove $[\xi.\zeta] \geq \sum_{p \in K} \nu_p e_p(\zeta) = [\xi'.\zeta] \geq \sum_{p \in K \cap Q} \nu_p \tau_p$).

4.15 Let ξ be a non-empty reduced germ of curve. Prove that if ξ goes through $\mathcal{K} = (K, \nu)$, then its adjoints go through $\mathcal{K}' = (K, \nu')$, $\nu'_p = \nu_p - 1$ for any $p \in K$.

4.16 Prove that if ξ is a non-empty reduced germ of curve, then $2\delta(\xi) = [\xi.\zeta]$ for a generic adjoint ζ of ξ.

Chapter 5

Analysis of branches

In the preceding chapters we have introduced Puiseux series and infinitely near points. The former give a local parametric representation of plane curves while the latter display the local geometry of curves and lead to the notion of equisingularity. Our main goal in this chapter is to relate Puiseux series and infinitely near points: we will show how the Puiseux series of an irreducible germ γ determine (and are in turn determined by) the infinitely near points on γ. In particular the Puiseux series of γ will provide a set of numerical equisingularity invariants, the characteristic exponents of γ, that determine its equisingularity class. Furthermore, the coefficients of the Puiseux series of irreducible germs will give rise to coordinates determining the position of the free infinitely near points they go through.

5.1 Characteristic exponents

Let $s = \sum_{j>0} a_j x^{j/n}$ be a Puiseux series and assume that it has polydromy order n, that is, that n and the integers i for which $a_i \neq 0$ have no common factor. We define a finite set of rational numbers $m_1/n, \ldots, m_k/n$, the *characteristic exponents* of s, in the following way: m_1/n is the first non-integral exponent that effectively appears in s, and, for each i, m_i/n is the first exponent effectively appearing in s that cannot be reduced to the minimal common denominator of $m_1/n, \ldots, m_{i-1}/n$. In other words, we have

$$m_1 = \min\{j \mid a_j \neq 0 \text{ and } j \notin (n)\},$$

and, inductively, provided that $n^{i-1} = \gcd(n, m_1, \ldots, m_{i-1}) \neq 1$,

$$m_i = \min\{j \mid a_j \neq 0 \text{ and } j \notin (n^{i-1})\}.$$

Since n is the polydromy order of s, it is clear that we will eventually reach an integer k for which $n^k = 1$: then the whole set of characteristic exponents of s, currently named the *system of characteristic exponents* of s, is $\{m_1/n, \ldots, m_k/n\}$.

Characteristic exponents did appear in the work of Smith ([76], 8), who call them *critical exponents*. The name *characteristic exponents* seems to be due to Halphen ([42], 64).

Notice that, by the definition, the system of characteristic exponents of an integral power series ($n = 1$) is the empty set. The terms $a_{m_i} x^{m_i/n}$, $i = 1, \ldots, k$, are called the *characteristic terms* or the *characteristic monomials* of s, while their coefficients a_{m_i} are often called the *characteristic coefficients*. Some authors prefer to deal with the pairs (m_i, n), $i = 1, \ldots, k$ and they call them *characteristic pairs*.

Sometimes the characteristic exponents of s are written in the form

$$\frac{m_1}{n} = \frac{\beta_1}{\alpha_1}, \quad \frac{m_2}{n} = \frac{\beta_2}{\alpha_1 \alpha_2}, \ldots, \quad \frac{m_k}{n} = \frac{\beta_k}{\alpha_1 \ldots \alpha_k},$$

where $n = \alpha_1 \ldots \alpha_k$ and $\alpha_i > 1$, $(\beta_i, \alpha_i) = (1)$ for $i = 1, \ldots, k$. In this way is shown, for each i, the factor α_i of n that prevents m_i/n from being reduced to the minimum common denominator of $m_1/n, \ldots, m_{i-1}/n$. Obviously we have

$$\beta_i = \frac{m_i}{n^i}, \quad \alpha_1 \ldots \alpha_i = \frac{n}{n^i}$$

and

$$\alpha_i = \frac{n^{i-1}}{n^i}$$

for $i > 1$.

After the definition of characteristic exponents, the series s may be written, more precisely, in the form

$$s = \sum_{\substack{j \in (n) \\ 1 \le j < m_1}} a_j x^{j/n} + \sum_{\substack{j \in (n^1) \\ m_1 \le j < m_2}} a_j x^{j/n} + \cdots + \sum_{\substack{j \in (n^{k-1}) \\ m_{k-1} \le j < m_k}} a_j x^{j/n} + \sum_{j \ge m_k} a_j x^{j/n}.$$

where all coefficients a_{m_i} are different from zero. Then the orders of the differences between s and its conjugates are easily computed:

Proposition 5.1.1 *For any n-th root of unity $\varepsilon \ne 1$, there is an $i \le k$ such that*

$$o_x(s - \sigma_\varepsilon(s)) = m_i/n.$$

Furthermore, for each such i there are just n^{i-1} conjugates $\sigma_\varepsilon(s)$ of s such that

$$o_x(s - \sigma_\varepsilon(s)) \ge m_i/n,$$

namely the series $\sigma_\varepsilon(s)$ for $\varepsilon^{n^{i-1}} = 1$, $s = \sigma_1(s)$ itself included.

PROOF: The former expression of s gives

$$\sigma_\varepsilon(s) = \sum_{\substack{j \in (n) \\ 1 \le j < m_1}} \varepsilon^j a_j x^{j/n} + \sum_{\substack{j \in (n^1) \\ m_1 \le j < m_2}} \varepsilon^j a_j x^{j/n} + \cdots$$

$$+ \sum_{\substack{j \in (n^{k-1}) \\ m_{k-1} \le j < m_k}} \varepsilon^j a_j x^{j/n} + \sum_{j \ge m_k} \varepsilon^j a_j x^{j/n}.$$

Assume that $\varepsilon^{n^i} \neq 1$, $\varepsilon^{n^{i-1}} = 1$, then, since $n^i = \gcd(m_i, n^{i-1})$, also $\varepsilon^{m_i} \neq 1$, so that

$$s - \sigma_\varepsilon(s) = a_{m_i}(1 - \varepsilon^{m_i})x^{m_i/n} + \cdots$$

has order m_i/n and the claim follows. $\diamond$

Remark 5.1.2 We have seen that the number of conjugates of s with $o_x(s - \sigma_\varepsilon(s)) = m_i/n$ is $n^{i-1} - n^i$.

Remark 5.1.3 Proposition 5.1.1 shows that the characteristic exponents of s may be equivalently defined as the orders of the differences between s and its conjugates.

Remark 5.1.4 Assume that s is convergent: then for each determination of its associated function $\bar{s}$ there are just $n^{i-1} - 1$ other determinations whose difference with the first one has infinitesimal order relative to x non-less than m_i/n.

Assume that local analytic coordinates x, y in a neighbourhood of O are fixed and let γ be an irreducible germ of curve at O. The characteristic exponents of all Puiseux series of γ relative to the coordinates x, y being the same, they will be called the *characteristic exponents of γ relative to the coordinates x, y*. Later on we will see that the characteristic exponents of γ do not depend on the coordinates as far as the second axis is non-tangent to γ. Each characteristic exponent has associated with it a finite sequence of consecutive free and satellite points on the germ. The first of these sequences is described in the next section.

5.2 The first characteristic exponent

Suppose that local coordinates x, y at O have been fixed, let γ be an irreducible germ with origin at O and s a Puiseux series of γ. Let us consider first the easier case, in which s is an integral power series

$$s = \sum_{j>0} a_j x^j.$$

Then, by 2.2.8 and 2.2.9, γ is smooth and non-tangent to the y-axis, and so all infinitely near points $\{p_j\}_{j>0}$ on γ are simple and free. The positions of these points depend on the coefficients a_j of the Puiseux series. To be precise, if dots $\ldots$ indicate terms of higher degree:

Proposition 5.2.1 *Take γ as above and fix $\ell \geq 0$. An irreducible germ $\bar{\gamma}$ with Puiseux series $\bar{s}$ goes through $O = p_0$ and the points $p_1, \ldots, p_\ell$ in its first, $\ldots$, ℓ-th neighbourhood on γ if and only if*

$$\bar{s} = \sum_{j=1}^{\ell} a_j x^j + \cdots.$$

Furthermore, there is an absolute coordinate a in the first neighbourhood of p_ℓ so that the satellite point (the point on the y-axis if $\ell = 0$) has coordinate ∞ and $\bar\gamma$ goes through the point of coordinate a $(a \neq \infty)$ if and only if it has a Puiseux series of the form

$$\bar s = \sum_{j=1}^{\ell} a_j x^j + a x^{\ell+1} + \cdots .$$

PROOF: The claim reduces to 3.2.4 if $\ell = 0$. If $\ell > 0$ the claim follows by induction on ℓ after blowing up O and taking local coordinates $x' = x$ and $y' = y/x - a_1$ at p_1, as then $x' = 0$ is the germ of the exceptional divisor. ◇

After 5.2.1 one may understand the coefficients of an integral power series $s \in \mathbb{C}\{x\}$ as coordinates of the infinitely near points on the smooth curve $y = s(x)$.

From now on in this section we assume that s is not an integral power series. Since we will make no use of terms beyond the first characteristic term, we write s in the form

$$s = a_1 x + \cdots + a_h x^h + a x^{m/n} + \cdots$$

where n is the polydromy order of s, $h = [m/n]$, $m/n \notin \mathbb{Z}$ and $a \neq 0$, so that in particular m/n is the first characteristic exponent of s.

Perform the successive divisions, as for computing $n_r = \gcd(m, n)$:

$$m = h_0 n + n_1 \quad (h_0 = h, \text{ thus})$$
$$n = h_1 n_1 + n_2$$

$$\cdot$$
$$\cdot$$
$$\cdot$$

$$n_{r-2} = h_{r-1} n_{r-1} + n_r$$
$$n_{r-1} = h_r n_r ,$$

where $r \geq 1$. In particular, using the notations of section 5.1, $n_r = n^1$. We will describe a first cluster of infinitely near points on γ under the assumption that γ is not tangent to the y-axis, that is, by 2.7.1, that $m/n > 1$ or, equivalently, $h > 0$.

Theorem 5.2.2 *If the irreducible germ γ is not tangent to the y-axis, using the notations introduced above, there are on γ in successive neighbourhoods*

(0) *h_0 n-fold points, namely the point O $(= p_{0,1})$ and for $i = 2, \ldots, h_0$, the point $p_{0,i}$ on the germ $\zeta : y = a_1 x + \cdots + a_h x^h$ in the first neighbourhood of $p_{0,i-1}$,*

(1) *h_1 n_1-fold points, namely the point $p_{1,1}$ on ζ in the first neighbourhood of p_{0,h_0} and, for $i = 2, \ldots, h_1$, the point $p_{1,i}$ proximate to $p_{0,h}$ in the first neighbourhood of $p_{1,i-1}$,*

and, inductively, for $j = 2, \ldots, r$,

    (j) h_j n_j-fold points, namely the point $p_{j,1}$ proximate to $p_{j-2,h_{j-2}}$ in the first neighbourhood of $p_{j-1,h_{j-1}}$ and, for $i = 2, \ldots, h_j$, the point $p_{j,i}$ proximate to $p_{j-1,h_{j-1}}$ in the first neighbourhood of $p_{j,i-1}$.

Furthermore the point q on γ in the first neighbourhood of p_{r,h_r} is a free point.

Notice that $p_{0,2}, \ldots, p_{0,h_0}$ and $p_{1,1}$ are free points and that their positions are determined by the coefficients $a_1, \ldots, a_h$ according to 5.2.1 above. All further points $p_{j,i}$ described in 5.2.2 are satellite points: figure 5.1 shows the corresponding Enriques diagram.

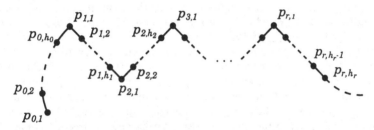

Figure 5.1: The points on a branch γ depending on the first characteristic exponent, as described in 5.2.2.

PROOF OF 5.2.2: First of all, after taking $y - a_1 x - \cdots - a_h x^h$ as a new second coordinate, there is no restriction if we assume that $a_1 = \cdots = a_h = 0$ and hence that ζ is the x-axis. Then we have

$$s = ax^{m/n} + \cdots$$

and, by hypothesis, $m > n$. Thus (2.2.8) $O = p_{0,1}$ is n-fold on γ. Now, perform $h - 1$ successive blowing-ups with centres on γ: it is clear, using 3.2.7, that the corresponding strict transforms of γ have their origins $p_{0,2}, \ldots, p_{0,h}$ on the x-axis and Puiseux series

$$s_1 = ax_1^{\frac{m}{n}-1} + \cdots$$
$$s_2 = ax_2^{\frac{m}{n}-2} + \cdots$$
$$\cdots$$
$$s_{h-1} = ax_{h-1}^{\frac{m}{n}-h+1} + \cdots.$$

In particular it is clear, again by 2.2.8, that $p_{0,2}, \ldots, p_{0,h}$ are n-fold on γ, as claimed.

    Blow up once more the origin of the last strict transform of γ, this time $p_{0,h}$: we find that the new strict transform still has its origin $p_{1,1}$ on the x-axis and that its Puiseux series has the form

$$s_h = ax_h^{\frac{m}{n}-h} + \cdots.$$

Now, since $m/n - h = n_1/n < 1$, 2.2.8 shows that $p_{1,1}$ is n_1-fold as claimed. Furthermore the strict transform of γ is now tangent to the y_h-axis (2.2.6) and so 3.2.7 no longer applies. Fortunately, we do not need to explicitly perform more blowing-ups: the satellite points that follow $p_{1,1}$ will be determined just by the fact that the multiplicity has dropped from n to n_1 after the last blowing-up.

Theorem 3.5.8 applied to $p_{0,h}, p_{1,1}$ shows that the next h_1 points on γ, call them $p_{1,2}, \ldots, p_{1,h_1}, p_{2,1}$, are all proximate to $p_{0,h}$ and have multiplicities $n_1, \ldots, n_1, n_2$, respectively. In the same way, 3.5.8 applied to $p_{1,h_1}, p_{2,1}$ shows that the next h_2 points, $p_{2,2}, \ldots, p_{2,h_2}, p_{3,1}$, are proximate to p_{1,h_1} and have multiplicities $n_2, \ldots, n_2, n_3$, and so on, until reaching the points $p_{r-1,h_{r-1}}, p_{r,1}$ which have multiplicities n_{r-1} and n_r: by using once more 3.5.8, we get $p_{r,2}, \ldots, p_{r,h_r}$, all proximate to $p_{r-1,h_{r-1}}$ and n_r-fold on γ.

Lastly, that the next point q on γ is a free point is easy to see. Indeed, assume that q is proximate to some point $p' \neq p_{r,h_r}$, then p_{r,h_r} itself should be proximate to p', and hence either $p' = p_{r-1,h_{r-1}}$ or $p' = p_{r,h_r-1}$ (notice that $h_r > 1$). Since $n_{r-1} = h_r n_r$, we have already found all points on γ proximate to $p_{r-1,h_{r-1}}$, by 3.5.3 or 3.5.8. By the same reason p_{r,h_r} is the only point on γ proximate to p_{r,h_r-1}, so the point p' does not exist and q is free as claimed. ◇

Notice that we have

$$\frac{m}{n} = h_0 + \cfrac{1}{h_1 + \cfrac{1}{\ddots \cfrac{1}{h_r}}}$$

so that the first characteristic exponent may be recovered from the proximity relations between the infinitely near points (or from an Enriques diagram representing them, the h_j being the lengths of its straight pieces, see figure 5.1) and in particular we obtain

Corollary 5.2.3 *The first characteristic exponent of the Puiseux series of an irreducible germ of curve γ does not depend on the coordinates as far as they are taken with the second axis non-tangent to γ.*

From now on this first characteristic exponent, shared by all Puiseux series of γ relative to coordinates whose second axis is non-tangent to γ, will be called the *first characteristic exponent of γ*.

We already know from 3.12.1 that on a non-smooth irreducible germ there is always a satellite point. Actually 5.2.2 shows $p_{1,2}$ to be the first satellite point on γ. Since smooth germs cannot go through it, there is a bound for the intersection multiplicities of γ with smooth germs. This bound provides an interesting geometrical interpretation of the numerator m of the first characteristic exponent of γ:

Corollary 5.2.4 *Assume that an irreducible germ γ with origin at O has first characteristic exponent m/n, n being the order of γ. Then there exists a smooth germ ζ such that $[\zeta.\gamma] = m$ and for any smooth germ ξ, $[\xi.\gamma] \leq m$. Furthermore if this inequality is strict, then necessarily $[\xi.\gamma] \in (n)$.*

PROOF: Use the notations of 5.2.2 and let ξ be a smooth germ. Since it cannot go through $p_{1,2}$, the Noether formula 3.3.1 and 5.2.2 give $[\xi.\gamma] = hn + n_1 = m$ if ξ goes through $p_{1,1}$ and $[\xi.\gamma] = jn$, $j \leq h$, otherwise. Since 5.2.2 assures us that the first possibility occurs for $\xi = \zeta$, the proof is complete. $\diamond$

Corollary 5.2.4 shows that the numerator m of the first characteristic exponent of γ is the highest intersection multiplicity of γ and a smooth germ: because of this, m is often called the *maximal contact of γ*. The maximal contact of a smooth germ is currently defined to be ∞. Clearly, the maximal contact is an intrinsic character of the irreducible germ. It may be used together with the multiplicity of the germ as a control function for desingularization. Indeed, the next corollary shows that the maximal contact controls the number of blowing-ups leaving invariant the multiplicity of the germ.

Corollary 5.2.5 *Let m and n be, respectively, the maximal contact and the multiplicity of an irreducible germ γ at O. Denote by γ' the strict transform of γ after blowing up O and by m' and n' its maximal contact and multiplicity. Then, either $n' < n$ or $n' = n$ and $m' < m$.*

PROOF: After 5.2.2 we have either $h = 1$ and then $n' = n_1 < n$ or $h > 1$ and thus $n' = n$ and $m' = m - n$. $\diamond$

Notice that 5.2.5 says nothing about m' if $n' < n$. The reader may show by suitable examples that in such a case m' is unbounded and may even be ∞.

Before closing this section we will pay some attention to the case in which the local coordinates have been taken so that the second axis is tangent to the germ γ. Then (any of) the Puiseux series of γ relative to these coordinates has order strictly less than one, and therefore equal to its first characteristic exponent. We keep for the germ γ the same notations as above, still assuming in particular that the first characteristic exponent of γ is m/n.

Proposition 5.2.6 *Let γ be an irreducible germ with first characteristic exponent m/n, n being the order of γ. Assume that we have local coordinates $\bar{x}, \bar{y}$ at the origin of γ with the $\bar{y}$-axis tangent to γ. Let $\bar{s}$ be a Puiseux series of γ relative to such coordinates, and $\bar{m}/\bar{n}$ be the first characteristic exponent of $\bar{s}$, $\bar{n} = \nu(\bar{s})$. Then we have $\bar{m} = n$ and there are two possibilities, namely either*

(a) $\bar{n}/\bar{m} \in \mathbb{Z}$ and then $\bar{n}/\bar{m} \leq h = [m/n]$, or

(b) $\bar{n}/\bar{m} \notin \mathbb{Z}$ and then $\bar{n}/\bar{m} = m/n$ and hence $\bar{n} = m$.

PROOF: The Puiseux series of γ has the form

$$\bar{s} = b\bar{x}^{\bar{m}/\bar{n}} + \cdots, \quad b \neq 0$$

and since by hypothesis $\bar{m} < \bar{n}$, $\bar{m} = e_O(\gamma) = n$ by 2.2.8.

As noticed in 2.6.6, $\bar{n}$ equals the intersection multiplicity of γ and the $\bar{y}$-axis: if the $\bar{y}$-axis has not maximal contact with γ, then, as seen in 5.2.4, $\bar{n} = jn$, $j \leq h$. If, otherwise, the $\bar{y}$-axis has maximal contact, then $\bar{n} = m$. $\diamond$

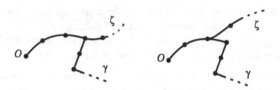

Figure 5.2: On the left ζ has maximal contact with γ, while on the right it has not. If ζ is the y-axis, we have cases (a) (right) and (b) (left) of 5.2.6.

Remark 5.2.7 One may perform successive divisions from $\bar{m}/\bar{n}$: we obviously obtain $\bar{m} = 0\bar{n} + \bar{m}$ first, and then say $\bar{n} = \bar{h}\bar{m} + n'$.

Assume that $n' = 0$ so that we are in case (a) of 5.2.6. Then $\bar{h} \leq h$, $\bar{m} = n$ and the intersection multiplicity of γ and the $\bar{y}$-axis is $\bar{n} = \bar{h}\bar{m}$. The only information we get from $\bar{m}/\bar{n}$ in this case is that γ contains, in successive neighbourhoods, the origin O and $\bar{h} - 1$ (necessarily free) points lying on the y-axis, all of multiplicity $\bar{m} = n$, followed by a further free point not on the y-axis.

If $n' \neq 0$ we are in case (b) of 5.2.6: $\bar{n} = m$ and $\bar{m} = n$, hence the last division and the successive ones are the same as we have performed from m/n, and therefore they still give the numbers h_j and multiplicities n_j of the points following O, just as described in 5.2.2. Now all free points lie on the $\bar{y}$-axis, as the intersection multiplicity of γ and the y-axis is $\bar{n} = m$.

5.3   Beyond the first cluster of satellite points

In this section we will study the first neighbourhood of the last satellite point p_{r,h_r} on γ we have so far determined. First of all we summarize the results we have already obtained in a single claim covering both the cases $m/n > 1$ and $m/n < 1$.

Let γ be an irreducible germ with Puiseux series

$$s = a_1 x + \cdots + a_h x^h + a x^{m/n} + \cdots \tag{5.1}$$

where $n = \nu(s)$, $a \neq 0$ and $h < m/n < h + 1$. In particular, even for $m/n < 1$, we denote the first characteristic exponent of s by m/n.

As in the preceding section, write

$$n_{i-1} = h_i n_i + n_{i+1},$$

$i = 0, \ldots, r$, $n_{-1} = m$, $n_0 = n$, $h_0 = h$ and $n_{r+1} = 0$, for the successive divisions leading to $\gcd(m, n)$. Furthermore, in the case $m/n > 1$, denote by ζ the smooth germ $\zeta : y = a_1 x + \cdots + a_h x^h$. We have seen in 5.2.2, 5.2.6 and 5.2.7:

Proposition 5.3.1 *There are on the germ γ, in successive neighbourhoods:*

- $h_0 + 1$ *points $p_{0,1}, \ldots, p_{0,h_0}, p_{1,1}$ on ζ, the first of them being the origin O of γ,*

- $h_1 - 1$ *points $p_{1,2}, \ldots, p_{1,h_1}$ and, if $r > 1$, a further $p_{2,1}$, all of which either are proximate to p_{0,h_0} if $h_0 > 0$, or belong to the y-axis if $h_0 = 0$, and, inductively, for $1 < i < r$,*

- h_i *points $p_{i,2}, \ldots, p_{i,h_1}, p_{i+1,1}$ all proximate to $p_{i-1,h_{i-1}}$, and, finally,*

- $h_r - 1$ *points $p_{r,2}, \ldots, p_{r,h_r}$ all proximate to $p_{r-1,h_{r-1}}$.*

The multiplicities of these points are $e_{p_{i,j}}(\gamma) = n_i$ for $i = 1, \ldots, r$, $j = 1, \ldots, h_i$. Furthermore the point on γ in the first neighbourhood of p_{r,h_r} is a free point and (just for the case $m/n = 1/h_1$) it does not belong to the y-axis.

In the sequel the points $p_{0,1}, \ldots, p_{r,h_r}$ will be called the *points depending on the first characteristic exponent of s.* If γ is tangent to the y-axis (i.e. $m/n < 1$), they are the points on γ that precede the first free point on γ not on the y-axis. Otherwise $(m/n > 1)$ $p_{0,1}, \ldots, p_{r,h_r}$ are all points on γ up to the first satellite point that precedes a free point on γ, and so they do not depend on the coordinates. In this case m/n is the first characteristic exponent of γ, and we will call $p_{0,1}, \ldots, p_{r,h_r}$ the *points depending on the first characteristic exponent of γ.*

Remark 5.3.2 In any case, the characteristic exponent m/n is determined by $p_{0,1}, \ldots, p_{r,h_r}$, or, more precisely, by their proximity relations and the number of them that lie on the y-axis, as

$$\frac{m}{n} = h_0 + \cfrac{1}{h_1 + \cfrac{1}{\ddots \cfrac{1}{h_r}}}.$$

The set of points depending on the first characteristic exponent of s is a cluster we will denote by $K^1(s)$, or even by $K^1(\gamma)$ in the case $m/n > 1$. The effective multiplicities of γ may be taken as a system ν_s of virtual multiplicities for $K^1(s)$:

$$\nu_s(p_{i,j}) = e_{p_{i,j}}(\gamma) = n_i,$$

thus giving rise to an obviously consistent weighted cluster $\mathcal{K}^1(s) = (K^1(s), \nu_s)$.

The reader may notice that, as shown in the proof of 5.2.2, the somewhat complex structure of satellite points in $K^1(s)$ is just the consequence of the multiplicity dropping at $p_{1,1}$, by the laws of proximity.

In the sequel we will often consider the case in which $a_1 = \cdots = a_h = 0$ (i.e., ζ is the x-axis or $m/n < 1$). Then, once the coordinates are fixed, $K^1(s)$, ν_s and $\mathcal{K}^1(s)$ depend only on the integers m and n and so they will be denoted by $K(m, n)$, $\nu_{m,n}$ and $\mathcal{K}(m, n)$, respectively. The reader may notice from 5.3.1 that $K(m, n) = K(m', n')$ if $m/n = m'/n'$.

Write $m' = m/n_r$ and $n' = n/n_r$ and consider the pencil $\mathcal{P} = \{\xi_{\lambda_0,\lambda_1} : \lambda_0 y^{n'} - \lambda_1 x^{m'} = 0\}$, $(\lambda_0, \lambda_1) \neq (0, 0,)$. Since n', m' are coprime, the germs $\xi_{\lambda_0,\lambda_1}$ with both $\lambda_0 \neq 0$ and $\lambda_1 \neq 0$ are irreducible and have Puiseux series $s = \lambda_1 \lambda_0^{-1} x^{m'/n'}$, so that they go through $K(m, n) = K(m', n')$ with effective multiplicities equal to the virtual multiplicities $\nu_{m',n'}$. It follows in particular that all germs in $\mathcal{P}$ actually go through $\mathcal{K}(m', n')$ (by 4.1.1). Denote by $\check{\xi}_{\lambda_0,\lambda_1}$ the virtual transform of $\xi_{\lambda_0,\lambda_1}$ with origin at p_{r,h_r}, relative to the virtual multiplicities $\nu_{m',n'}$.

The next lemma allows us to handle the projective structure of the first neighbourhood of the last point p_{r,h_r} in $K(n, m)$. Write $p_{r,h_r} = p$ in order to simplify the notations.

Lemma 5.3.3 *By mapping the germ $\xi_{\lambda_0,\lambda_1}$ to the point in the first neighbourhood of p on its virtual transform $\check{\xi}_{\lambda_0,\lambda_1}$, we get a linear projectivity between $\mathcal{P}$ and the first neighbourhood of p.*

PROOF: By 4.1.1 we may choose local equations f and g of $\check{\xi}_{1,0}$ and $\check{\xi}_{0,1}$ at p, so that, for any $(\lambda_0, \lambda_1) \neq (0, 0)$, $\lambda_0 f + \lambda_1 g$ is a local equation of $\check{\xi}_{\lambda_0,\lambda_1}$. If both $\lambda_0, \lambda_1 \neq 0$, then $\check{\xi}_{\lambda_0,\lambda_1}$ is the strict transform of $\xi_{\lambda_0,\lambda_1}$: it is thus smooth at p, and so are in particular $\check{\xi}_{1,1}$ and $\check{\xi}_{1,-1}$. It follows that the initial forms of both $f + g$ and $f - g$ have degree one. On the other hand, an easy computation using 5.3.1 shows that

$$[\xi_{1,1}.\xi_{1,-1}] = m'n' = \sum_{i=0}^{r} h_i(n_i')^2 = \sum_{q \in K(m,n)} e_q(\xi_{1,1}) e_q(\xi_{1,-1}).$$

Then, by the Noether formula, $\xi_{1,1}$ and $\xi_{1,-1}$ share no points beyond p and thus $\check{\xi}_{1,1}$ and $\check{\xi}_{1,-1}$ do not have the same tangent. It follows that the initial forms of $f + g$ and $f - g$ are linearly independent and hence that the initial forms of f and g have degree one and are linearly independent too. We obtain thus a linear projectivity by mapping $\xi_{\lambda_0,\lambda_1}$ to the tangent line to $\check{\xi}_{\lambda_0,\lambda_1}$, after which the claim follows from 3.2.2. $\diamond$

Now we may introduce an absolute coordinate in the first neighbourhood of the last point p_{r,h_r} in $K^1(s)$ which is closely related to the characteristic coefficient.

Theorem 5.3.4 *Let γ be an irreducible germ whose Puiseux series s has first characteristic exponent m/n, say*

$$s = a_1 x + \cdots + a_h x^h + a x^{m/n} + \cdots,$$

with $a \neq 0$. Denote by p the last point on γ depending on m/n. Then a second irreducible germ γ' goes through all points on γ depending on m/n and has a free point not on the y-axis in the first neighbourhood of p if and only if γ' has a Puiseux series of the form

$$s' = a_1 x + \cdots + a_h x^h + \bar{a} x^{m/n} + \cdots, \quad \bar{a} \neq 0.$$

Furthermore, there is an absolute projective coordinate in the first neighbourhood of p such that for all branches γ' above, the absolute coordinate of the point on γ' in the first neighbourhood of p is $\bar{a}^{n'}$, $n' = n/\gcd(m,n)$.

PROOF: Let s' be a Puiseux series of γ' and assume that γ' goes through p and has a free point not on the y-axis in its first neighbourhood. Then s and s' have the same set of points depending on the first characteristic exponent and thus, by 5.3.2, s' has first characteristic exponent m/n. Furthermore, if $m/n > 1$, both series have the same terms preceding the first characteristic one, as their coefficients are the coordinates of the free points preceding p, by 5.2.1 and 5.3.1. The converse is direct from 5.3.1.

For the second part of the claim take $y - a_1 x + \cdots + a_h x^h$ as a new second coordinate, after which we may assume without restriction that $a_1 = \cdots = a_h = 0$. Assume furthermore that we take n to be the polydromy order of s' and write $m/n = m'/n'$, $(m,n) = (1)$. Then we know from 5.3.3 that one may take projective coordinates in the first neighbourhood of p so that, for $\lambda \neq 0$, the point on $\xi_\lambda : y^{n'} - \lambda x^{m'} = 0$ has absolute coordinate λ. An easy direct computation shows that $[\xi_\lambda . \gamma'] \geq mn'$, the inequality being strict if and only if $\bar{a}^{n'} = \lambda$. Since on the other hand, by an easy computation using 5.3.1, we get

$$mn' = \sum_{q \in K^1(s)} e_q(\xi_\lambda) e_q(\gamma'),$$

it follows from the Noether formula that ξ_λ and γ' have the same point in the first neighbourhood of p if and only if $\bar{a}^{n'} = \lambda$, hence the claim. ⋄

Remark 5.3.5 Assume that $n/m \notin \mathbb{Z}$, then p is a satellite point, and therefore there are two satellite points in its first neighbourhood. Since the branches γ' above have a free point in the first neighbourhood of p, the coordinates of these satellite points are 0 and ∞. The reader may see as an exercise, using 5.3.3 and effective computation of the virtual transforms of $y^{n'} = 0$ and $x^{m'} = 0$, that the coordinate of the satellite point proximate to p_{r,h_r-1} is 0 if r is odd, and is ∞ if r is even. Consequently the coordinate of the satellite point proximate to $p_{r-1,h_{r-1}}$ is ∞ if r is odd, and is 0 if r is even.

The case $n/m \in \mathbb{Z}$ is obvious: the point in the first neighbourhood of p on the y-axis has coordinate ∞ and the satellite point has coordinate 0.

5.4 Composing blowing-ups

Let m, n be positive integers. Our aim in this section is to get equations for the blowing-up of $K(m, n)$. In fact the equations we will find are not new, as they have already appeared in chapter 1, when describing the Newton–Puiseux algorithm. These equations will acquire now a deeper geometrical sense, as equations of a composition of blowing-ups, and will allow us to extend the analysis of section 5.2 to further characteristic exponents.

Let p be the last point in $K(m, n)$, p' any free point in its first neighbourhood and (just for the case $n/m \in \mathbb{Z}$) not on the y-axis, and $\varphi : S_{p'} \longrightarrow S$ the blowing-up of $K(m, n)$, that is, the composition of the blowing-ups giving rise to p'. As above, put $m' = m/\gcd(n, m)$, $n' = n/\gcd(n, m)$. Denote by E the germ at p' of the exceptional divisor of φ: since p' is free, E is just the germ of the exceptional divisor of blowing up p and is in particular smooth. Assume that p' has the coordinate of 5.3.4 equal to b, so that $b \neq 0$ and p' belongs to the germ $y^{n'} - b x^{m'} = 0$. For any $f \in \mathcal{O}_O$, write $\varphi^*(f) = \bar{f}$ for the pull-back of f. We have

Lemma 5.4.1 *The germs at p' defined by the equations $\bar{x}$, $\bar{y}$ and $\bar{y}^{n'} - b\bar{x}^{m'}$ are, respectively, $n'E$, $m'E$ and $n'm'E + \zeta$ where ζ is a smooth germ with origin at p' and non-tangent to E.*

PROOF: Write m'/n' as a continued fraction

$$\frac{m'}{n'} = h_0 + \cfrac{1}{h_1 + \cfrac{1}{\ddots \cfrac{1}{h_r}}}$$

and define positive integers u_i, v_i, $i = 0, \ldots, r$ so that

$$\frac{u_i}{v_i} = h_0 + \cfrac{1}{h_1 + \cfrac{1}{\ddots \cfrac{1}{h_i}}}$$

and $\gcd(u_i, v_i) = 1$. Notice that in particular, $u_r = m'$ and $v_r = n'$. As is well known (see for instance [78], Chap. X):

$$u_i = h_i u_{i-1} + u_{i-2} \tag{5.2}$$
$$v_i = h_i v_{i-1} + v_{i-2} \tag{5.3}$$

where $i = 1, \ldots, r$ and $u_{-1} = 1$, $v_{-1} = 0$.

Take any $g \in \mathcal{O}_O$ and let ξ be the germ at O defined by g: obviously $\bar{g}$ defines the total transform $\bar{\xi}_{p'}$ of ξ at p' and $\bar{\xi}_{p'} = \rho E + \tilde{\xi}_{p'}$ where ρ is a non-negative integer and $\tilde{\xi}_{p'}$ the strict transform of ξ at p'.

Since x and y define germs at p which do not (effectively) go through p', we have $\tilde{\xi} = \emptyset$ if either $g = x$ or $g = y$, whereas if $g = y^{n'} - bx^{m'}$, then $\tilde{\xi}_{p'}$ is smooth and non-tangent to E because p' and the points infinitely near to p' on ξ are all simple and free, by 5.3.1.

We complete the proof by computing the values of ρ. Write

$$e_{i,j} = e_{p_{i,j}}(\xi), \quad \rho_{i,j} = v_{p_{i,j}}(\xi),$$

$i = 1, \ldots, r$, $j = 1, \ldots, h_i$, for the effective multiplicities and values of ξ. In particular, $\rho = \rho_{r,h_r}$. Let us write $\rho_{i,j} = 0$ for $i < 0$. From the proximity matrix of $K(m, n)$ it is easy to get, for $i = 1, \ldots, r$, the relations

$$e_{i,1} = \rho_{i,1} - \rho_{i-1,h_{i-1}} - \rho_{i-2,h_{i-2}}$$

and

$$e_{i,j} = \rho_{i,j} - \rho_{i-1,h_{i-1}} - \rho_{i,j-1}$$

for $j = 2, \ldots, h_i$. It turns out to be

$$\rho_{i,h_i} = h_i \rho_{i-1,h_{i-1}} + \rho_{i-2,h_{i-2}} + \sum_{j=1}^{h_i} e_{i,j}, \quad i = 1, \ldots, r. \qquad (5.4)$$

For $g = x$ we have $e_{i,j} = 1$ if $(i, j) = (0, 1)$ and $e_{i,j} = 0$ otherwise. Then relations 5.3 and 5.4 give $\rho_{i,h_i} = v_i$, $i = 0, \ldots, s$ and in particular $\rho_{r,h_r} = n'$ as claimed.

Similarly for $g = y$ we have $e_{i,j} = 1$ if either $i = 0$ or $(i, j) = (1, 1)$, and $e_{i,j} = 0$ otherwise, which gives $\rho_{r,h_r} = m'$.

For $g = y^{n'} - bx^{m'}$ we have $e_{i,j} = n_i/n_r = n_i'$ (say), so that we may write in equality 5.4

$$\sum_{j=1}^{h_i} e_{i,j} = h_i n_i'$$

for i even, and

$$\sum_{j=1}^{h_i} e_{i,j} = n_{i-1}' - n_{i+1}'$$

for i odd, n_{r+1}' being taken equal to zero. Using the equalities 5.2 we obtain $\rho_{i,h_i} = n' u_i$ for i even and $\rho_{i,h_i} = n' u_i - n_{i+1}'$ for i odd. In any case, since $n_{r+1}' = 0$, $\rho_{r,h_r} = n'm'$ as claimed. $\diamond$

Proposition 5.4.2 *Fix any n'-th root $b^{1/n'}$ of the coordinate b of p'. Then there exist local coordinates $\tilde{x}, \tilde{y}$ at p' related to x, y by the formulas*

$$\bar{x} = \tilde{x}^{n'}$$

$$\bar{y} = (b^{1/n'} + \tilde{y})\tilde{x}^{m'}$$

and such that $\tilde{x}$ is an equation of the germ at p' of the exceptional divisor.

PROOF: Since from 5.4.1 we know that $\bar{x}$ is an equation of $n'E$, we have $\bar{x} = ug^{n'}$ where g is an equation of E and u is invertible. Then there are n'-th roots of u in $\mathcal{O}_{p'}$ and we define $\tilde{x}$ as being one of the n'-th roots of $\bar{x}$, $\tilde{x} = \bar{x}^{1/n'}$ after which $\tilde{x}$ is an equation of E.

On the other hand, since $\bar{y}$ defines $m'E$ (5.4.1), we may take $\tilde{y} = \bar{y}\tilde{x}^{-m'} - b^{1/n'} \in \mathcal{O}_{p'}$. We have

$$\bar{y}^{n'} - b\bar{x}^{m'} = \left((\tilde{y} + b^{1/n'})^{n'} - b\right) \tilde{x}^{n'm'}$$

so that, again using 5.4.1, $(\tilde{y} + b^{1/n'})^{n'} - b$ defines a smooth germ non-tangent to E. The same germ being defined by $\tilde{y}$, the proof is complete. ◇

Note that the formulas in 5.4.2 give only a local representation of the composite blowing-up φ near p'. The reader may easily see that if the same formulas are considered for all values of $\tilde{x}, \tilde{y}$, then they define a rational transformation of degree n' (so in general not a birational one) between complex affine planes.

5.5   Enriques' theorem

In this section we will determine the multiplicities of the infinitely near points on an irreducible germ γ, as well as their proximity relations, from one of the Puiseux series of γ. It will result in particular in the determination of the equisingularity class of γ from its characteristic exponents. This analysis was performed by Enriques ([35] IV.I). Further results he obtained on the way, relative to the conditions on a variable branch in order to go through points on a fixed one (*position of points*) will be presented separately in section 5.7.

In his original treatment, Enriques used the abstract definition of infinitely near points (see section 3.4), by means of a close and delicate examination of the intersections of a fixed branch with variable ones. The method is interesting, as it builds its own tools on the way, using for the variable branches the results just obtained for the fixed one, but unfortunately it becomes hard to follow beyond the first characteristic exponent. In fact, in [35] IV.I, the proof is written in detail for the first characteristic exponent only, and just sketched for the further ones.

But for a little bit of complicated notation not easy to avoid, the structure of the points on an irreducible germ is not difficult to describe, as they repeat, for each characteristic exponent, the structure we have already described in section 5.2 for the points depending on the first one. The proof we present here is quite simple as it profits from this fact: the points depending on the first characteristic exponent being already determined, we blow up all of them. Then 5.4.2 allow us to easily relate the Puiseux series s of γ with the Puiseux series $\tilde{s}$ of its strict transform and, in particular, to compute the characteristic exponents of $\tilde{s}$ from that of s. It turns out that $\tilde{s}$ has one fewer characteristic exponent than s, and then the claim easily follows by using the relationship between the characteristic exponents of both series and induction on the number of characteristic exponents.

After proving Enriques' theorem, we will see that the characteristic exponents do not depend on the coordinates, as far as the coordinates are chosen with the y-axis non-tangent to the germ. Even if this is the most important case, we also include in the claim the case of germs tangent to the y-axis, as it is needed for the induction to work. Furthermore, this case has its own interest and will allow an easy proof of the inversion formula in section 5.6.

Assume as before that γ is an irreducible germ with origin at O, that s is (one of) its Puiseux series relative to fixed local coordinates x, y and that s has polydromy order n and characteristic exponents $\{m_1/n, \ldots, m_k/n\}$.

Put $m_0 = 0$ and $n^i = \gcd(n, m_1, \ldots, m_i)$ so that, in particular, $n^0 = n$ and $n^k = 1$. For each $i = 1, \ldots, k$, perform the successive Euclidean divisions leading to $n^i = \gcd(n^{i-1}, m_i) = n^i_{r(i)}$,

$$m_i - m_{i-1} = h^i_0 n^{i-1} + n^i_1$$
$$n^{i-1} = h^i_1 n^i_1 + n^i_2$$

$$\cdot$$
$$\cdot$$
$$\cdot$$

$$n^i_{r(i)-2} = h^i_{r(i)-1} n^i_{r(i)-1} + n^i_{r(i)}$$
$$n^i_{r(i)-1} = h^i_{r(i)} n^i_{r(i)},$$

and notice that $r(i) \geq 1$, $h^i_0 \geq 0$ and $h^i_j > 0$ for $j = 1, \ldots, r(i)$. Then we have:

Theorem 5.5.1 (Enriques) *There are on the irreducible germ γ in successive neighbourhoods, corresponding to the i-th characteristic exponent of s,*

h^i_0 n^i_0-*fold points*

$$p^i_{0,1}, \ldots, p^i_{0,h^i_0},$$

h^i_1 n^i_1-*fold points*

$$p^i_{1,1}, \ldots, p^i_{1,h^i_1},$$

$\ldots$

$h^i_{r(i)-1}$ $n^i_{r(i)-1}$-*fold points*

$$p^i_{r(i)-1,1}, \ldots, p^i_{r(i)-1,h^i_{r(i)-1}}$$

and $h^i_{r(i)}$ $n^i_{r(i)}$-fold points

$$p^i_{r(i),1}, \ldots, p^i_{r(i),h^i_{r(i)}}.$$

The first of these points (namely either $p^i_{0,1}$ if $h^i_0 \neq 0$ or $p^i_{1,1}$ if $h^i_0 = 0$) is the origin O if $i = 1$ or a free point in the first neighbourhood of $p^{i-1}_{r(i-1),h^{i-1}_{r(i-1)}}$ if $i > 1$. Furthermore all points on γ after $p^k_{r(k),h^k_{r(k)}}$ are simple and free.

To describe how the above points $p^i_{j,\ell}$ are related by proximity, let us write $p^i_{0,0} = p^{i-1}_{r(i-1),h^{i-1}_{r(i-1)}}$ for $i > 1$.

We exclude first the case $i = 1$ and $m_1/n < 1$: then all points

$$p^i_{0,1}, \ldots, p^i_{0,h^i_0}, p^i_{1,1},$$

$i = 1, \ldots, k$, but $p^i_{0,1} = O$ are free points. The remaining ones are satellite points: more precisely, for $j = 1, \ldots, r(i) - 1$,

$$p^i_{j,1}, \ldots, p^i_{j,h^i_j}, p^i_{j+1,1}$$

are proximate to $p^i_{j-1,h^i_{j-1}}$, and

$$p^i_{r(i),1}, \ldots, p^i_{r(i),h^i_{r(i)}}$$

are proximate to $p^i_{r(i)-1,h^i_{r(i)-1}}$

In the case $i = 1$ and $m_1/n < 1$, we have $h^1_0 = 0$, then $p^1_{1,1} = O$,

$$p^1_{1,2}, \ldots, p^1_{1,h^1_1},$$

and also $p^1_{2,1}$ if $r(1) > 1$, are free points on the y-axis. The remaining ones are all satellite and proximity between them is as above.

It is clear that 5.5.1 determines the multiplicities of all points on γ as well as their proximity relations. In particular the equisingularity class and the Enriques diagram of γ are determined, see figure 5.3. Let us add some further comments before proving 5.5.1.

The reader may notice that 5.5.1 splits the set points on γ in $k + 1$ pairwise disjoint sets of consecutive points. Indeed, for each $i = 1, \ldots, k$ we have the finite set $\{p^i_{0,1}, \ldots, p^i_{r(i),h^i_{r(i)}}\}$ whose points will be called the *points depending on the i-th characteristic exponent*. The $(k + 1)$-th and last set is infinite and consists of all points after $p^k_{r(k),h^k_{r(k)}}$, which are all simple and free.

Remark 5.5.2 Each set of points depending on a characteristic exponent consists of a certain number of consecutive free points followed by satellite points. There is at least one free point depending on each characteristic exponent as

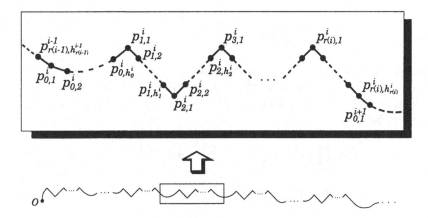

Figure 5.3: The Enriques diagram of an irreducible germ, with an enlarged view of the section corresponding to a characteristic exponent.

$r(i) > 0$ for all i. By the same reason there are satellite points depending on each characteristic exponent but for $i = 1$ and $m_1/n < 1$: in this case the points $p^1_{1,\ell}$ are all free and therefore there are no satellite points if and only if $r(1) = 1$ (i.e., m_1 divides n). In particular all points depending on the characteristic exponents are singular points but for the trivial case $m_1 = 1$ which clearly corresponds to a smooth branch tangent to the y-axis.

PROOF OF 5.5.1: We use induction on the number k of characteristic exponents of γ. The case $k = 0$ being obvious we assume $k > 0$ and the theorem verified for branches having a Puiseux series with $k - 1$ characteristic exponents. The part of the claim corresponding to the first characteristic exponent has been proved either in 5.2.2 or in 5.2.7 (see 5.3.1), so only the part of the claim relative to the points after $K^1(s)$ needs to be proved. For this we will blow up $K^1(s)$.

Assume that the Puiseux series of γ is

$$s = \sum_{j \geq n} a_j x^{j/n}.$$

Once again, by taking $y - a_n x - \cdots - a_{h^1_0 n} x^{h^1_0}$ as a new second coordinate we may assume that the Puiseux series of γ has its first characteristic term as initial term, all terms from the first characteristic one onwards being not modified by the change of coordinates. So we take

$$s = \sum_{j \geq m_1} a_j x^{j/n}$$

and, as equation of γ

$$f = \prod_{\varepsilon^n = 1} \left(y - \sum_{j \geq m_1} \varepsilon^j a_j x^{j/n} \right).$$

Write $m_1/n^1 = m'$, $n/n^1 = n'$. The coordinate of the first point p' on γ after $K^1(s)$ (if $k > 1$, then either $p' = p_{0,1}^2$ or $p' = p_{1,1}^2$) is $a_{m_1}^{n'}$ (5.3.4), so we take the local coordinates at p' of 5.4.2 with $b^{1/n'} = a_{m_1}$. The equation of the total transform $\bar{\gamma}_{p'}$ is then

$$\bar{f} = \prod_{\varepsilon^n = 1} \left(\tilde{x}^{m'} (a_{m_1} + \tilde{y}) - \sum_{j \geq m_1} \varepsilon^j a_j \tilde{x}^{j/n^1} \right)$$

$$= \tilde{x}^{nm'} \prod_{\varepsilon^n = 1} \left(a_{m_1} + \tilde{y} - \sum_{j \geq m_1} \varepsilon^j a_j \tilde{x}^{(j - m_1)/n^1} \right).$$

An equation of the strict transform at p', $\tilde{\gamma}_{p'}$, is thus

$$\prod_{\varepsilon^n = 1} \left(a_{m_1} + \tilde{y} - \sum_{j \geq m_1} \varepsilon^j a_j \tilde{x}^{(j - m_1)/n^1} \right).$$

If $\varepsilon^{n^1} \neq 1$, then the corresponding factor in the above equation is invertible. After dropping invertible factors, we may take as equation of $\tilde{\gamma}_{p'}$

$$\prod_{\varepsilon^{n^1} = 1} \left(\tilde{y} - \sum_{j > m_1} \varepsilon^j a_j \tilde{x}^{(j - m_1)/n^1} \right),$$

from which it is clear that

$$\tilde{s} = \sum_{j > m_1} a_j \tilde{x}^{(j - m_1)/n^1}$$

is a Puiseux series of $\tilde{\gamma}_{p'}$. Since its characteristic exponents clearly are $(m_2 - m_1)/n^1, \ldots, (m_k - m_1)/n^1$, it is enough to use the induction hypothesis on $\tilde{\gamma}_{p'}, \tilde{s}$ to reach the claim. ⋄

There are many important consequences of the Enriques theorem, the first one is the intrinsic character of the characteristic exponents, as far as the y-axis is non-tangent to the germ:

Corollary 5.5.3 *Assume that the y-axis is not tangent to γ. Then the characteristic exponents of the Puiseux series of γ are determined by the multiplicities of the points on γ and their proximity relations (or by the Enriques diagram of γ). In particular the characteristic exponents of a Puiseux series of γ do not depend on the coordinates, as far as the coordinates are chosen with the second axis non-tangent to γ (i.e., as far as $m_1/n > 1$).*

PROOF: If γ is not tangent to the y-axis then, by 2.2.9, n is the order of γ, $n = e_O(\gamma)$. On the other hand, as noticed in 5.5.2, there are on γ just k non-empty groups of successive satellite points separated by free points, and it is clear from 5.5.1 that m_1 is the sum of the multiplicities of the points preceding the first group of satellite points while, for $i > 1$, $m_i - m_{i-1}$ is the sum of the multiplicities of the free points between the $(i-1)$-th and the i-th groups of satellite points. ◇

From now on, the characteristic exponents of any Puiseux series of γ, relative to coordinates whose y-axis is not tangent to γ, will be called *the characteristic exponents of* γ. They determine and are in turn determined by the equisingularity class of the irreducible germ:

Corollary 5.5.4 *Irreducible germs* γ *and* $\bar{\gamma}$ *are equisingular if and only if they have the same characteristic exponents.*

PROOF: An equisingularity between γ and $\bar{\gamma}$ preserving multiplicities and proximity, it is clear from 5.5.3 that the characteristic exponents of γ and $\bar{\gamma}$ are the same. The converse follows from 5.5.1. ◇

Equisingularity of reduced germs is also easily characterized:

Corollary 5.5.5 *Two reduced germs* ξ *and* ζ *are equisingular if and only if there is a one to one correspondence between branches of* ξ *and branches of* ζ *such that*

(a) corresponding branches have the same characteristic exponents, and

(b) any two branches of ξ *have the same intersection multiplicity as their corresponding branches of* ζ.

PROOF: Follows from 5.5.4 and 3.8.6. ◇

We close the present section with two corollaries of 5.5.1 that are stated for further reference. The first one is just a particular case of 5.5.2.

Corollary 5.5.6 *If an irreducible germ* γ *has* k *characteristic exponents, then the satellite points on* γ *are distributed in* k *maximal sets of consecutive satellite points.*

Corollary 5.5.7 *If an irreducible germ* γ *has a Puiseux series with characteristic exponents*

$$\{m_1/n, \ldots, m_k/n\}$$

and we still call $n^i = \gcd(n, m_1, \ldots, m_i)$, *then*

$$2\delta(\gamma) = \sum_{i=1}^{k} m_i(n^{i-1} - n^i) - n + 1.$$

PROOF: Make a direct computation from 3.11.12 using 5.5.1. See exercise 5.6 for details. ◇

5.6   The inversion formula

In this section we will relate characteristic exponents of an irreducible germ γ and those of one of its Puiseux series in the only non-obvious case, namely when the y-axis is tangent to the germ. The result is called the *inversion formula* because it in particular relates the characteristic exponents of the Puiseux series $y = s(x)$, relative to coordinates x, y, and those of the Puiseux series $x = s'(y)$, relative to coordinates y, x, when one of the axis is tangent to the germ.

Proposition 5.6.1 (Inversion formula) *Let γ be an irreducible germ with characteristic exponents $m_1/n, \ldots, m_k/n$, n the order of γ. If x, y are local coordinates with the y-axis tangent to γ, s is a Puiseux series of γ relative to such coordinates, $\nu(s) = \bar{n}$ and $\bar{m}_1/\bar{n}, \ldots, \bar{m}_{\bar{k}}/\bar{n}$ are the characteristic exponents of s, then $\bar{n}$ is the intersection multiplicity of γ and the y-axis and one of the following (clearly incompatible) claims is true:*

(a) $\bar{n} \in (n)$, $n < \bar{n} < m_1$, $\bar{k} = k + 1$, $\bar{m}_1 = n$ and $\bar{m}_{i+1} + \bar{n} = m_i + n$ for $i = 1, \ldots, k$.

(b) $\bar{n} = m_1$, $\bar{k} = k$, $\bar{m}_1 = n$ and $\bar{m}_i + \bar{n} = m_i + n$ for $i = 2, \ldots, k$.

PROOF: That $\bar{n}$ is the intersection multiplicity of γ and the y-axis has been seen in 2.6.6. We will compare the results of applying 5.5.1 to γ twice, by using its own characteristic exponents and the characteristic exponents of s, as the results need to be the same. In particular we will use that there are k well determined maximal sets of consecutive satellite points on γ (5.5.6): we call them *groups of satellite points* for short. Notice first that, by the hypothesis, $\bar{m}_1/\bar{n} < 1 < m_1/n$. Then, we know from 5.2.6 (or even from 2.2.8) that $n = e_O(\gamma) = \bar{m}_1$, which is claimed in both (a) and (b) and gives $n = \bar{m}_1 < \bar{n}$, as claimed in (a).

Assume first that there are no satellite points depending on $\bar{m}_1/\bar{n}$ (i.e., $\bar{n}/\bar{m}_1 \in \mathbb{Z}$, the case (a) of 5.2.6). Since we know from 5.5.2 that each further characteristic exponent $\bar{m}_i/\bar{n}$, $i = 2, \ldots, \bar{k}$ gives rise to a group of satellite points, we have $\bar{k} - 1 = k$. Furthermore, the points preceding the first group of satellite points are O and the free points depending on m_1/n on one hand, and O and the free points depending on either $\bar{m}_1/\bar{n}$ or $\bar{m}_2/\bar{n}$ on the other. By adding up their multiplicities using 5.5.1 in both ways we get

$$m_1 = \bar{n} + \bar{m}_2 - \bar{m}_1,$$

which gives in particular

$$\bar{n} < m_1$$

and

$$\bar{m}_2 + \bar{n} = m_1 + n,$$

both claimed in part (a).

Similarly, for $1 < i \le k$, the set of free points depending on m_i/n and that of free points depending on $\bar{m}_{i+1}/\bar{n}$ agree, as both are the set of points between the $(i-1)$-th and the i-th groups of satellite points. By adding up their multiplicities computed in both ways we get

$$m_i - m_{i-1} = \bar{m}_{i+1} - \bar{m}_i$$

and an easy induction gives the remaining equalities of part (a).

Assume now that there are satellite points depending on $\bar{m}_1/\bar{n}$: this time each characteristic exponent of s gives rise to a group of satellite points and so, necessarily, $\bar{k} = k$. The same argument used above shows that the free points depending on m_i/n and those depending on $\bar{m}_i/\bar{n}$ are the same, for $i = 1, \ldots, k$. By adding up the multiplicities of the free points depending on each characteristic exponent we get

$$m_1 = \bar{n}$$

and

$$m_i - m_{i-1} = \bar{m}_i - \bar{m}_{i-1},$$

for $1 < i \leq k$, from which the remaining equalities of part (b) easily follow. $\diamond$

Corollary 5.6.2 *The characteristic exponents of a Puiseux series s of an irreducible germ γ are determined by the characteristic exponents of γ (or its equisingularity type) and the intersection multiplicity of γ with the y-axis.*

PROOF: Write $\bar{n}$ for the intersection multiplicity of γ and the y-axis. If $\bar{n} = e_O(\gamma)$, then γ is not tangent to the y-axis and hence the characteristic exponents of s are those of γ, by definition. If $\bar{n} > e_O(\gamma)$, γ is tangent to the y-axis and then the claim follows from the inversion formula 5.6.1. $\diamond$

Corollary 5.6.3 (of 5.6.2 and the proof of 5.5.1) *Let γ be an irreducible germ, s a Puiseux series of γ relative to coordinates x, y, and $m_1/n, \ldots, m_k/n$ (n the polydromy order of s) the characteristic exponents of s. If p is the first point on γ not depending on m_1/n, $\tilde{x}, \tilde{y}$ local coordinates at p with $\tilde{x} = 0$ the germ of the exceptional divisor at p and $\tilde{s}$ a Puiseux series of the strict transform $\tilde{\gamma}_p$ relative to these coordinates, then $\tilde{s}$ has characteristic exponents $(m_2 - m_1)/n^1$, $\ldots, (m_k - m_1)/n^1$, $n^1 = \gcd(n, m_1)$ being the polydromy order of $\tilde{s}$.*

PROOF: By 5.6.2 it is not restrictive to assume that both coordinates x, y and $\tilde{x}, \tilde{y}$ are those of 5.4.2, in which case the claim has been already obtained when proving 5.5.1. $\diamond$

5.7   The position of points

Fix local coordinates x, y at O and assume that p is any point infinitely near to O. In this section we will give necessary and sufficient conditions on the Puiseux series of an irreducible germ ζ for it to go through p. To explain these conditions a little bit more, assume that p lies in the first neighbourhood of a point q and that ζ is already going through q. Then if p either lies on the y-axis or is a satellite point, ζ goes or goes not through p depending on certain exponent of its Puiseux series. By contrast, if p is a free point and does not lie on the y-axis, then the condition of going through it depends on a coefficient of the Puiseux series. We will even see that either this coefficient or a suitable power of it may be taken as the projective coordinate of p (in the first neighbourhood of q), thus extending 5.2.1 and 5.3.4.

Let us fix an irreducible germ γ and one of its Puiseux series s that we assume has characteristic exponents $\{m_1/n, \ldots, m_k/n\}$, and hence the form

$$
s = \sum_{\substack{j \in (n) \\ 1 \le j < m_1}} a_j x^{j/n} + \sum_{\substack{j \in (n^1) \\ m_1 \le j < m_2}} a_j x^{j/n} + \cdots + \sum_{\substack{j \in (n^{k-1}) \\ m_{k-1} \le j < m_k}} a_j x^{j/n} + \sum_{j \ge m_k} a_j x^{j/n}.
$$

We still use the notations of section 5.5 for the points on γ. It will be useful to introduce a double notation for the points at the corners of the Enriques diagram of γ by setting $p^i_{j,h^i_j+1} = p^i_{j+1,1}$ for $i = 1, \ldots, k$ and $j = 0, \ldots, r(j) - 1$. We also denote by $p^{k+1}_{0,j}$, $j \ge 1$, the point on γ in the j-th neighbourhood of $p^k_{r(k),h^k_{r(k)}}$, the last point depending on the last characteristic exponent, and set $h^{k+1}_0 = \infty$.

In order to lighten the notations a little bit, we denote by $[s]_\tau$ the partial sum of degree τ of any power series s, and by $[s]_{<\tau}$ the sum of the monomials of s of degree strictly less than τ.

Proposition 5.7.1 (going through a free point) *Fix a point $p^i_{0,\ell}$ on γ, $1 \le i \le k+1$, $1 \le \ell \le h^i_0 + 1$: $p^i_{0,\ell} = O$ if $i = \ell = 1$, otherwise it is a free point. An irreducible germ ζ goes through $p^i_{0,\ell}$ if and only if it has a Puiseux series $\bar{s}$ such that*

$$
[\bar{s}]_{(m_{i-1}+(\ell-1)n^{i-1})/n} = [s]_{(m_{i-1}+(\ell-1)n^{i-1})/n}.
$$

Furthermore, there is a projective absolute coordinate in the first neighbourhood of $p^i_{0,\ell}$ such that the satellite point (the point on the y-axis if $p^i_{0,\ell} = O$) has coordinate ∞ and, for any $a \in \mathbb{C}$, ζ goes through the point of coordinate a if and only if

$$
\bar{s} = [s]_{(m_{i-1}+(\ell-1)n^{i-1})/n} + a x^{(m_{i-1}+\ell n^{i-1})/n} + \cdots.
$$

PROOF: For $i = 1$ the claim is just that of 5.2.1. Thus one may assume $i > 1$ and make induction on i. In this case the proof is achieved using 5.4.2 and arguing as in the proof of 5.5.1, details are left to the reader. ◇

Remark 5.7.2 The order of ζ in 5.7.1 need not be n, but just a multiple of n/n^i. The reader may notice that if the y-axis is tangent to γ, then the only point $p^1_{0,\ell}$ is $p^1_{0,1} = p^1_{1,1} = O$ and then nothing is said in the claim about the free points $p^1_{1,\ell}$ on the y-axis (see 5.5.1).

Since we have taken $1 \leq \ell$, proposition 5.7.1 interprets the coefficients of the non-characteristic terms as coordinates of points. The next proposition relates the coefficient of a characteristic term and the coordinate of the point immediately after those depending on the corresponding characteristic exponent. Fix i as above:

Proposition 5.7.3 (still going through a free point) *An irreducible germ ζ, with Puiseux series $\bar{s}$, goes through the last point depending on m_i/n, $p^i_{r(i),h^i_r(i)}$, and has a free point not on the y-axis in its first neighbourhood if and only if*

$$\bar{s} = [s]_{<m_i/n} + ax^{m_i/n} + \cdots$$

for some $a \in \mathbb{C}$, $a \neq 0$. Furthermore, one may choose an absolute projective coordinate z in the first neighbourhood of the last point depending on the i-th characteristic exponent, $p^i_{r(i),h^i_{r(i)}}$, so that, for any $a \neq 0$, ζ goes through the point of coordinate z in the first neighbourhood of $p^i_{r(i),h^i_{r(i)}}$ if and only if $a^{n^{i-1}/n^i} = z$.

PROOF: The case $i = 1$ has been already proved in 5.3.4. Then the general one follows by induction on i, once again using 5.4.2 as in the proof of 5.5.1. ⋄

Again the series $\bar{s}$ need not have polydromy order n if $i < k$. The reader may notice that in the case $i = 1$ and $m_1 = n^1$, $p^i_{r(i),h^i_{r(i)}}$ belongs to the y-axis. Otherwise it is a satellite point and no point in its first neighbourhood may belong to the y-axis. Then, as in the case of the first characteristic exponent (5.2.3), the satellite points in the first neighbourhood of $p^i_{r(i),h^i_{r(i)}}$ have absolute coordinates 0 and ∞, depending on the parity of $r(i)$ and on which points each of them is proximate to.

Before dealing with the conditions for going through a satellite point let us make some further comments on propositions 5.7.1 and 5.7.3. Assume that the y-axis is not tangent to γ, that is, that $m_1/n > 1$. Let us call $\mathcal{J} = \mathcal{J}(m_1/n, \ldots, m_k/n)$ the set of indices allowed in a Puiseux series with characteristic exponents $\{m_1/n, \ldots, m_k/n\}$, that is,

$$\mathcal{J} = \{j \in \mathbb{Z} | j > 0 \text{ and } j \in (n^{i-1}) \text{ if } j < m_i, i = 1, \ldots, k\}.$$

Then our Puiseux series s may be written as

$$s = \sum_{j \in \mathcal{J}} a_j x^{j/n}.$$

On the other hand, let us call $\mathcal{F} = \mathcal{F}(\gamma)$ the set of all free points on γ, ordered by the natural order of infinitely near points. By mapping the r-th element of $\mathcal{J}$ to the r-th element of $\mathcal{F}$ we get a one to one map from $\mathcal{J}$ onto $\mathcal{F}$. The image p of $j \in \mathcal{J}$ by this map will be called the free point *corresponding* to j (or to the coefficient a_j, or to the monomial $a_j x^{j/n}$) and, conversely, j (resp.

a_j, $a_j x^{j/n}$) will be the index (resp. coefficient, monomial) corresponding to the point p. Let us write $m_0 = n$ and $m_{k+1} = \infty$. Then it easily follows from 5.5.1 that if $j \in \mathcal{J}$ and $m_{i-1} \leq j < m_i$, then the free point corresponding to j is just $p^i_{0,\ell}$, $\ell = (j - m_{i-1})/n^{i-1} + 1$. One has thus $j = m_{i-1} + (\ell - 1)n^{i-1}$, after which it follows from 5.7.1 and 5.7.3 that each coefficient of the Puiseux series s determines the position of its corresponding free point. The next lemma gives further information on the correspondence between indices and free points and will be useful later on. Keep all notations as above and still assume that the y-axis is not tangent to γ.

Lemma 5.7.4 *If p is a free point on γ, then its corresponding index j equals the sum of the multiplicities on γ of O and all free points preceding p.*

PROOF: Assume that $p = p^i_{0,\ell}$, $1 \leq i \leq k+1$ and $1 \leq \ell \leq h^i_0 + 1$. If $i = 1$, by 5.5.1 (or 5.2.2), there are before $p^i_{0,\ell}$ just O and $\ell - 1$ free points, and all of them have multiplicity n. So the sum of their multiplicities is $n\ell = m_0 + (\ell - 1)n = j$ as wanted.

If $i > 1$, by 5.5.1 the sum of the multiplicities of O and the free points depending on the first, ... ,$(i-1)$-th characteristic exponents is

$$m_1 + (m_2 - m_1) + \cdots + (m_{i-1} - m_{i-2}) = m_{i-1}.$$

Furthermore and still by 5.5.1, there are $\ell - 1$ free points depending on the i-th characteristic exponent and preceding $p^i_{0,\ell}$, and all of them have multiplicity n^{i-1}. It is enough to add up all multiplicities to get $m_{i-1} + (\ell - 1)n^{i-1} = j$ as claimed. $\diamond$

Now we will consider the conditions for going through a satellite point or a point on the y-axis. Again allow γ to be tangent to the y-axis and keep all notations as before. Fix i, $1 \leq i \leq k$, and a point $p^i_{t,j}$ with either $0 < t < r(i)$, $2 \leq j \leq h^i_t + 1$ or $t = r(i)$, $2 \leq j \leq h^i_{r(i)}$. Notice that it is a satellite point but for $i = t = 1$ and $m_1/n < 1$, in which case it belongs to the y-axis (see 5.5.1). Consider any irreducible germ ζ going through $p^i_{1,1}$: according to 5.7.1, ζ has a Puiseux series of the form

$$\bar{s} = [s]_{<m_i/n} + bx^{\bar{m}/\bar{n}} + \cdots$$

where $\bar{n}$ is the polydromy order of $\bar{s}$ and we show the first term $bx^{\bar{m}/\bar{n}}$, $b \neq 0$, after those prescribed by 5.7.1. Then $\bar{n}$ needs to be a multiple of the polydromy order of $[s]_{<m_i/n}$, that is, $\bar{n} = dn/n^{i-1}$, $d \in \mathbb{N}$.

The first $j - 1$ characteristic exponents of $\bar{s}$ are of course those of s: they may be written

$$\frac{dm_1/n^{i-1}}{\bar{n}}, \ldots, \frac{dm_{i-1}/n^{i-1}}{\bar{n}},$$

after which $\gcd(dm_1/n^{i-1}, \ldots, dm_{i-1}/n^{i-1}, \bar{n}) = d$. After writing $\bar{m}/d$ as a continued fraction in the form

$$\frac{\bar{m}}{d} = \frac{m_{i-1}}{n^{i-1}} + \bar{h}_0 + \cfrac{1}{\bar{h}_1 + \cfrac{1}{\ddots \cfrac{1}{\bar{h}_{\bar{r}}}}}$$

we have:

Proposition 5.7.5 (going through a satellite point) *The irreducible germ* ζ *goes through* $p^i_{t,j}$ *if and only if either* $\bar{h}_0 = h^i_0, \ldots, \bar{h}_{t-1} = h^i_{t-1}, \bar{h}_t \geq j - 1$ *if* $\bar{r} > t$, *or* $\bar{h}_0 = h^i_0, \ldots, \bar{h}_{t-1} = h^i_{t-1}, \bar{h}_t \geq j$ *if* $\bar{r} = t$.

The reader may notice that the arithmetic conditions in the claim just say that the part of the Enriques diagram of ζ depending on $\bar{m}/\bar{n}$ fits onto the part of the Enriques diagram of γ depending on m_i/n, up to the point $p^i_{t,j}$.

PROOF OF 5.7.5: Since, by hypothesis, $\bar{m}/\bar{n} > (m_{i-1} + h^i_0 n^{i-1})/n$, necessarily $\bar{h}_0 \geq h^i_0$. First of all we use theorem 5.5.1 in order to see that there are on ζ, after the points depending on the $(i-1)$-th characteristic exponent, just $\bar{h}_0 + 1$ free points ($p^i_{1,1}$ being the $(h^i_0 + 1)$-th of them), then $\bar{h}_1$ points proximate to the last but one of them, then $\bar{h}_2$ points proximate to the last but one of the former ones, and so on. After this it is enough to compare, point by point, these points and the points on γ depending on m_i/n, as described in 5.5.1: each satellite point being uniquely determined by the points it is proximate to, the claim follows. $\diamond$

To close this section we will pay some attention to totally ordered sequences of consecutive infinitely near points in order to characterize those that occur as the sequences of points on irreducible germs. Assume that $P = \{p_i\}$, $i \geq 0$ is a sequence of points so that $p_0 = O$ and p_i is in the first neighbourhood of p_{i-1} for $i > 0$. Clearly from 3.7.7, if P is the sequence of points on an irreducible germ, then P contains at most finitely many satellite points. However, this condition is not a sufficient one for P to be the sequence of points on an irreducible germ. In order to see this let us in assume in the sequel that P is not the sequence of points on the y-axis and also that P contains at most finitely many satellite points. We will associate with P a formal Puiseux series in the following way. Fix any index j so that all points p_i, $i \geq j$ are free points and no one of them lies on the y-axis. By 5.7.1 there are a positive integer n and a polynomial $s_j \in \mathbb{C}[x^{1/n}]$ so that an irreducible germ goes through p_j if and only if it has a Puiseux series with partial sum s_j. Let d/n be the degree of s_j. By repeated application of 5.7.1, one may inductively define polynomials $s_i \in \mathbb{C}[x^{1/n}]$, $i > j$, of degree $(d + i - j)/n$ so that

$$s_i = s_{i-1} + a_i x^{(d+i-j)/n},$$

$a_i \in \mathbb{C}$, and such that an irreducible germ γ goes through p_i if and only if it has a Puiseux series with partial sum s_i. The reader may notice that a_i is the absolute coordinate of p_i in the first neighbourhood of p_{i-1} according to 5.7.1. Then the polynomials s_i are the partial sums of a formal series $s \in \mathbb{C}[[x^{1/n}]]$ and we have:

Proposition 5.7.6 *The sequence P is the sequence of the points on an irreducible germ if and only if the series s is a convergent one.*

PROOF: If a branch γ goes through all points p_i, then one of its Puiseux series needs to be s: otherwise one may find $i \geq j$ so that no Puiseux series of γ has partial sum s_i against 5.7.1 and the hypothesis of γ to be going through p_i. Then s is convergent as it is a Puiseux series of an irreducible germ.

Conversely, if s is convergent, so are its conjugate series $\sigma_\varepsilon(s) = s(\varepsilon x^{1/n})$, $\varepsilon^n = 1$, n the polydromy order of s, so that $\prod_\varepsilon (y - \sigma_\varepsilon(s))$ is convergent too and defines an irreducible germ with Puiseux series s and that goes through all points p_i just because s has partial sums s_i, by 5.7.1. ⋄

Remark 5.7.7 The coefficients a_i being the coordinates of the points p_i, they may be freely chosen in order to define a sequence P. Hence, in particular, sequences P giving rise to a non-convergent series s may be easily constructed.

From a more algebraic viewpoint, one may consider formal, non-necessarily analytic, germs of curve: A *formal germ of curve* is by definition associated with a non-zero principal ideal of $\mathbb{C}[[x,y]]$ whose generators are called the equations of the formal germ. Even if their equations need not be convergent, for most aspects formal germs behave as the ordinary analytic ones. It is not difficult to see that sequences P as above, with finitely many satellite points, are in one to one correspondence with irreducible formal germs, an equation of the formal germ associated with the sequence P being constructed from the series s just as in the convergent case.

5.8   The semigroup of a branch

Let γ be, once again, an irreducible germ with origin at O and equation f. Throughout this section we will assume that the local coordinates x, y have been chosen so that the y-axis is not tangent to γ and we still use for the points on γ and its Puiseux series the same notations as above. Let us denote by $\bar{\mathbb{N}}$ the semigroup of all non-negative integers, $\bar{\mathbb{N}} = \mathbb{N} \cup \{0\}$.

Define the *semigroup of γ*, $\Sigma(\gamma)$, as

$$\Sigma(\gamma) = \{j \in \bar{\mathbb{N}} | j = [\gamma.\xi] \text{ for a germ } \xi \text{ with origin at } O\}$$
$$= \{j \in \bar{\mathbb{N}} | j = o_\gamma(g), \ g \in \mathcal{O}_{S,O}, \ g \notin (f)\}.$$

Since intersection multiplicity is additive and $0 \in \Sigma(\gamma)$, corresponding to the empty germ at O, $\Sigma(\gamma)$ is a semigroup. Furthermore, it is clear from its definition that $\Sigma(\gamma)$ is intrinsically associated with the germ γ. The positive integers not in $\Sigma(\gamma)$ are the gaps of the intersection multiplicity with γ, i.e., the positive integers that cannot be obtained as the intersection multiplicity of γ and a germ of curve at O. In this section we will compute a minimal set of generators for $\Sigma(\gamma)$, after which it will turn out that $\Sigma(\gamma)$ is an equisingularity invariant that determines in turn the equisingularity type of γ.

First of all we introduce the integers that generate $\Sigma(\gamma)$. Choose, for each $i = 1, \ldots, s$, a branch γ_i going through all points on γ up to $p_{1,1}^i$ (the last free point corresponding to the i-th characteristic exponent) and having $p_{1,1}^i$ as a non-singular point.

As it clear from theorem 5.5.1 and the results of the last section, this is equivalent to taking γ_i defined by any Puiseux series of the form

$$s_i = \sum_{\substack{j\in(n) \\ 1\leq j<m_1}} a_j x^{j/n} + \cdots + \sum_{\substack{j\in(n^{i-2}) \\ m_{i-2}\leq j<m_{i-1}}} a_j x^{j/n} + \sum_{\substack{j\in(n^{i-1}) \\ m_{i-1}\leq j<m_i}} a_j x^{j/n} + \cdots$$

where the non-explicit terms are assumed not to increase the polydromy order n/n^{i-1} of s_i.

It is clear from the Noether formula (3.3.1) that the intersection multiplicities $[\gamma_i.\gamma]$ do not depend on the way the germs γ_i have been chosen. Anyway, this will follow from the next lemma, as it computes these intersection multiplicities from the characteristic exponents of γ. Write $\breve{m}_i = [\gamma_i.\gamma]$, $i = 1, \ldots, k$:

Lemma 5.8.1 *We have*

$$\breve{m}_i = \frac{(n - n^1)m_1}{n^{i-1}} + \frac{(n^1 - n^2)m_2}{n^{i-1}} + \cdots + \frac{(n^{i-2} - n^{i-1})m_{i-1}}{n^{i-1}} + m_i.$$

PROOF: Either make a direct computation of the orders of the differences between the n conjugates of s and the n/n_{i-1} conjugates of s_i (5.1.2 may help) or use 5.5.1 and the Noether formula 3.3.1 to get

$$\breve{m}_i = \sum_{\ell=1}^{i} \sum_{j=1}^{r(\ell)} h_j^\ell (n_j^\ell)^2 / n^{i-1} + m_i - m_{i-1},$$

from which a straightforward computation gives the claim. $\diamond$

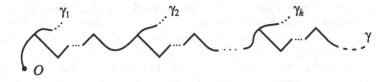

Figure 5.4: The Enriques diagram of γ and the branches γ_i that provide the generators of $\Sigma(\gamma)$ according to 5.8.2.

Now, our main result in this section is

Theorem 5.8.2 *The semigroup of γ is generated by the order n of γ together with the integers $\bar{m}_i$, $i = 1, \ldots, k$.*

PROOF: It is clear that $n \in \Sigma(\gamma)$ and also that, by their own definition, $\bar{m}_i \in \Sigma(\gamma)$, $i = 1, \ldots, k$. Thus the claim we need to prove is that for any non-empty germ ξ at O, one has $[\xi.\gamma] = b_0 n + b_1 \bar{m}_1 + \cdots + b_k \bar{m}_k$, $b_i \in \bar{\mathbb{N}}$.

Denote by p the last point belonging to both γ and ξ, and let j be the order of the neighbourhood of O the point p is belonging to. We will make the proof using induction on j. The case $j = 0$ ($p = O$) is obvious, as then we have

$$[\xi.\gamma] = e_O(\xi)e_O(\gamma) = e_O(\xi)n.$$

Hence, assume $j > 0$ and the claim verified for all germs missing p. Denote by K the cluster consisting of p and all points preceding it and let p' be the point on γ in the first neighbourhood of p. Take the effective multiplicities of ξ as a system of virtual multiplicities ν for K:

$$\nu_q = e_q(\xi), \quad q \in K.$$

It is clear that $\mathcal{K} = (K, \nu)$ is consistent, so (4.2.2) we may choose a germ $\bar{\xi}$ going through $\mathcal{K}$ with effective multiplicities equal to the virtual ones, missing p' and having all points after p simple and free. Then it follows from the Noether formula that $[\xi.\gamma] = [\bar{\xi}.\gamma]$ and thus, intersection multiplicity being additive, it is enough to make the proof for all branches of $\bar{\xi}$. From them, those missing p satisfy the claim because of the induction hypothesis. Hence, it is enough to prove the claim for the branches of $\bar{\xi}$ through p, that is, for any irreducible germ through p with no singular points after p.

Thus we will assume in the sequel that ξ is an irreducible germ that goes through p and has no singular point infinitely near to p. The proximity equality at p (3.5.3) forces the point p itself to be simple on ξ. Put $m = [\xi.\gamma]$.

We consider first the case in which p has a single satellite point in its first neighbourhood (i.e., p is free) and furthermore such a satellite point belongs to γ: then $p = p_{1,1}^i$ for some i and, by the definition of $\bar{m}_i$, $m = \bar{m}_i$, after which there is nothing to prove.

In the remaining cases there is a satellite point in the first neighbourhood of p, say q_0, that does not belong to γ. Denote by q the point on ξ in the first neighbourhood of p and still call $\mathcal{K} = (K, \nu)$ the cluster of all points p' on γ, $p' \leq p$ taken with virtual multiplicities $\nu_{p'} = e_{p'}(\xi)$. It follows from 4.2.2 that there are germs ξ' going through $\mathcal{K}$ with effective multiplicities equal to the virtual ones. Any such ξ' must have $e_p(\xi') = e_p(\xi) = 1$ and therefore it has a single point, necessarily simple, say q', in the first neighbourhood of p. Using 4.2.2, we choose the germ ξ' with $q' \neq q$ and then, by the Noether formula, $m \leq [\xi'.\gamma]$. The reader may equivalently obtain ξ' by a generic modification of the suitable coefficient of the Puiseux series of ξ, according to the results of section 5.7 above.

Fix equations g and g' for ξ and ξ', respectively, and consider the pencil $\mathcal{P} = \{\lambda g + \lambda' g' = 0\}$, $(\lambda, \lambda') \in \mathbb{C}^2 - \{(0,0)\}$. Notice first that since $m = [\xi.\gamma] \leq$

$[\xi'.\gamma]$, by 2.7.3 we have $m \leq [\zeta.\gamma]$ for all $\zeta \in \mathcal{P}$ and furthermore at most one germ $\zeta \in \mathcal{P}$ may have $m < [\zeta.\gamma]$.

Consider the virtual transforms $\check{\zeta}$, with origin at p and relative to the virtual multiplicities ν, of all germs $\zeta \in \mathcal{P}$. Such virtual transforms still describe a pencil $\check{\mathcal{P}}$, their equations being of the form $\lambda \check{g} + \lambda' \check{g}' = 0$ where $\check{g}$ and $\check{g}'$ are equations of the virtual transforms of ξ and ξ', respectively. By the definitions of K and ξ', the virtual transforms of ξ and ξ' agree with the strict ones. Thus both virtual transforms are smooth and, since we have chosen $q' \neq q$, they have different principal tangents at p. It follows that all germs in $\check{\mathcal{P}}$ are smooth and that their principal tangents are all different and describe the whole pencil of tangent lines at p. Then denote by ζ_0 the germ in $\mathcal{P}$ whose virtual transform $\check{\zeta}_0$ has its principal tangent corresponding to the satellite point q_0 (see 3.2.2), and, similarly, by ζ_1 the one whose virtual transform $\check{\zeta}_1$ has its principal tangent corresponding to the point p' on γ. Since $p' \neq q_0$, $\check{\zeta}_1 \neq \check{\zeta}_0$ and therefore $\zeta_1 \neq \zeta_0$.

From the virtual Noether formula (4.1.2) we easily obtain $m < [\zeta_1.\gamma]$, the point p' being on both γ and $\check{\zeta}_1$. Then since $\zeta_0 \neq \zeta_1$, necessarily $m = [\zeta_0.\gamma]$ (2.7.3).

Add the satellite point q_0 to K in order to obtain a second cluster K', extend the virtual multiplicities ν to a system ν' for K' by taking virtual multiplicity one for q_0 and call $\mathcal{K}' = (K', \nu')$. Denote by p_0 the point other than p the point q_0 is proximate to. Of course p is also proximate to p_0. The point q being free, it is not proximate to p_0 and therefore all points on ξ proximate to p_0 are in K: it follows that

$$\nu_{p_0} = e_{p_0}(\xi) = \sum_{\bar{q}} e_{\bar{q}}(\xi) = \sum_{\bar{q}} \nu_{\bar{q}},$$

the summations running on the points $\bar{q} \in K$ proximate to p_0. Thus the excess of $\mathcal{K}$ at p_0 is 0 and hence, after adding q_0 counted once, $\mathcal{K}'$ fails to be consistent at p_0, the excess there being -1.

When unloading multiplicities on p_0, the virtual multiplicities of q_0 and p drop from 1 to 0, after which they remain unmodified by further unloadings because p and q_0 are the two last points in K'. Thus generic germs through $\mathcal{K}'$ do not effectively go through p.

On the other hand we have already found a germ through $\mathcal{K}'$, namely ζ_0, whose intersection with γ is just m. Since any germ through $\mathcal{K}'$ has intersection multiplicity with γ non-less than m, again by the virtual Noether formula, all elements in $H_{\mathcal{K}'}$ but those in a certain linear subspace $F \subset H_{\mathcal{K}'}$, $F \neq H_{\mathcal{K}'}$, define a germ whose intersection multiplicity with γ is m.

It follows thus that there exists a germ ξ'' that does not effectively go through p and such that $[\xi''.\gamma] = m$: by the induction hypothesis the claim is true for ξ'' and so the proof is done. $\diamond$

The reader may notice that the proof of 5.8.2 is geometric and uses the definition of the generators $\check{m}_i$ rather than their explicit expression from the characteristic exponents 5.8.1. In particular the reason for defining the germs γ_i as above should be clear after the proof. For different (and more algebraic) proofs of 5.8.1, the reader is referred to [11] or [94], II.3.

Remark 5.8.3 It follows from 5.8.1 that $n^i = \gcd(n, \bar{m}_1, \ldots, \bar{m}_i)$, $i = 0, \ldots, k$, and also that n^i does not divide $\bar{m}_{i+1}$ if $i < k$. Thus, it is clear that $n, \bar{m}_1, \ldots, \bar{m}_k$ is a minimal system of generators of $\Sigma(\gamma)$.

Remark 5.8.4 The generators $n, \bar{m}_1, \ldots, \bar{m}_k$ are in turn determined by the semigroup $\Sigma(\gamma)$: since it is clear from 5.8.1 that $n < \bar{m}_1 < \cdots < \bar{m}_k$, n is the least positive element of $\Sigma(\gamma)$ and $\bar{m}_i$ is the least element of $\Sigma(\gamma)$ not divided by $n^{i-1} = \gcd(n, \bar{m}_1, \ldots, \bar{m}_{i-1})$ for $i = 1, \ldots, k$.

Now it is easy to see that the semigroup of an irreducible germ determines and is in turn determined by the equisingularity class of the germ:

Corollary 5.8.5 *Irreducible germs* γ *and* γ' *are equisingular if and only if* $\Sigma(\gamma) = \Sigma(\gamma')$.

PROOF: After 5.5.4 we know that γ and γ' are equisingular if and only if they have the same characteristic exponents. Then the equality of semigroups is clearly necessary for the equisingularity of the germs as their generators may be computed from the characteristic exponents, by 5.8.1 and 5.8.2 (generators may also be computed from the Enriques diagram of the germ, by the definition of the generators). The converse is clear, as, by 5.8.4, $n, \bar{m}_1, \ldots, \bar{m}_k$ are determined by the semigroup and $m_1, \ldots, m_k$ may in turn be determined from them using 5.8.1. ◇

We will devote the rest of this section to establishing some algebraic properties of $\Sigma(\gamma)$. In the sequel we will write $\Sigma(\gamma) = \Sigma$.

We take the notations as in section 3.10, so in particular we still denote by R the local ring of γ, $R = \mathcal{O}/(f)$, f being an equation of γ. Since γ is irreducible, we know from 3.11.6 that the integral closure of R in its quotient field is $\bar{R} = \mathbb{C}\{t\}$, the inclusion morphism $R \subset \mathbb{C}\{t\}$ being a uniformization morphism of γ. Therefore the semigroup of γ may be viewed as

$$\Sigma = \{j \in \bar{\mathbb{N}} | j = o_t(\mathbf{g}), \ \mathbf{g} \in R \ , \ g \neq 0\}.$$

Recall that the conductor of γ is an ideal of both R and $\mathbb{C}\{t\}$, so in particular we may write it as $(t^c) = t^c \mathbb{C}\{t\}$. It is clear that $c = \dim_{\mathbb{C}} \mathbb{C}\{t\}/(t^c)$ so that, by 4.8.2, $c = 2\delta$ if δ is the order of singularity of γ. The integer c may be characterized as follows:

Proposition 5.8.6 *We have* $j \in \Sigma$ *for all integers* j *with* $j \geq c$. *Furthermore* c *is the least integer satisfying this property.*

PROOF: If $j \geq c$, write $t^j = t^{j-c}t^c$, from which, since t^c is in the conductor, it is clear that $t^j \in R$ and therefore $j \in \Sigma$. The proof will be complete after showing that $c - 1 \notin \Sigma$. Assume $c - 1 \in \Sigma$ in order to get a contradiction: then there exists $\mathbf{g} \in (t^{c-1}) - (t^c)$, $\mathbf{g} \in R$. If u is any element of $\mathbb{C}\{t\}$, write $u = u_0 + u_1$ where $u_0 \in \mathbb{C}$ and $o_t u_1 > 0$. We have

$$u\mathbf{g} = u_0\mathbf{g} + u_1\mathbf{g}$$

where obviously $u_0 g \in R$, and also $u_1 g \in R$ because $u_1 g$ has order non-less than c and therefore it belongs to the conductor. It follows that $ug \in R$ for any $u \in \mathbb{C}\{t\}$ and hence that g is in the conductor against the hypothesis $g \notin (t^c)$. ⋄

If Σ is any semigroup contained in $\bar{\mathbb{N}}$ and there is a $c \in \Sigma$ such that $j \in \Sigma$ if $j \geq c$ while $c - 1 \notin \Sigma$, then c is uniquely determined by these properties and is currently called the *conductor* of Σ.

Back to the case $\Sigma = \Sigma(\gamma)$, we compute the number of elements in the complement of Σ:

Proposition 5.8.7 *We have* $\sharp(\bar{\mathbb{N}} - \Sigma) = \delta$.

PROOF: Write $I_j = (t^j) \cap R$. Obviously one has $I_j = I_{j+1}$ if and only if $j \notin \Sigma$, or, since in any case $\dim_{\mathbb{C}}(t^j)/(t^{j+1}) = 1$,

$$\dim_{\mathbb{C}} I_j/I_{j+1} = \begin{cases} 1 \text{ if } j \in \Sigma \\ 0 \text{ if } j \notin \Sigma. \end{cases}$$

Since all elements in $\bar{\mathbb{N}} - \Sigma$ are less than c, by 5.8.6, it is clear that $R/I_c = R/(t^c)$ has dimension equal to $c - \sharp(\bar{\mathbb{N}} - \Sigma)$. Then the exact sequence

$$0 \longrightarrow R/(t^c) \longrightarrow \mathbb{C}\{t\}/(t^c) \longrightarrow \mathbb{C}\{t\}/R \longrightarrow 0,$$

already used in the proof of 4.8.3, gives

$$\delta = \dim_{\mathbb{C}} \mathbb{C}\{t\}/R = \sharp(\bar{\mathbb{N}} - \Sigma)$$

and hence the claim. ⋄

Let us write $[0, c-1] = \{j \in \mathbb{Z} | 0 \leq j \leq c-1\}$. By 5.8.6, $[0, c-1]$ contains the complement of Σ, so that, by 5.8.7, from the $c = 2\delta$ elements of $[0, c-1]$, δ elements are in Σ and the remaining δ are the whole of $\bar{\mathbb{N}} - \Sigma$. The next property is often called the *symmetry* of the semigroup Σ.

Proposition 5.8.8 *For any* $j \in [0, c-1]$, $j \in \Sigma$ *if and only if* $c - 1 - j \notin \Sigma$.

PROOF: Consider the reflection on $c - 1$:

$$\sigma : [0, c-1] \longrightarrow [0, c-1]$$
$$j \longmapsto c - 1 - j.$$

Clearly, $\sigma^2(j) = j$ for any $j \in [0, c-1]$, so in particular σ is bijective. Since $c - 1 \notin \Sigma$, by 5.8.6, it is clear that $\sigma(j) \notin \Sigma$ if $j \in \Sigma$. Then since $[0, c-1] \cap \Sigma$ and $\bar{\mathbb{N}} - \Sigma$ have the same cardinal, one has $\sigma([0, c-1] \cap \Sigma) = \bar{\mathbb{N}} - \Sigma$ and hence $\sigma(\bar{\mathbb{N}} - \Sigma) = [0, c-1] \cap \Sigma$ which proves the claim. ⋄

An arbitrary semigroup Σ in $\bar{\mathbb{N}}$ with finite complement obviously has a conductor c, but in general it does not satisfy the claim of 5.8.8, i.e., it is not symmetric. Irreducible germs of curve in $\mathbb{C}^n$, $n > 2$, give rise to semigroups just as in the plane case, but these semigroups are in general non-symmetric: in fact such a semigroup is symmetric if and only if the local ring of the germ satisfies a special algebraic property called the *Gorenstein property*. This is in particular the case if the germ is a complete intersection, i.e., it can be defined by a set of $n - 1$ equations. See [5] or [16] for further information.

To close this section let us just mention that the definition of the semigroup $\Sigma(\gamma)$ may be extended to reduced (non-necessary irreducible) germs: if a reduced germ ξ has branches γ_i, $i = 1, \ldots, r$, one takes

$$\Sigma(\xi) = \{([\zeta \cdot \gamma_1], \ldots, [\zeta \cdot \gamma_r])\} \subset \bar{\mathbb{N}}^r$$

where ζ runs on all germs of curve at O sharing no branch with ξ. For $r > 1$ this still gives rise to a semigroup which is, however, far more difficult to handle than in the irreducible case. The interested reader is referred to [27].

5.9   Approximate roots of polynomials in $\mathbb{C}\{x\}[y]$

Let $P \in \mathbb{C}\{x\}[y]$ be a monic polynomial of degree n, $P = \sum_{i=0}^{n} b_i y^i$, $b_n = 1$. For each divisor d of n there is a well determined monic polynomial $Q \in \mathbb{C}\{x\}[y]$ so that $\deg(P - Q^d) < n - n/d$. Such a Q is called the *d-th approximate root of P*. In this section we are interested in the case in which P is the Weierstrass equation of an irreducible germ γ. If the Puiseux series of γ (relative to the local coordinates x, y) have characteristic exponents $m_1/n, \ldots, m_k/n$, our main goal is to prove that for each $i = 1, \ldots, k$ there is an approximate root of P which is irreducible and defines a germ whose Puiseux series agree with those of γ up to the degree m_i/n (this one excluded). This result, as well as the introduction of the approximate roots, is due to Abhyankar and Moh ([3] and [4]). Some of the arguments we use below come from [68].

Let us begin by proving the existence and uniqueness of approximate roots for non-necessarily irreducible polynomials.

Proposition 5.9.1 *If $P \in \mathbb{C}\{x\}[y]$ is a monic polynomial of degree n and d divides n there is a uniquely determined monic polynomial $Q \in \mathbb{C}\{x\}[y]$ such that $\deg(P - Q^d) < n - n/d$.*

PROOF: Let us write $P = \sum_{i=0}^{n} b_i y^i$, $b_n = 1$, It is clear that Q needs to have degree just n/d and so we may write $Q = \sum_{i=0}^{n/d} c_i y^i$, $c_{n/d} = 1$. Then $c_{(n/d)-1}$ is uniquely determined by equating the terms of degree $n - 1$ in P and Q^d and, inductively, by equating the monomials of degree $n - j$, $j = 2, \ldots, n/d$, one gets $c_{(n/d)-j}$ as the unique solution of a linear equation involving the $c_{(n/d)-\ell}$, $\ell < j$ already computed. ◇

Remark 5.9.2 If an arbitrary entire ring A such that $1/d \in A$ is taken instead of $\mathbb{C}\{x\}$, the claim of 5.9.1 still holds true, and the same proof applies. It is also worth noticing that the coefficients c_i of the approximate root belong to the subalgebra $\mathbb{C}[b_0, \ldots, b_{n-1}]$ of $\mathbb{C}\{x\}$ generated by the coefficients of P.

The next lemma will be useful in the sequel.

Lemma 5.9.3 *If d divides $n = \deg P$, ℓ divides n/d, G is a d-th approximate root of P and H is a ℓ-th approximate root of G, then H is a $d\ell$-approximate root of P.*

PROOF: follows from the equality

$$P - H^{d\ell} = P - G^d + G^d - H^{d\ell} = P - G^d + (G - H^d)(G^{d-1} + \cdots + H^{\ell(d-1)}),$$

as, by hypothesis $\deg(P - G^d) < n - n/d \le n - n/d\ell$ and $\deg(P - H^d) + \deg(G^{d-1} + \cdots + H^{\ell(d-1)}) < n/d - n/d\ell + n - n/d = n - n/d\ell$. ⋄

Assume now that F is any monic polynomial of degree n/d. We will write $C_P(F)$ for the quotient of the Euclidean division of $P - F^d$ by F^{d-1}. The reader may notice that $\deg C_P(F) < n/d$ and prove the next lemma directly from the definitions.

Lemma 5.9.4 *The polynomial F is the d-th approximate root of P if and only if $C_P(F) = 0$.*

Define $T_P(F) = F + d^{-1}C_P(F)$, the *Tschirnhausen transform* of F relative to P. Notice that still $\deg T_P(F) = n/d$. The next proposition provides an alternative way of getting the d-th approximate root of P.

Proposition 5.9.5 *If $C_P(F) = 0$, then $T_P(F) = F$. Otherwise, $\deg C_P(F) > \deg C_P(T_P(F))$.*

PROOF: The first claim is obvious. Regarding the second one, we have

$$P - T_P(F)^d = P - F^d - F^{d-1}C_P(F) + P'$$

where P' is the sum of the remaining terms in the expansion of $T_P(F)^d$. If $\rho = \deg C_P(F) \ge 0$, then clearly $\deg P' < \rho + (d-1)n/d$, while by the definition of $C_P(F)$, $\deg(P - F^d - F^{d-1}C_P(F)) < (d-1)n/d$. All together we obtain $\deg(P - T_P(F)^d) < \rho + (d-1)n/d$ and so, after dividing by $T_P(F)^{d-1}$, the quotient $C_P(T_P(F))$ has degree less than ρ, as claimed. ⋄

Corollary 5.9.6 *For $j \ge n/d$, $(T_P)^j(F)$ is the d-th root of P.*

PROOF: Since $j \ge n/d > \deg C_P(F)$, by 5.9.5, $C_P((T_P)^j(F)) = 0$ and the claim follows from 5.9.4. ⋄

From now on assume that x, y are local coordinates at a point O on a smooth surface S and that P is the Weierstrass equation of an irreducible germ γ at O. The reader may easily see that this is always the case for any monic and irreducible polynomial $P \in \mathbb{C}\{x\}[y]$ after a suitable substitution $y = y' + c$, $c \in \mathbb{C}$. As in the preceding sections, let s be a Puiseux series of γ, n its polydromy order and $m_1/n, \ldots, m_k/n$ its characteristic exponents. Write $n^i = \gcd(n, m^1, \ldots, m^i)$. The main result of this section reads as follows:

Theorem 5.9.7 *For any i, $0 \leq i < k$, the n^i-th approximate root G_i of P is the Weierstrass equation of an irreducible germ one of whose Puiseux series is congruent with s mod $x^{m_{i+1}/n}$.*

The reader may notice that the irreducibility of the germ $G_i = 0$ forces the polydromy order of its Puiseux series $\bar{s}$ to be $\deg G_i = n/n^i$, the minimal value allowed by the condition $\bar{s} \equiv s$ mod $x^{m_{i+1}/n}$.

PROOF OF 5.9.7: It is clearly enough to deal with the case $i = k - 1$, as the remaining cases follow from this one by iteration using 5.9.3. In the sequel we will write $G = G_{k-1}$, $d = n_{k-1}$ and $p = p_{1,1}^k$. We will need two auxiliary lemmas which are stated and proved next. In both of them we assume that $\zeta : F = 0$ is a germ of curve at O, $F \in \mathbb{C}\{x\}[y]$.

Lemma 5.9.8 *If ζ goes through p, then $\deg F \geq n/d$.*

PROOF OF 5.9.8: By 5.7.1, if ζ goes through p, then one of its branches needs to have a Puiseux series with polydromy order non-less than n/d. This series and its conjugates being roots of F, the claim follows. $\diamond$

Lemma 5.9.9 *Assume that $\deg F = n/d$. Then $[\zeta.\gamma] \equiv 1$ mod d if and only if ζ goes through p. In such a case ζ is irreducible and if furthermore F is monic, then F is the Weierstrass equation of ζ.*

PROOF OF 5.9.9: By 5.5.1, the multiplicities on γ of all points preceding p are multiples of d. Since $[\zeta.\gamma] \equiv 1$ mod d, by the Noether formula 3.3.1, ζ needs to go through p.

Conversely, assume that ζ goes through p. Then if F' is the Weierstrass equation of a branch ζ' of ζ going through p, by 5.9.8, $n/d \leq \deg F' \leq \deg F = n/d$. Thus, $\zeta = \zeta'$ is irreducible, and even, if F is monic, $F = F'$.

So it remains to prove that $[\zeta.\gamma] \equiv 1$ mod d. By 5.7.1, ζ has a Puiseux series of the form $\bar{s} = [s]_{<m_k/n} + \ldots$, where the non-explicit terms cannot increase the polydromy order n/d of the explicit ones just because $\deg F = n/d$. Then, by 5.5.1, p and all points infinitely near to p on ζ are simple and free. Since, by hypothesis, the point on γ in the first neighbourhood of p is a satellite one, ζ and γ do not share any point infinitely near to p. By 5.5.1, the multiplicities on γ of all points preceding p are multiples of d, so the Noether formula gives $[\zeta.\gamma] \equiv e_p(\gamma)$ mod d. Now, since again by 5.5.1, $1 = n^k = \gcd(d, m_k) = \gcd(d, e_p(\gamma))$, the claim follows. $\diamond$

END OF THE PROOF OF 5.9.7: Let ζ be any germ that goes through p and whose Weierstrass equation F has degree n/d. For instance we may take for ζ the germ with Puiseux series $\bar{s} = [s]_{<m_k/n}$ which goes through p by 5.7.1. By 5.9.9, $o_\gamma(F) = [\zeta.\gamma] \equiv 1$ mod d. By dividing P and the successive remainders by decreasing powers of F we get an expression

$$P = F^d + C_1 F^{d-1} + \cdots + C_d \tag{5.5}$$

where $C_1 = C_P(F)$, $C_i \in \mathbb{C}\{x\}[y]$ and $\deg C_i < n/d$, for $i = 1, \ldots, d$. By 5.9.8 the germ $C_i = 0$ cannot go through p and so, in particular, $o_\gamma(C_i) < \infty$. Furthermore, since all multiplicities on γ of points preceding p are multiples of d, by the Noether formula 3.3.1, $o_\gamma(C_i) \equiv 0 \mod d$. It follows that $o_\gamma(C_i F^i) \equiv i \mod d$ for $i = 1, \ldots, d - 1$, while $o_\gamma(F^d) \equiv o_\gamma(C_0) \equiv 0 \mod d$.

The left hand side of the equality 5.5 identically vanishes after substituting $x = t^n$ $y = s(t^n)$. We have just seen that all summands on the right side have finite order along γ: after substitution, the initial terms of the summands of minimal order must cancel. Because of the above congruences, cancellation of initial terms may be done by the first and last summands only, and thus they are the only summands having minimal order along γ. We get $d[\zeta.\gamma] = o_\gamma(F^d) < o_\gamma(C_i F^{d-i})$, $i > 0$, and so, in particular, for $i = 1$, $[\zeta.\gamma] < o_\gamma(C_P(F))$.

Recall that we have already defined $T_P(F) = F + d^{-1}C_P(F)$, so that, clearly, it has degree n/d. Furthermore, by the above inequality, $o_\gamma(T_P(F)) = o_\gamma(F) \equiv 1 \mod d$ and hence, by 5.9.9, $T_P(F)$ is the Weierstrass equation of an irreducible germ ζ_1 going through p.

The whole argument may now be repeated by taking $T_P(F)$ instead of F, then $T_P^2(F)$, and so on. By 5.9.6, after at most n/d steps, the d-th approximate root G of P will be reached: thus also G is the Weierstrass equation of an irreducible germ going through p, after which the claim follows by using once again 5.7.1. $\diamond$

As it follows from 5.7.1, and has been in fact seen while proving it, 5.9.7 may be equivalently stated in the form:

Corollary 5.9.10 *For each $i = 0, \ldots, k - 1$, the germ $\gamma_i : G_i = 0$ is irreducible and $e_{p_{1,1}^{i+1}}(\gamma_i) = 1$.*

Corollary 5.9.11 *Keep the notations and hypothesis as in 5.9.7 and assume furthermore that γ is not tangent to the y-axis. Then $n = e_O(\gamma)$ and the $o_\gamma(G_i)$, $i = 0, \ldots, k - 1$, are a minimal system of generators of the semigroup of γ.*

PROOF: Follows from 5.9.10, 5.8.2 and the definition of the integers $\bar{m}_i$ in section 5.8. $\diamond$

5.10 Exercises

5.1 Relate the characteristic exponents of an irreducible germ with origin at O and those of its strict transform after blowing up O.

5.2 Let γ be an irreducible germ of curve with origin at a point O of the affine plane $\mathbb{A}_2$. Identify $\mathbb{A}_2$ with its own tangent space at O so that, in particular, the principal tangent ℓ to γ is a line in $\mathbb{A}_2$. Define the *class* of γ as the difference $[\gamma.\ell] - e_O(\gamma)$. Prove that the class is not an analytic invariant of the germ.

Let x, y be affine coordinates with origin O, the x-axis being the principal tangent ℓ and identify each line $y = mx + b$ with the point of affine coordinates m, b in a second affine plane $\mathbb{A}_2^*$. Prove that there is an open neighbourhood U of O so that for any representative γ' of γ in an open neighbourhood U' of O, $U' \subset U$, the tangents to γ' are the points of an analytic curve in $\mathbb{A}_2^*$ whose germ γ^* at $(0,0)$ (the *dual germ* of γ) depends only on γ and not on γ'.

Prove that the order and class of γ^* are, respectively, the class and order of γ.

5.3 Prove that the germ $\gamma : (x^4 - y^5)^3 - x^{14}y = 0$ is irreducible. Compute the characteristic exponents of its Puiseux series as well as those of γ itself, and draw its Enriques diagram.

5.4 Let $\{n_i\}$, $i \in \mathbb{N}$, be a non-increasing sequence of positive integers, all but finitely many of them being equal to one and satisfying the conditions of 3.5.9. Prove that there is a system of characteristic exponents so that all irreducible germs with these characteristic exponents have sequence of multiplicities $\{n_i\}$.

5.5 Prove that the sum of multiplicities of all satellite points on an irreducible germ γ with origin at O is just $e_O(\gamma) - 1$.

5.6 Let γ be an irreducible germ whose Puiseux series has characteristic exponents $m_1/n, \ldots, m_k/n$ and write $\gcd(n, m_1, \ldots, m_i) = n^i$ $(n^k = 1)$. Denote by T_i the set of points on γ depending on the i-th characteristic exponent. Put $T = \bigcup_{i=1}^{k} T_i$ and take $m_0 = 0$. Check the following equalities, the last one being that already claimed in 5.5.7:

$$\sum_{p \in T_i} e_p(\gamma) = m_i - m_{i-1} - n^i + n^{i+1}$$

$$\sum_{p \in T_i} e_p(\gamma)^2 = (m_i - m_{i-1})n^{i-1}$$

$$\sum_{p \in T} e_p(\gamma) = m_k + n - 1$$

$$\sum_{p \in T} e_p(\gamma)^2 = \sum_{i=1}^{k} (m_i - m_{i-1})n^{i-1}$$

$$2\delta(\gamma) = \sum_{i=1}^{k} m_i(n^{i-1} - n^i) - \dot{n} + 1.$$

5.7 Describe the proximity matrix of the cluster of all singular points of an irreducible germ of curve with characteristic exponents $m_1/n, \ldots, m_k/n$.

5.8 Notations for the irreducible germ γ being as in section 5.8, prove that if a germ ξ does not go through the last free point on γ depending on the i-th characteristic exponent, then $[\xi.\gamma]$ belongs to the semigroup generated by $n, \bar{m}_1, \ldots, \bar{m}_{i-1}$.

5.9 Let γ be an irreducible germ of curve with origin at O and $\tilde{\gamma}$ its strict transform after blowing up O. Relate the generators of the semigroups $\Sigma(\gamma)$ and $\Sigma(\gamma')$.

5.10 Take all notations as in section 5.8. Prove that

$$\frac{n^{i-1}}{n^i} \bar{m}_i \in\, < n, \bar{m}_1, \ldots, \bar{m}_{i-1} >$$

for $i = 1, \ldots, k$.

5.11 Take all notations as in section 5.8. Prove that the generators $\breve{m}_i$ of the semigroup of the branch γ satisfy the relations

$$n^{i-1}\breve{m}_i < n^i \breve{m}_{i+1}$$

for $i = 1, \ldots, k-1$.

Conversely, assume that there are given positive integers $\alpha_0, \ldots, \alpha_k$ so that if one takes $n^i = \gcd(\alpha_0, \ldots, \alpha_i)$, $i = 0, \ldots, k$, then

$$n^0 > n^1 > \cdots > n^k = 1$$

and

$$n^{i-1}\alpha_i < n^i \alpha_{i+1}$$

for $i = 1, \ldots, k-1$. Prove that there is an irreducible germ of curve whose semigroup is generated by $\alpha_0, \ldots, \alpha_k$.

5.12 Let γ be an irreducible germ, assume that its Puiseux series has characteristic exponents $m_1/n, \ldots, m_k/n$. Still write $n^i = \gcd(n, m_1, \ldots, m_i)$ and assume $n^k = 1$. Let ξ be any germ so that $[\eta.\xi] \leq n/n^{i-1}$ where η denotes the germ of the y-axis. Prove that

$$[\gamma.\xi] \leq \frac{(n-n^1)m_1}{n^{i-1}} + \frac{(n^1-n^2)m_2}{n^{i-1}} + \cdots + \frac{(n^{i-2}-n^{i-1})m_{i-1}}{n^{i-1}} + m_i.$$

Prove also that if the equality holds, then ξ is an irreducible germ whose Puiseux series has polydromy order n/n^i and characteristic exponents $m_1/n, \ldots, m_{i-1}/n$.

Chapter 6

Polar germs and related invariants

Polar germs are germs of curve that are associated with each non-empty germ of plane curve and carry very deep information on it. They will be studied in this chapter and used for introducing and computing some new equisingularity invariants of a reduced germ of curve, such as its Milnor number and polar quotients, for giving examples of non-isomorphic equisingular germs and for constructing analytic invariants that are not invariant by equisingularity.

6.1   Definition and first properties of polar germs

As in the preceding chapters, let us fix a point O on a smooth surface S. Throughout this chapter we will denote by g an analytic function (or germ of function) defined near O such that the curve (or germ) $\eta : g = 0$ is smooth at O. Then, if x, y are local coordinates at O, at least one of the derivatives $\partial g / \partial x$, $\partial g / \partial y$ is non-zero at O.

Let $\xi : f = 0$ be a non empty-germ of curve at O and fix a germ of analytic function g as above, that is, with $\eta : g = 0$ smooth at O. Take local coordinates x, y in a neighbourhood U of O and denote by $\partial(f, g)/\partial(x, y)$ the jacobian determinant

$$\frac{\partial(f, g)}{\partial(x, y)} = \begin{vmatrix} \partial f / \partial x & \partial f / \partial y \\ \partial g / \partial x & \partial g / \partial y \end{vmatrix}.$$

Definition: If $\partial(f, g)/\partial(x, y) \neq 0$, it defines a germ of curve at O which is called the *g-polar germ* (or just the *g-polar*) of ξ relative to its equation f. It will be denoted by $P_g(f)$ in the sequel. A polar or a polar germ of ξ is just a g-polar germ of ξ for some g defining a smooth curve at O.

It makes no sense to define polar germs of the empty germ, as any g-polar germ $P_g(f)$, $f(O) = 0$, would appear as the g-polar germ of the empty germ relative to its equation $1 + f$. In the sequel, when speaking about polar germs of a germ ξ, it will be always implicitly assumed that $\xi \neq \emptyset$.

Of course, if f' and g' are representatives of f and g defined in a neighbourhood V of O where the local coordinates are also defined, the curve ζ : $\partial(f', g')/\partial(x, y) = 0$ in V is a representative of $P_g(f)$. Note that the locus of ζ is the set of the points $p \in V$ at which $f' - f'(p) = 0$ and $g' - g'(p) = 0$ are not transverse.

Remark 6.1.1 The polar germ $P_g(f)$ does not depend on the coordinates used in its definition. Indeed, if x', y' also are local coordinates at O, then $\partial(x', y')/\partial(x, y)$ is invertible at O and

$$\frac{\partial(f, g)}{\partial(x', y')} \frac{\partial(x', y')}{\partial(x, y)} = \frac{\partial(f, g)}{\partial(x, y)}.$$

Remark 6.1.2 If φ is an analytic isomorphism defined in a neighbourhood of a point O' in a surface S' and $\varphi(O') = O$, then the elementary derivation rules show that if the polar germ $P_g(f)$ is defined, then so is $P_{\varphi^*(g)}(\varphi^*(f))$ and

$$\varphi^*(P_g(f)) = P_{\varphi^*(g)}(\varphi^*(f)).$$

In this sense, the notion of polar germ is invariant under analytic isomorphisms.

Remark 6.1.3 If the germ g is linear in the coordinates, say $g = ay - bx$, $a, b \in \mathbb{C}$, then the equation of the polar germ has the form

$$a\frac{\partial f}{\partial x} + b\frac{\partial f}{\partial y}.$$

Conversely, any polar has an equation of this form if suitable local coordinates are used: one may even choose g as one of the coordinates, say $x = g$, and then the equation of the polar is just a derivative, namely $\partial f/\partial y = 0$.

Remark 6.1.4 The polar germ remains undefined (because its equation identically vanishes) only if $\xi = r\eta$. Indeed, arguing as in 6.1.3 one may assume that the polar has equation $\partial f/\partial y$. If $\partial f/\partial y = 0$, then f depends only on x and hence $f = x^r u$ with u invertible, as claimed. Unless explicitly noticed, this case will be excluded in the sequel and thus, when speaking about a g-polar of ξ, we will always assume not only that ξ is non-empty, but also that it has some irreducible component different from $\eta : g = 0$.

Remark 6.1.5 Also the case in which η is a component of ξ has little interest: from $f = g^r f'$ an easy computation gives $P_g(f) = r\eta + P_g(f')$, i. e., the polar is composed of η and the polar of the germ $\xi' : f' = 0$ which may be assumed not to have component η.

Remark 6.1.6 In general, the polar $P_g(f)$ depends on the equation f and not only on the germ ξ: according to the derivation rules, by modifying the equation of ξ one adds to the equation of the polar a multiple of the equation of ξ:

$$\frac{\partial(uf, g)}{\partial(x, y)} = u\frac{\partial(f, g)}{\partial(x, y)} + f\frac{\partial(u, g)}{\partial(x, y)}.$$

However, we will often use the notation $P_g(\xi)$ for $P_g(f)$, f an equation of ξ, if no confusion may result, as in most cases we will be interested in aspects of $P_g(f)$ that do not depend on the equation f but only on the germ ξ itself.

In the frame of projective geometry, the *polar curve* of the algebraic plane curve $C : F(X_0, X_1, X_2) = 0$ relative to the point $p = [a_0, a_1, a_2]$ (often also called the polar of p relative to C) is defined as the projective curve with equation

$$a_0\frac{\partial F}{\partial X_0} + a_1\frac{\partial F}{\partial X_1} + a_2\frac{\partial F}{\partial X_2} = 0.$$

This is a very classical notion on which the local definition of polar germs is modelled. The reader may easily see that if O belongs to C and $p \neq O$, then the germs of the (projective) polar curves are polar germs of C_O in our local sense. In particular the results on polar germs we will obtain in this chapter may be applied to the projective polars, see, for instance, exercises 6.8 and 6.9.

Back to the local case, let us begin by showing the easiest properties of the polar germs:

Proposition 6.1.7 *Let ξ be a non-empty germ at O, $\zeta = P_g(\xi)$ any g-polar of ξ and $\eta : g = 0$. Assume $\xi \neq r\eta$, then we have:*

(a) $[\zeta.\eta] = [\xi.\eta] - 1$.

(b) $e_O(\zeta) \geq e_O(\xi) - 1$ and the inequality is strict if and only if all branches of ξ are tangent to η.

(c) If no branch of ξ is tangent to η, then no branch of the polar ζ is tangent to η either.

PROOF: Take local coordinates x, y with $x = g$ so that the polar ζ has equation $\partial f/\partial y$, by 6.1.3. Then, since $\xi \neq \emptyset$, $o_y f(0, y) > 0$ and so

$$[\xi.\eta] = o_y f(0, y) = 1 + o_y \frac{\partial f}{\partial y}(0, y) = 1 + [\zeta.\eta]$$

and claim (a) is proved.

The Newton diagram of $\partial f/\partial y$, $\Delta(\partial f/\partial y)$, comes from that of f by deleting the points that lie on the α-axis and moving all other points one step downwards. Since $e_O(\xi) = \min\{\alpha + \beta|(\alpha, \beta) \in \Delta(f)\}$ and $e_O(\zeta) = \min\{\alpha + \beta|(\alpha, \beta) \in \Delta(\partial f/\partial y)\}$, clearly $e_O(\zeta) \geq e_O(\xi) - 1$. Furthermore, the inequality is strict if and only if all points $(\alpha, \beta) \in \Delta(f)$ for which $\alpha + \beta = e_O(\xi)$ lie on the α-axis.

Clearly, this occurs if and only if all sides of $\mathbf{N}(\xi)$ have slope strictly less than -1, that is, if and only if all branches of ξ are tangent to the y-axis (2.2.7).

Lastly, for part (c), one has

$$e_O(\zeta) \leq [\zeta \cdot \eta] = [\xi \cdot \eta] - 1 = e_O(\xi) - 1,$$

the last equality coming from the hypothesis that no branch of ξ is tangent to η. Using part (b) above one obtains

$$e_O(\zeta) = [\zeta \cdot \eta]$$

which in turn implies that no branch of ζ is tangent to η. ◇

In the sequel we will say that a polar $P_g(\xi)$ is a *transverse polar* if and only if no branch of ξ is tangent to $\eta : g = 0$. Otherwise we will say that the polar is a *non-transverse* one.

6.2 Transforming polar germs

As is the preceding section, assume that f is an equation of a non-empty germ of curve ξ at O and consider the ideal $\mathbf{J}(\xi) = \mathbf{J}(f) = (\partial f/\partial x, \partial f/\partial y, f)$, often called the *jacobian ideal* of ξ. An easy computation shows that the jacobian ideal does not depend on the equation f of ξ or the local coordinates x, y. The linear system of germs it defines will be called the *jacobian system* of ξ. It will be denoted by $\mathcal{J}(\xi)$ or also by $\mathcal{J}(f)$. The next lemma shows that the jacobian ideal is closely related to the polar germs.

Lemma 6.2.1 *The equations of the polars of a germ ξ generate its jacobian ideal. Furthermore any $h \in \mathbf{J}(f)$,*

$$h = u_1 \frac{\partial f}{\partial x} + u_2 \frac{\partial f}{\partial y} + u_3 f = 0,$$

with either u_1 or u_2 invertible at O defines a polar of ξ.

PROOF: Clearly $\partial f/\partial x$ and $\partial f/\partial y$ are equations of polar germs. It is enough to add an equation of, say, $P_x((1 + y)f)$ to generate the ideal. Since it follows from their definition and 6.1.6 that all polars of ξ have their equations in the jacobian ideal, the first half of the claim is proved.

For the second one, note first that the vector field

$$u_1 \frac{\partial}{\partial x} + u_2 \frac{\partial}{\partial y}$$

is, by the hypothesis, regular at O. Thus, after a suitable change of local coordinates one may assume without restriction that $h = \partial f/\partial y + uf$. Then, if v is any analytic function defined in a neighbourhood of O and such that $\partial v/\partial y = u$, the germ $h = 0$ is the x-polar of ξ relative to its equation $e^v f$. ◇

Remark 6.2.2 It is clear from 6.1.7 and 6.2.1 that for any $\zeta \in \mathcal{J}(\xi)$, $e_O(\zeta) \geq e_O(\xi) - 1$. It follows also from 6.1.7 and 6.2.1 that all germs $\zeta \in \mathcal{J}(\xi)$ with $e_O(\zeta) = e_O(\xi) - 1$ are polar germs, and hence there is a non-empty Zariski-open subset of $\mathbf{J}(\xi)$ all of whose elements define polar germs, namely $\mathbf{J}(\xi) - \mathcal{M}_{S,O}^{e_O(\xi)}$. Thus generic germs in the jacobian system can always be assumed to be polar germs and consequently we will often say *generic polar* instead of *generic germ in the jacobian system*.

In the forthcoming section 8.5 we will come back to polar germs in order to obtain further properties of them related to the linear structure of the jacobian system. Most important among these properties and worth noting here is the fact (8.5.1) that germs defined by the elements of a certain non-empty Zariski-open subset of $\mathbf{J}(\xi)$ (i.e., generic polars of ξ) all have the same equisingularity type.

Next we will obtain some information about the behaviour of polar germs under blowing up.

Proposition 6.2.3 *Fix any point p on ξ in the first neighbourhood of O and take O with virtual multiplicity $e_O(\xi) - 1$. Then the equations of the virtual transform of any polar of ξ belong to the jacobian ideal of the strict transform of ξ, $\tilde{\xi}_p$.*

PROOF: By 6.2.1 the claim may be equivalently stated for all germs defined by elements of the jacobian ideal of ξ, and then it is enough to prove it for the germs defined by a system of generators of the jacobian ideal. Thus we choose local coordinates x, y so that p is on the x-axis and prove the claim for the polars

$$\frac{\partial f}{\partial x} = 0, \quad \frac{\partial f}{\partial y} = 0,$$

the case of the germ ξ itself, corresponding to the generator f, being obvious. Local coordinates at p, $\tilde{x}, \tilde{y}$ are related to x, y by the formulas $\bar{x} = \tilde{x}$, $\bar{y} = \tilde{x}\tilde{y}$, where $\tilde{x}$ is an equation of the exceptional divisor and the bar means pull-back by the blowing-up (3.1.2). Take $\tilde{f} = \tilde{x}^{-e}\bar{f}$, $e = e_O(\xi)$, as equation of the strict transform of ξ. Since $\bar{f} = \tilde{x}^e \tilde{f}$, clearly,

$$\frac{\partial \bar{f}}{\partial \tilde{x}} = \tilde{x}^e \frac{\partial \tilde{f}}{\partial \tilde{x}} + e\tilde{x}^{e-1}\tilde{f}$$

$$\frac{\partial \bar{f}}{\partial \tilde{y}} = \tilde{x}^e \frac{\partial \tilde{f}}{\partial \tilde{y}}.$$

On the other hand $\bar{f} = f(\bar{x}, \bar{y}) = f(\tilde{x}, \tilde{x}\tilde{y})$, from which

$$\frac{\partial \bar{f}}{\partial \tilde{x}} = \overline{\frac{\partial f}{\partial x}} + \overline{\frac{\partial f}{\partial y}}\tilde{y}$$

$$\frac{\partial \bar{f}}{\partial \tilde{y}} = \overline{\frac{\partial f}{\partial y}}\tilde{x}.$$

Then, an easy computation gives

$$\overline{\frac{\partial f}{\partial x}} = \tilde{x}^e \frac{\partial \tilde{f}}{\partial \tilde{x}} - \tilde{x}^{e-1}\tilde{y}\frac{\partial \tilde{f}}{\partial \tilde{y}} + e\tilde{x}^{e-1}\tilde{f}$$

$$\overline{\frac{\partial f}{\partial y}} = \tilde{x}^{e-1}\frac{\partial \tilde{f}}{\partial \tilde{y}}$$

and hence the claim. ◇

More precise information may be found after blowing up points until reaching a free point not belonging to η:

Proposition 6.2.4 *Assume that ξ is a germ of curve at O, γ is a branch of ξ and p is the first free point on γ not on $\eta : g = 0$. Then the strict transform at p of any g-polar of ξ is a $\tilde{g}$-polar of the strict transform of ξ at p, $\tilde{g}$ being an equation of the exceptional divisor at p.*

PROOF: Take local coordinates at O, x, y, with $x = g$ and fix an equation f of ξ so that the polar we consider has equation $\partial f/\partial y$ (6.1.3). The Puiseux series of γ may be written in the form

$$s = ax^{m/n} + \ldots$$

with $m/n \leq 1$ and $a \neq 0$ if $m/n < 1$. Assume furthermore that m and n are coprime. Then the point p is the first point on γ after the points depending on the characteristic exponent m/n if $m/n < 1$, or just the point on γ in the first neighbourhood of O if $m/n = 1$ and hence $m = n = 1$. In any case it has coordinate a^n (5.3.4 or 3.1.2) and we know from 5.4.2 or 3.1.2 that one may choose local coordinates $\tilde{x}, \tilde{y}$ at p so that $\tilde{x}$ is a local equation of the exceptional divisor and

$$\bar{x} = \tilde{x}^n$$
$$\bar{y} = \tilde{x}^m(a + \tilde{y}),$$

the bar meaning pull-back by the composition of the blowing-ups giving rise to p. Write the equation of ξ in the form

$$f = \sum_{n\alpha+m\beta=k} a_{\alpha,\beta}x^\alpha y^\beta + \sum_{n\alpha+m\beta>k} a_{\alpha,\beta}x^\alpha y^\beta$$

where at least one of the $a_{\alpha,\beta}$ with $n\alpha + m\beta = k$ is non-zero. Then we have

$$\bar{f} = \tilde{x}^k\left(\sum_{n\alpha+m\beta=k} a_{\alpha,\beta}(a+\tilde{y})^\beta\right) + \tilde{x}^{k+1}(\cdots)$$

from which it is clear that $\tilde{f} = \tilde{x}^{-k}\bar{f}$ cannot be divided by $\tilde{x}$ and is therefore an equation of the strict transform of ξ with origin at p. Computing derivatives

$$\overline{\frac{\partial f}{\partial y}} = \frac{\partial \bar{f}}{\partial \bar{y}} = \frac{\partial \bar{f}}{\partial \tilde{y}}\left(\frac{\partial \bar{y}}{\partial \tilde{y}}\right)^{-1} = \tilde{x}^{k-m}\frac{\partial \tilde{f}}{\partial \tilde{y}}.$$

Since $\tilde{x}$ does not divide $\tilde{f}$, it does not divide $\partial \tilde{f}/\partial \tilde{y}$ either. Therefore the former equalities prove that $\partial \tilde{f}/\partial \tilde{y}$ is an equation of the strict transform of $P_x(f)$ just as claimed. $\diamond$

Remark 6.2.5 We have proved the equality

$$\widetilde{P_x(f)}_p = P_{\tilde{x}}(\tilde{f})$$

where $\tilde{f}$ is the equation of $\tilde{\xi}_p$ obtained from the pull-back of f: $\tilde{f} = \tilde{x}^{-k}\bar{f}$.

Remark 6.2.6 Proposition 6.2.4 applies in particular to the points p in the first neighbourhood of O that belong to ξ but not to η. It is a relevant fact that the first neighbouring point on η is excluded from the claim. Just because of it, iterated use of 6.2.4 on the strict transforms of ξ obtained by successively blowing up the points on a non-smooth branch necessarily stops as soon as the first satellite point is reached. Thus, using 6.2.4 in this way is quite useless for a general analysis of the behaviour of polar germs at infinitely near points. By contrast, 6.2.4 allows us to jump over the satellite points and still control the polar germ at the next free ones, which, as it will turn out in forthcoming sections, is a very successful strategy. We make use of it in the proof of the next corollary.

Corollary 6.2.7 *Let ξ be a non-empty germ of curve with origin at O. Assume that g defines a smooth germ $\eta : g = 0$ at O. If p is any free point on ξ not on η and E_p denotes the germ at p of the exceptional divisor, then $[\tilde{\zeta}_p.E_p] = [\tilde{\xi}_p.E_p]-1$ for any g-polar ζ of ξ.*

The reader may notice that the same claim holds true for $p = O$ and $E_O = \eta$, as proved in 6.1.7.

PROOF: By induction on the number r of free points not on η and preceding p. The claim follows from 6.2.4 and 6.1.7 if $r = 0$. Otherwise, still using 6.2.4, the claim follows from the induction hypothesis applied to the strict transforms of ξ and ζ with origin at the first free point not on η and preceding p. $\diamond$

Keeping the hypothesis of 6.2.7 we have:

Corollary 6.2.8 *The polar ζ goes through p if and only if p is a singular point of ξ.*

PROOF: Just recall that p is singular for ξ if and only if $[\tilde{\xi}_p.E_p] > 1$ (3.8.1) and then apply 6.2.7. $\diamond$

6.3   The Plücker formula

In this section we compute the intersection multiplicities of the branches of a (non-empty) germ ξ and any of its polar germs, say $P_g(\xi)$. As a consequence we will get a formula for the intersection multiplicity of ξ and $P_g(\xi)$ which is closely related to the classical first Plücker formula. Still denote by η the germ defined by g.

Theorem 6.3.1 *For any branch γ of a non-empty germ ξ and any g-polar $P_g(\xi)$ of ξ,*

$$[\gamma.P_g(\xi)] = \sum_{p \in \gamma} e_p(\gamma)(e_p(\xi) - 1) + [\gamma.\eta] - 1$$

in the sense that one side is finite if and only if the other is so, and then they agree.

PROOF: If γ is a multiple component of ξ, then an easy computation shows that γ is also a component of $P_g(\xi)$: then both sides of the equality are infinity. Also if $\gamma = \eta$ both sides of the equality are infinity, this time by 6.1.5. Thus we will assume in the sequel that $\gamma \neq \eta$ and that γ is a simple component of ξ. The claim will be proved in two steps:

  Step 1: Assume ξ irreducible, that is, $\xi = \gamma$. First we will see that if the claimed formula is true for all transverse polars, then it is true for the non-transverse ones too. Indeed, assume that γ is tangent to η. Choose local coordinates with $x = g$ and let f be the equation of γ for which $P_x(\gamma) = P_x(f)$. Then η is the y-axis, the x-axis $\tau : y = 0$ is non-tangent to γ and therefore $P_y(f)$ is a transverse polar. Let $t \mapsto (x(t), y(t))$, $x(t), y(t) \in \mathbb{C}\{t\}$ be a uniformizing map of a representative of γ. We have $o_t x(t) = [\gamma.\eta]$ and $o_t y(t) = [\gamma.\tau] = e_O(\gamma)$. Since $f(x(t), y(t)) = 0$ for t close enough to 0,

$$0 = \frac{\partial f}{\partial x}(x(t), y(t))\frac{dx(t)}{dt} + \frac{\partial f}{\partial y}(x(t), y(t))\frac{dy(t)}{dt}.$$

Then, by equating the orders in t of both summands, we get

$$[\gamma.P_x(f)] = [\gamma.P_y(f)] + [\gamma.\eta] - [\gamma.\tau]$$

and it is enough to compute $[\gamma.P_y(f)]$ using the formula we are assuming to be true for transverse polars.

  Now we need to prove the claim just for transverse polars. We use induction on the number of multiple points on γ, the case of γ smooth being obvious.

  Since γ is not tangent to η, $P_x(\gamma)$ has multiplicity at O equal to $e_O(\gamma) - 1$, by (6.1.7). From 3.3.4 and 6.2.6 we get

$$[\gamma.P_g(\gamma)] = e_O(\gamma)(e_O(\gamma) - 1) + [\tilde{\gamma}.P_{\tilde{x}}(\tilde{\gamma})],$$

where q is the point on γ in the first neighbourhood of O, $\tilde{\gamma}$ the strict transform of γ with origin at q and $\tilde{x}$ an equation of the germ E_q of the exceptional divisor at q. Since the induction hypothesis gives

$$[\tilde{\gamma}.P_{\tilde{x}}(\tilde{\gamma})] = \sum_{p \in \tilde{\gamma}} e_p(\tilde{\gamma})(e_p(\tilde{\gamma}) - 1) + [\tilde{\gamma}.E_q] - 1,$$

the claim follows after using that $e_p(\tilde{\gamma}) = e_p(\gamma)$ for all $p \in \tilde{\gamma}$ and $[\tilde{\gamma}.E_q] = e_O(\gamma) = [\gamma.\eta]$, the latter equality being true because γ is not tangent to η.

Step 2: ξ is non-irreducible. Denote by $\gamma_1, \ldots, \gamma_r$ the branches of ξ and assume, after a suitable reordering, that $\gamma = \gamma_1$. As before, take coordinates x, y with $x = g$ and let f be the equation of ξ for which $P_x(\xi) = P_x(f)$. Write $f = \prod_{i=1}^r f_i^{\alpha_i}$, each f_i being an equation of γ_i, and, by hypothesis, $\alpha_1 = 1$. Still take $t \mapsto (x(t), y(t))$ to be a uniformizing map of a representative of γ. Since $f_1(x(t), y(t)) = 0$, we easily get

$$\frac{\partial f}{\partial y}(x(t), y(t)) = \prod_{i=2}^r f_i(x(t), y(t))^{\alpha_i} \frac{\partial f_1}{\partial y}(x(t), y(t)),$$

so that, by taking orders in t and using the claim already proved in step 1,

$$[\gamma.P_x(\xi)] = \sum_{i=2}^r \alpha_i[\gamma.\gamma_i] + [\gamma.P_x(\gamma)]$$

$$= \sum_{i=2}^r \sum_{p\in\gamma} \alpha_i e_p(\gamma)e_p(\gamma_i) + \sum_{p\in\gamma} e_p(\gamma)(e_p(\gamma) - 1) + [\gamma.\eta] - 1$$

$$= \sum_{p\in\gamma} e_p(\gamma)(e_p(\xi) - 1) + [\gamma.\eta] - 1$$

which ends the proof. $\diamond$

Corollary 6.3.2 (Plücker formula) *If r denotes the number of branches of ξ, using the same conventions as in 6.3.1,*

$$[\xi.P_g(\xi)] = \sum_{p\in\xi} e_p(\xi)(e_p(\xi) - 1) + [\xi.\eta] - r.$$

PROOF: Just add up the equalities 6.3.1 gives rise to when applied to the branches of ξ. $\diamond$

Remark 6.3.3 The Plücker formula looks shorter using the order of singularity (3.11.12), namely
$$[\xi.P_g(\xi)] = 2\delta(\xi) + [\xi.\eta] - r.$$

In fact 6.3.2 corresponds to the local part of the classical Plücker formula. Assume that $C : F(X_0, X_1, X_2) = 0$ is a reduced plane algebraic curve in $\mathbb{P}_2$ and $Q = [a_0, a_1, a_2]$ a point. If P is a simple point of C, the equality

$$a_0\left(\frac{\partial F}{\partial X_0}\right)(P) + a_1\left(\frac{\partial F}{\partial X_1}\right)(P) + a_2\left(\frac{\partial F}{\partial X_2}\right)(P) = 0$$

may be equivalently read either as saying that the line tangent to C at P goes through Q, or that the polar of C relative to Q goes through P. Thus the number of tangents to C through a generic point Q may be computed as the number of intersections of C and its polar relative to Q that are simple points of C, the lines from Q to the singular points of ξ being not considered as proper tangent lines if Q is generic. This number of intersections equals the total intersection number $(\ell(\ell - 1), \ell$ the degree of ξ) minus the sum of the intersection multiplicities at the singular points, given by 6.3.2. For a more precise claim, see exercise 6.9.

As is clear from 6.3.2, the integer $d_g(\xi) = [\xi.P_g(\xi)]$ does not depend on g and is an equisingularity invariant of ξ as far as $\eta : g = 0$ is not tangent to ξ. In such a case we will just write $d(\xi) = d_g(\xi)$. Smith, in his remarkable paper [76], named $d(\xi)$ the *discriminantal index* of ξ. The reader may have noticed that the discriminantal index is just the control function we have already used in the proof of 3.7.1. Under no assumption about the relative position of ξ and $\eta : g = 0$, we will call $d_g(\xi)$ the *discriminantal index* of ξ *relative to* g (or to η).

Corollary 6.3.4 *A branch γ of ξ is also a branch of $P_g(\xi)$ if and only if either $\gamma = \eta$ or γ is a multiple branch of ξ. In particular, if η is not a branch of ξ, then ξ is reduced if and only if it shares no branch with $P_g(\xi)$.*

PROOF: By 6.3.1, $[\gamma.P_g(\xi)] = \infty$ if and only if either $[\gamma.\eta] = \infty$ or $e_p(\xi) > 1$ for all p on γ, hence the claim. $\diamond$

6.4   The Milnor number

In this section we will use the intersection multiplicity of two polar germs to introduce the Milnor number, one of the most interesting local invariants of curves.

Theorem 6.4.1 *Let $\xi : f = 0$ be a non-empty germ of curve and x, y local coordinates at O. If ξ has r branches, then it holds*

$$[P_x(f).P_y(f)] = \sum_{p \in \xi} e_p(\xi)(e_p(\xi) - 1) + 1 - r,$$

in the sense that if one side is finite, then so is the other and they agree.

PROOF: The claim is obvious if $P_x(f) = \emptyset$, as then ξ is smooth. Thus assume $P_x(f) \neq \emptyset$, denote by γ any branch of the polar $P_x(f)$ and assume that $(x(t), y(t))$ is a uniformizing map of γ. Then we have

$$\frac{\partial f}{\partial y}(x(t), y(t)) = 0$$

identically in t, and hence

$$\frac{d}{dt}f(x(t), y(t)) = \frac{\partial f}{\partial x}(x(t), y(t))\frac{dx}{dt}.$$

By equating the orders in t of both sides we get

$$[\xi \cdot \gamma] = [P_y(f) \cdot \gamma] + [\eta \cdot \gamma],$$

η still denoting the y-axis. Thus, after adding up the above equalities for all branches γ of ξ, we obtain

$$[\xi \cdot P_x(f)] = [P_y(f) \cdot P_x(f)] + [\eta \cdot P_x(f)].$$

By 6.1.7, $[\eta \cdot P_x(f)] = [\eta \cdot \xi] - 1$, after which it is enough to use 6.3.2 in order to reach the claim. $\diamond$

It follows from 6.4.1 that the intersection multiplicity $[P_x(f) \cdot P_y(f)]$ does not depend on the coordinates x, y or the equation f of the germ ξ. Thus we can set the following definition following Milnor [60]:

Definition: The *Milnor number* (often called Milnor's μ) of a non-empty germ of curve ξ is

$$\mu(\xi) = [P_x(f) \cdot P_y(f)] = \sum_{p \in \xi} e_p(\xi)(e_p(\xi) - 1) + 1 - r,$$

where f is any equation of ξ, r its number of branches and x, y are local coordinates at the origin of ξ.

Remark 6.4.2 Clearly $\mu(\xi)$ is an equisingularity invariant. It is worth noticing that $\mu(\xi)$ is finite (that is, $P_x(f)$ and $P_x(f)$ share no branch) if and only if ξ is reduced (by 6.4.1, 3.7.1 and 3.7.10). Of course, by 3.11.10, one may equivalently take, in a more algebraic form,

$$\mu(\xi) = \dim_{\mathbb{C}} \mathbb{C}\{x, y\}/(\partial f/\partial x, \partial f/\partial y).$$

As for other local characters, if ξ' is a curve and O a point of ξ', we will write $\mu_O(\xi')$ for the Milnor number of the germ of ξ' at O.

By introducing $\delta(\xi)$ in the formula of 6.4.1, via 3.11.12, we get a shorter formula, often named the *Milnor formula*, that relates the Milnor number, the order of the singularity and the number of branches:

Corollary 6.4.3 *For any non-empty reduced germ of curve ξ,*

$$\mu(\xi) = 2\delta(\xi) - r + 1$$

where r is the number of branches of ξ.

The additivity formula for δ directly gives a similar one for the Milnor number, namely:

Proposition 6.4.4 *If ξ and ζ are non-empty germs of curve, then*

$$\mu(\xi + \zeta) = \mu(\xi) + \mu(\zeta) + 2[\xi \cdot \zeta] - 1.$$

PROOF: If either both germs share a branch or one of them is non-reduced, then both sides are clearly non-finite. Otherwise the claim follows from 6.4.3 and 3.11.13. ◇

Remark 6.4.5 As follows from 6.4.1, the Milnor number is related to the older discriminantal index by the formula

$$d(\xi) = \mu(\xi) + e_O(\xi) - 1,$$

and so both invariants carry similar information. Nevertheless $\mu(\xi)$ provides a sharper control of equisingularity (see section 7.3). For this reason, and also because of its nice topological meaning (see [60] and exercise 6.12), the use of the Milnor number is today largely preferred.

6.5   Virtual behaviour of polar germs

There are many claims in classical literature that are intended to describe the equisingularity type of the (in some sense) general polars $P_g(\xi)$ from the equisingularity type of the germ ξ. Nevertheless, one of the most important facts about polars is that the equisingularity type of ξ does not determine, in general, the equisingularity types of its polar germs, nor even those of the generic or transverse ones. This fact, that proves all these classical claims to be wrong, was pointed out by Pham ([67]), who gave the first example of equisingular germs with non-equisingular polars.

Example 6.5.1 (Pham) All germs

$$\xi_\lambda : f = y^3 - x^{11} + \lambda x^8 y = 0, \quad \lambda \in \mathbb{C}$$

are equisingular, as all have a one-sided Newton polygon with ends $(0, 3)$ and $(11, 0)$ and therefore are irreducible with single characteristic exponent $11/3$. Nevertheless, the reader may easily plot the Newton polygons of $\partial f/\partial x, \partial f/\partial y$ in order to draw the Newton polygon of a generic polar of ξ_λ. It has a single side and first end $(0, 2)$ regardless of the value of λ, but its last end is $(8, 0)$ if $\lambda \neq 0$ and $(10, 0)$ if $\lambda = 0$, which clearly give rise to non-equisingular polar germs. The reader may notice that the monomial $\lambda x^8 y$ is irrelevant to the equisingularity type of ξ_λ. However, after taking the y-derivative, it gives rise to λx^8 in the equation of the polar germ, which is critical to its equisingularity type.

It is clear that 6.5.1 shows a phenomenon that may be reproduced in many other cases: taking y-derivative (resp. x-derivative) deletes the points in the Newton diagram of ξ which lie on the first (resp. second) axis and moves the other points one step downwards (resp. leftwards). Then points that do not lie on the Newton polygon of ξ may give rise to points on the Newton polygon of a polar and therefore, by annihilating the corresponding monomial in the equation of ξ, the equisingularity type of the polar may be modified without modifying that of the germ ξ itself.

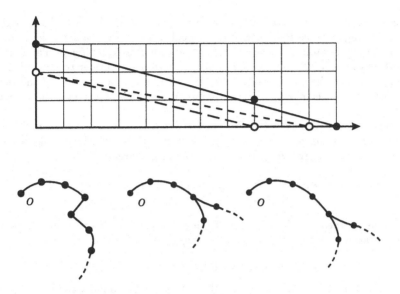

Figure 6.1: Newton polygons and Enriques diagrams of the germs $\xi_\lambda : y^3 - x^{11} + \lambda x^8 y = 0$ and their generic polars for $\lambda = 0$ and $\lambda \neq 0$ (6.5.1).

Thus one cannot expect to get a precise description of the equisingularity types of the polars of ξ from that of ξ itself. In this section we will just describe the virtual behaviour of the polar germs by showing that they go through a certain weighted cluster that depends on the multiple points on ξ and on the intersections of the branches of ξ with $\eta : g = 0$.

Fix $g \in \mathcal{O}_O$ defining an smooth germ η at O, as in preceding sections. For each non-empty reduced germ ξ at O with no component η define a weighted cluster $\partial_g(\xi)$ by taking:

(a) All multiple points p of ξ, each with virtual multiplicity $e_p(\xi) - 1$.

(b) On each branch γ of ξ, the first $[\gamma.\eta] - 1$ points on γ which are simple on ξ, each virtually counted once.

Notice that if no branch of ξ is tangent to η, then the weighted cluster does not depend on g because the intersections $[\gamma.\eta]$ are the orders of the branches, $e_O(\gamma)$. In such a case we may write $\partial(\xi)$ instead of $\partial_g(\xi)$.

Theorem 6.5.2 *If the non-empty germ ξ is reduced and does not contain η : $g = 0$, then all g-polars of ξ go through $\partial_g(\xi)$.*

PROOF: We will deal first with the multiple points of ξ by proving that all germs in the jacobian system of ξ, and hence in particular all polar germs of ξ, go through the multiple points p of ξ with the virtual multiplicities $e_p(\xi) - 1$. We will use induction on the number of multiple points on ξ, the case of a smooth germ ξ being obvious. By linearity, it is enough to prove this claim for the germs defined by generators of the jacobian ideal, namely $\partial f/\partial x, \partial f/\partial y, f$. The claim is obvious for ξ itself, defined by f. Then we need only consider the case of the x- and y-polars: both have multiplicity at least $e_O(\xi) - 1$ at O, by 6.1.7 and the claim for the other multiple points follows from the induction hypothesis and 6.2.3.

Now fix a branch γ of ξ and write $\ell = [\gamma.\eta]$. Let p_0 be the last multiple point of ξ on γ and $p_1, \ldots, p_{\ell-1}$ the points in its successive neighbourhoods on γ. Repeated use of the virtual Noether formula 4.1.2 easily gives

$$[\gamma.P_g(\xi)] = \sum_{O \le p \le p_0} e_p(\gamma)(e_p(\xi) - 1) + [\tilde{\gamma}_{p_1}.\hat{P}_g(\xi)],$$

$\hat{P}_g(\xi)$ being the virtual transform of the polar $P_g(\xi)$ with origin at p_1, each $p < p_1$ taken with the virtual multiplicity $e_p(\xi) - 1$. Using 6.3.1 we obtain

$$[\tilde{\gamma}_{p_1}.\hat{P}_g(\xi)] = \ell - 1.$$

If $\ell = 1$ there is nothing to prove. If $\ell > 1$ the former equality and the similar ones that come from it by using $\ell - 2$ times the virtual Noether formula 4.1.2 show that $\hat{P}_g(\xi)$ goes through $p_1, \ldots, p_{\ell-1}$, all taken with virtual multiplicities one, as wanted. $\diamond$

Direct consequences of 6.5.2 are:

Corollary 6.5.3 (Noether–Enriques) *All polar germs of a reduced germ ξ go through $\partial(\xi)$.*

Corollary 6.5.4 *All polars of a reduced germ ξ are adjoints of ξ.*

The classical claim about the behaviour of the polar germs by M. Noether ([63]) says that the polars germs have multiplicity at least $e_p(\xi) - 1$ at each multiple point p of ξ. This claim is false unless multiplicities are understood as virtual (and then it is just 6.5.4, actually the result Noether was interested in). An easy example is given by the germ

$$\gamma : y^{10} - x^{23} = 0 :$$

it is irreducible with sequence of multiplicities $10, 10, 3, 3, 3, 1 \ldots$ and an easy computation shows that its generic polars have effective multiplicity 1 at the point in the 4-th neighbourhood. Enriques gave the correct claim in [35], IV.II, but Noether's claim still appeared in later literature (see for instance [24], IV.1 or [72]).

6.6 On the effective behaviour of polar germs

In the preceding section we have seen that the polar germs of a reduced germ ξ go through the weighted cluster $\partial(\xi)$. However, but for the most trivial cases,

$\partial(\xi)$ is a non-consistent cluster and hence no polar of ξ can go through it with effective multiplicities equal to the virtual ones. By unloading one may get from $\partial(\xi)$ an equivalent consistent cluster $\widetilde{\partial(\xi)}$ with the same points. Then the polars of ξ still go through $\widetilde{\partial(\xi)}$ and, since it is consistent, it makes sense to ask if generic polars of ξ actually go sharply through $\widetilde{\partial(\xi)}$. This statement, which Enriques wrongly claimed to be true ([35], IV.II.18), is false, as one may expect after Pham's example 6.5.1: the reader may easily see that generic polars of $\xi_0 : y^3 - x^{11} = 0$ (6.5.1) do not go sharply through $\widetilde{\partial(\xi_0)}$, see figure 6.2.

Figure 6.2: From left to right, the Enriques diagrams of $\partial(\xi_0)$, $\widetilde{\partial(\xi_0)}$ and a generic polar of $\xi_0 : y^3 - x^{11} = 0$.

Anyway, Enriques was not so wrong, because his claim is true if the germ ξ is, in some sense, generic within its equisingularity class. In this section we shall give an easy proof of this fact for ξ irreducible with a single characteristic exponent. This particular case will be enough to show how the polars of a germ ξ may be used for unveiling properties of ξ that depend on its isomorphism class and not only on its equisingularity class. For the general case the reader is referred to [20] and [22].

Fix a characteristic exponent m/n, ($m/n > 1$, $\gcd(m,n) = 1$) and choose a system of local coordinates x, y. We consider the weighted cluster $\mathcal{K}(m,n) = (K(m,n), \nu_{m,n})$ as introduced in section 5.3 and keep the notations used there. After performing the Euclidean divisions

$$n_{i-1} = h_i n_i + n_{i+1}, \quad i = 0, \ldots, r,$$

where $n_{-1} = m$, $n_0 = n$, $n_r = 1$ and $n_{r+1} = 0$, the points of $K(m,n)$ are the points $p_{i,j}$, $i = 1, \ldots, r$, $j = 1, \ldots h_i$, depending on the first characteristic exponent of $y^n - x^m = 0$ and already defined in section 5.3. The system of virtual multiplicities $\nu_{n,m}$ assigns to each $p_{i,j}$ virtual multiplicity n_i.

The next lemma will allow us to easily handle all isomorphism classes of germs with single characteristic exponent m/n. It is worth noting that no similar statement is true for germs with many branches or many characteristic exponents.

Lemma 6.6.1 *Assume that local coordinates x, y at O are fixed. For a germ of curve γ at O the following conditions are equivalent:*

(i) γ has equation

$$f = a_{0,n}y^n + a_{m,0}x^m + \sum_{ni+mj>nm} a_{i,j}x^iy^j, \qquad (6.1)$$

with $a_{0,n} \neq 0$, $a_{m,0} \neq 0$.

(ii) γ is irreducible, has single characteristic exponent m/n and Puiseux series

$$s = ax^{m/n} + \cdots, \quad a \neq 0 \qquad (6.2)$$

(iii) γ goes sharply through $\mathcal{K}(m,n)$.

Furthermore any irreducible germ γ' with single characteristic exponent m/n is analytically isomorphic to a germ γ satisfying the above equivalent conditions.

PROOF: That (i) $\Rightarrow$ (ii) is easily proved using the Newton–Puiseux algorithm, while the converse was already seen in 2.2.5.

Theorem 5.2.2 (or 5.5.1) shows that (ii) $\Rightarrow$ (iii). To see the converse, assume that a germ ξ goes sharply through $\mathcal{K}(m,n)$: then one of its branches, say γ, goes through p_{r,h_r} and has a simple and free point in its first neighbourhood. By 5.3.4, γ has then single characteristic exponent m/n and a Puiseux series as (6.2) above: thus $n = e_O(\gamma) \leq e_O(\xi)$ which by hypothesis is equal to n. This forces $\gamma = \xi$ and proves (ii).

For the remaining part, note first that γ' has order n and maximal contact m. Thus, if one takes local coordinates $\bar{x}, \bar{y}$ at the origin of γ' with the $\bar{x}$-axis having maximal contact with γ', then γ' has an equation of the form

$$a_{0,n}\bar{y}^n + a_{m,0}\bar{x}^m + \sum_{ni+mj>nm} a_{i,j}\bar{x}^i\bar{y}^j = 0, \quad a_{0,n} \neq 0, \quad a_{m,0} \neq 0.$$

Then define an analytic isomorphism φ between suitable neighbourhoods of O by taking $\varphi(p)$ as the point whose coordinates (x,y) equal the coordinates $(\bar{x},\bar{y})$ of p. Clearly $\varphi^*x = \bar{x}$ and $\varphi^*y = \bar{y}$ and therefore φ transforms γ' into a germ γ as claimed. $\diamond$

After lemma 6.6.1 we may restrict ourselves to consider only germs going sharply through $\mathcal{K}(m,n)$, as they represent all the isomorphism classes of irreducible germs with single characteristic exponent m/n. Notice that

$$H_{\mathcal{K}(m,n)} = \{f = a_{0,n}y^n + a_{m,0}x^m + \sum_{ni+mj>nm} a_{i,j}x^iy^j\},$$

its open subset $U = \{f \in H_{\mathcal{K}(m,n)} | a_{0,n} \neq 0, \quad a_{m,0} \neq 0\}$ being, by 6.6.1, that of the equations of the germs going sharply through $\mathcal{K}(m,n)$. Notice also that the elements of U generate $H_{\mathcal{K}(m,n)}$ as a $\mathbb{C}$-vector space.

Define a second system of virtual multiplicities $\nu' = \nu'_{m,n}$ for $K(m,n)$ by the following rule, sometimes called the *Enriques alternance rule*:

$$\nu'(p_{i,j}) = \begin{cases} n_i & \text{for } i \text{ odd and } (i,j) \neq (r, h_r) \\ n_i - 1 & \text{for } i \text{ even or } (i,j) = (r, h_r). \end{cases}$$

Define $\mathcal{K}'(n, m) = (K(m, n), \nu')$. The next lemma will allow us to take it instead of $\widetilde{\partial(\gamma)}$:

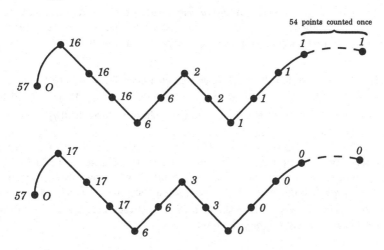

Figure 6.3: Weighted clusters $\partial(\gamma)$ and $\widetilde{\partial(\gamma)}$ for an irreducible germ γ with single characteristic exponent 75/58.

Lemma 6.6.2 *If γ goes sharply through $\mathcal{K}(m, n)$, the weighted cluster $\widetilde{\partial(\gamma)}$ consists of all points in $\mathcal{K}'(m, n)$ with the same virtual multiplicities plus further points, all of which lie on γ and have virtual multiplicity zero.*

PROOF: According to its definition $\partial(\gamma)$ consists of the points $p_{i,j} \in K(m, n)$ taken with virtual multiplicities $n_i - 1$ together with the first $n - 1$ simple points on γ, all taken with virtual multiplicity one. We will refer to the latter points as the simple points of $\partial(\gamma)$. All we need to do is an easy exercise of unloading: we successively unload all the virtual multiplicities of the simple points onto the points in $K(m, n)$. The reader may see that increasing by one the virtual multiplicity of $p_{i,j}$ by successive unloadings takes n_i unities from the whole amount of virtual multiplicities of the simple points. We start with $n - 1$ unities of multiplicity on the simple points, not enough to increase the multiplicity of p_{0,h_0}. Thus we spend $h_1 n_1$ unities from them to increase the virtual multiplicities of the points $p_{1,j}$, $j = 1, \ldots, h_1$. Then there remain $n_2 - 1$ unities of virtual multiplicity on the simple points, from which $h_3 n_3$ are used for increasing the multiplicities of the $p_{3,j}$, $j = 1, \ldots, h_3$ and so on until spending either the last $n_{r-2} - 1 = h_{r-1} n_{r-1}$ unities on $p_{r-1,1}, \ldots, p_{r-1,h_{r-1}}$ if r is even, or the last $n_{r-1} - 1 = h_r - 1$ unities on $p_{r,1}, \ldots, p_{r,h_r-1}$ if r is odd. At this stage the system of virtual multiplicities being consistent, the proof is complete. ◇

In the sequel we will use $\mathcal{K}'(m,n)$ instead of $\widetilde{\partial(\gamma)}$. Notice that $\mathcal{K}'(m,n)$ does not depend on γ. Our main result in this section is:

Theorem 6.6.3 *There is a non-empty Zariski-open set $V \subset H_{\mathcal{K}(m,n)}$ such that for any $f \in V$ the germ $\gamma : f = 0$ is irreducible, has single characteristic exponent m/n and its generic polars go sharply through $\widetilde{\partial(\gamma)}$.*

PROOF: First notice that, by 4.2.7, once a single polar of γ goes sharply through $\widetilde{\partial(\gamma)}$, then generic polars of γ go sharply through $\widetilde{\partial(\gamma)}$ too. Hence it will be enough to show that there exists a polar going sharply through $\widetilde{\partial(\gamma)}$.

In fact, it will be enough to show that there is a polar ζ going sharply through $\mathcal{K}'(m,n)$. Indeed, if this is the case, ζ is a transverse polar by 6.1.7. The intersection multiplicity $[\gamma.\zeta]$ being given by the Plücker formula (6.3.2), an easy computation using the Noether formula (3.3.1) shows that ζ does not (effectively) go through points on γ other than those in $\mathcal{K}'(m,n)$, after which 6.6.2 shows that ζ goes sharply through $\widetilde{\partial(\gamma)}$. All together, the proof will be complete after proving the next lemma:

Lemma 6.6.4 *There is a non-empty Zariski-open set $V \subset H_{\mathcal{K}(m,n)}$ such that for any $f \in V$ the germ $\gamma : f = 0$ is irreducible, has single characteristic exponent m/n and its polar $P_x(f)$ goes sharply through $\mathcal{K}'(m,n)$.*

PROOF: Write $H = H_{\mathcal{K}(m,n)}$ and $H' = H_{\mathcal{K}'(m,n)}$. Still denote by U the Zariski-open subset of H of the equations of the germs going sharply through $\mathcal{K}(m,n)$. Taking the y-derivative defines a $\mathbb{C}$-linear map $\bar{\Delta} : \mathbb{C}\{x,y\} \longrightarrow \mathbb{C}\{x,y\}$ that, as the reader may easily check, is continuous for the Zariski topology. We know from 6.6.2 that all elements in U have their y-derivatives in H'. Since U generates H, the same is true for an arbitrary element of H and so $\bar{\Delta}$ restricts to

$$\Delta : H \longrightarrow H',$$

which maps equations of germs to equations of their x-polars. Since $\mathcal{K}'(m,n)$ is consistent, by 4.2.7, there is a non-empty Zariski-open set $V' \subset H'$ such that the germs $f' = 0$, $f' \in V'$, go sharply through $\mathcal{K}'(m,n)$. Then it is clearly enough to prove that Δ is onto, because in such a case one may take $V = \Delta^{-1}(V') \cap U$ and the claim is satisfied.

In order to prove that Δ is an epimorphism, note first that the kernel of Δ consists of all series in H that do not depend on y, which easily gives $\ker \Delta = x^m \mathbb{C}\{x\}$. Thus we have the commutative diagram of $\mathbb{C}$-vector spaces

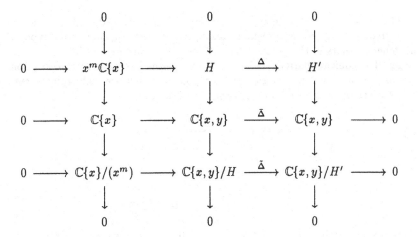

We obviously have $\dim \mathbb{C}\{x\}/x^m = m$, while from 4.7.1 we have

$$\dim \mathbb{C}\{x,y\}/H = \sum_{i=1}^{r} \frac{h_i n_i (n_i+1)}{2} = \frac{nm+n+m-1}{2}.$$

In the same way, it easily follows from the definition of ν' that

$$\dim \mathbb{C}\{x,y\}/H' = \frac{nm+n-m-1}{2},$$

and so

$$\dim \mathbb{C}\{x,y\}/H = \dim \mathbb{C}\{x\}/x^m + \dim \mathbb{C}\{x,y\}/H'.$$

Using this equality it is elementary to check that all rows and columns in the above diagram are exact, which in turn, by a well known algebraic argument, proves that Δ is onto, as wanted. $\diamond$

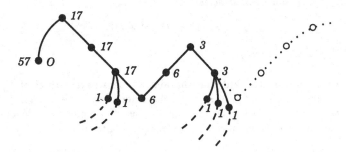

Figure 6.4: The generic polars of a generic germ γ through $\mathcal{K}(75, 58)$ (6.6.3 – 6.6.5).

The weighted cluster $\widetilde{\partial(\gamma)}$ has been explicitly described in 6.6.2 and so 6.6.3 gives a complete and explicit description of the equisingularity type of the generic polars of the germs $\gamma : f = 0$ for $f \in V$, in terms of infinitely near points. The equisingularity type of these polar germs also has a nice description in terms of branches and characteristic exponents which we will explain next. Let us write m/n as an even-order continued fraction:

$$\frac{m}{n} = h'_0 + \cfrac{1}{h'_1 + \cfrac{1}{\ddots \quad h'_{2\ell-1} + \cfrac{1}{h'_{2\ell}}}}$$

where we have either $2\ell = r$ and $h'_i = h_i$, $i = 0, \ldots, r$ if r is even, or $2\ell = r + 1$, $h'_i = h_i$, $i = 0, \ldots, r - 1$, $h'_{2\ell-1} = h_r - 1$ and $h'_{2\ell} = 1$ if r is odd.

Denote by u_i/v_i the i-th reduced fraction of the former continued fraction, that is, for $i = 1, \ldots, 2\ell$,

$$\frac{u_i}{v_i} = h'_0 + \cfrac{1}{h'_1 + \cfrac{1}{\ddots \quad h'_{i-1} + \cfrac{1}{h'_i}}}$$

with $\gcd(u_i, v_i) = 1$. Then theorem 6.6.3 may be equivalently stated as:

Theorem 6.6.5 *There is a non-empty Zariski-open set $V \subset H_{\mathcal{K}(m,n)}$ such that for any $f \in V$ the germ $\gamma : f = 0$ is irreducible, has single characteristic exponent m/n and its generic polars are composed of branches $\zeta_{k,j}$, $k = 1, \ldots, \ell$, $j = 1, \ldots, h_{2k}$, each branch $\zeta_{k,j}$ having single characteristic exponent u_{2k-1}/v_{2k-1} and Puiseux series*

$$s_{k,j} = a_{k,j} x^{u_{2k-1}/v_{2k-1}} + \cdots$$

with $a_{k,j}^{v_{2k-1}} \neq a_{k,j'}^{v_{2k-1}}$ if $j \neq j'$.

It may be equivalently said that the above polars have a total of $\sum_1^\ell h_{2k}$ branches, from which just h_{2k} have single characteristic exponent u_{2k-1}/v_{2k-1}, for $k = 1, \ldots, \ell$, all branches go through the last free point on γ before the satellite ones and no two of them share a free point after it.

PROOF OF 6.6.5: Take V as in 6.6.3, so that the generic polars of $\gamma : f = 0$, $f \in V$ go sharply through $\mathcal{K}'(m,n)$. From the definition of $\mathcal{K}'(m,n)$, we easily compute the excesses ρ_p for $p \in K(m,n)$ to get $\rho_p = 0$ for all $p \in K(n,m)$ but for $p = p_{2k-1,h_{2k-1}}$, $k = 1,\ldots,\ell$, in which case $\rho_p = h'_{2k}$. Then, by 4.2.6, a generic polar of γ has a total of $\sum_{k=1}^{\ell} h'_{2k}$ branches, for each $k = 1,\ldots,\ell$, exactly h_{2k} of these branches have a non-singular point in the first neighbourhood of $p_{2k-1,h_{2k-1}}$, and all these non-singular points are different. Once we know the position of a non-singular point on each branch of the polar, the characteristic exponent of the branch may be recovered as a continued fraction, as done when proving 5.2.3, after which the claim follows using 5.3.4 for the inequalities between the powers of the coefficients of the characteristic monomials. ◇

A direct proof of 6.6.5 using Newton polygons may be seen in [19].

Once we know the behaviour of generic polars of generic germs with given characteristic exponent, the reader may easily get examples of germs γ whose generic polars fail to go through $\widetilde{\partial(\gamma)}$ with effective multiplicities equal to the virtual ones. The easiest ones are given by the so called *quasihomogeneous germs*, that is, the germs of curve that in a suitable system of local coordinates have equation $x^m - y^n = 0$. One of them was already used in Pham's example 6.4.1.

6.7 Base points of polars and analytic invariants

It has been pointed out in 6.1.2 that isomorphic germs have isomorphic polars while, as repeatedly said in the last two sections, equisingular germs need not have equisingular polars. This is one of the most interesting features of polar germs: their equisingularity types depend not only on the equisingularity classes of the original germs but also on their isomorphism classes. Hence, polars may be used for unveiling information on the isomorphism classes of germs that is not determined by their equisingularity classes (analytic information). The most obvious of such uses consists of giving examples of non-isomorphic equisingular germs: indeed, by 6.1.2, isomorphic germs must have isomorphic and hence equisingular polars so that germs with non-equisingular generic polar germs are non-isomorphic. Pham's example 6.4.1 gives thus an example of non-isomorphic equisingular germs. The reader may build many others from the results of the last section.

In this section we will fix an irreducible germ γ with a single characteristic exponent and assume that its generic polars go sharply through $\widetilde{\partial(\gamma)}$. Then we will show that generic polars of γ effectively go through a finite set of (necessarily simple and free) infinitely near points that do not lie on γ. We will say that such points, together with the points of positive virtual multiplicity in $\widetilde{\partial(\gamma)}$, are the *base points of the polar germs* of γ. They are intrinsically related to γ and provide analytic information on it. Base points of polar germs will be used in this section for constructing some examples of continuous analytic invariants of germs of curve and families of equisingular germs with a continuous (i.e., not finite or countable) range of analytic types. Results similar to those obtained in this section hold for irreducible germs with an arbitrary number of characteristic exponents, see [21]. We will deal with base points of polar germs of arbitrary reduced germs in the forthcoming sections 8.5 and 8.6.

We use all notations introduced in the preceding section. Still assume that γ is irreducible, has single characteristic exponent m/n, and, which is not restrictive by 6.6.1, that it goes sharply through $\mathcal{K}(m, n)$. Assume furthermore that its generic polars go sharply through $\widetilde{\partial(\gamma)}$. Fix ζ to be one of such polars and, as in 6.6.5, denote its branches by $\zeta_{k,j}$, $k = 1, \ldots, \ell$, $j = 1, \ldots, h'_{2k}$, each $\zeta_{k,j}$ having single characteristic exponent u_{2k-1}/v_{2k-1}. Then we have

Proposition 6.7.1 *If ζ' is any polar of γ going sharply through $\widetilde{\partial(\gamma)}$, then for each (k, j), $k = 1, \ldots, \ell$, $j = 1, \ldots, h'_{2k}$, ζ' effectively goes through the first*

$$u_{2k-1} - v_{2k-1} - 1$$

infinitely near points on $\zeta_{k,j}$ not in $K(m, n)$.

PROOF: Modify the coordinates if needed, so that $\zeta = P_x(f)$ where f is an equation of γ and take $\xi = P_y(f)$. Fix k and perform the Euclidean algorithm for $\gcd(u_{2k-1}, v_{2k-1})$,

$$e_{i-1} = h'_i e_i + e_{i+1}, \quad i = 0, \ldots, 2k - 1,$$

where $e_{-1} = u_{2k-1}$, $e_0 = v_{2k-1}$, $e_{2k-1} = 1$ and $e_{2k} = 0$. By 5.2.2, $e_{p_{i,s}}(\zeta_{k,j}) = e_i$ for $i = 0, \ldots, 2k-1$ and $s = 1, \ldots, h'_i$. Then one may compute, using the Noether formula,

$$[\gamma.\zeta_{k,j}] = \sum_{i=0}^{2k-1} h'_i n_i e_i. \tag{6.3}$$

Let $x(t) = t^{v_{2k-1}}$, $y(t) = at^{u_{2k-1}} + \ldots$ be a uniformizing map of a representative of $\zeta_{k,j}$. By computing just as in the proof of 6.4.1 we get

$$\frac{d}{dt} f(x(t), y(t)) = v_{2k-1} \frac{\partial f}{\partial x}(x(t), y(t)) t^{v_{2k-1}-1}.$$

By equating the orders in t and using equality 6.3 above we obtain

$$[\xi.\zeta_{k,j}] = [\gamma.\zeta_{k,j}] - v_{2k-1} = \sum_{i=0}^{2k-1} h'_i n_i e_i - v_{2k-1}. \tag{6.4}$$

Since the equation of ζ' is a linear combination of the equations of ζ, ξ and γ, it is clear from equations 6.3 and 6.4 that

$$[\zeta'.\zeta_{k,j}] \geq \sum_{i=0}^{2k-1} h'_i n_i e_i - v_{2k-1}.$$

On the other hand, since ζ' goes sharply through $\mathcal{K}'(m, n)$, the sum of products of multiplicities at the points of $K(m, n)$ is

$$\sum_{p \in K(m,n)} e_p(\zeta') e_p(\zeta_{k,j}) = \sum_{\substack{i=0,\dots,2k-1 \\ i \text{ even}}} h_i'(n_i - 1)e_i + \sum_{\substack{i=0,\dots,2k-1 \\ i \text{ odd}}} h_i' n_i e_i$$

$$= \sum_{i=0}^{2k-1} h_i' n_i e_i - \sum_{\substack{i=0,\dots,2k-1 \\ i \text{ even}}} h_i' e_i$$

$$\leq [\zeta'.\zeta_{k,j}] + v_{2k-1} - u_{2k-1} + 1.$$

Now, since the points not in $K(m, n)$ are simple for both ζ' and $\zeta_{k,j}$, the Noether formula and the last computation show that ζ' and $\zeta_{k,j}$ share at least $u_{2k-1} - v_{2k-1} - 1$ points besides those in $K(m, n)$, which gives the claim. ◇

Figure 6.5: An Enriques diagram showing the base points of the polar germs of figure 6.4. There are no base points outside of $K(75, 58)$ on the branches $\zeta_{1,j}$, while there are four on each $\zeta_{2,j}$, as claimed in 6.7.1.

Obviously, the base points we have found do not depend on the polar ζ, as they are on any polar going sharply through $\mathcal{K}'(m, n)$. As exercise the reader may check that there are no more base points for the generic polar germs besides the points in $K(m, n)$ and those described in 6.7.1, by computing twice the intersection multiplicity of ζ and ζ', namely as $\mu(\gamma)$ and by the Noether formula.

We will devote the rest of this section to obtaining some analytic invariants from the base points of the polar germs. First of all let us show how to get some analytic invariants of many-branched germs. We know (3.1.4) that an analytic isomorphism φ induces a linear projectivity between the first neighbourhoods of any (ordinary or infinitely near) point p and its corresponding $\varphi(p)$. Furthermore, this projectivity sends satellite points to satellite points and preserves their proximity relationship with former points, as proximity is invariant under analytic isomorphisms (3.5.1). Thus, if ξ is any germ through p, consider the set of points in the first neighbourhood of p that either lie on ξ or are satellite points. Any projective invariant of this set of points is an analytic invariant of the germ. We will give two examples. The easier one is that of a germ at O that has four smooth branches with different principal tangents; it has four distinct points p_1, p_2, p_3, p_4 in the first neighbourhood of O and so any function of the cross-ratio of p_1, p_2, p_3, p_4, invariant under permutations of the points, is an analytic invariant of the germ. For the second example, consider the germ $\xi : y^4 - ax^3y^2 + x^6$, $a^2 \neq 4$: it has two branches, both with characteristic exponent $3/2$ and going through the same satellite points, let p be the last of them. In the first neighbourhood of p there are two distinct points p_1, p_2 on ξ and two satellite points, enough to give rise to an analytic invariant: p_1, p_2 have coordinates (5.3.4) a_1, a_2 with $a_1.a_2 = 1$, $a_1 + a_2 = a$, while the satellite points have coordinates 0 and ∞. The four points have cross-ratio equal to a_1/a_2 and hence the symmetric function

$$a^2 = \frac{a_1}{a_2} + \frac{a_2}{a_1} + 2$$

is an analytic invariant of ξ.

Of course there are no analytic invariants of such kind for irreducible germs γ, as fewer than four points on a projective line give rise to no non-trivial projective invariant. Nevertheless, any analytic isomorphism φ transforms polars of γ into polars of $\varphi(\gamma)$ so that the image by φ of the base points of the polars of γ are base points of the polars of $\varphi(\gamma)$. Then, each time we have two or more base points of the polar germs in the first neighbourhood of a satellite point in K (or three or more base points in the first neighbourhood of a free point), we may construct an analytic invariant of the germ by taking any projective invariant of the set of such base points together with the satellite points in the same first neighbourhood. We close this section with an explicit example.

Example 6.7.2 Consider the germs

$$\gamma_a : 3y^5 - x^{12} + 15x^{10}y - 5ax^5y^3, \quad a \in \mathbb{C}.$$

An easy computation shows that generic polars of γ_a go sharply through $\mathcal{K}(12, 5)$ if $a^2 \neq 4$, which we assume in the sequel. The generic polars of γ_a have two branches, both with single characteristic exponent $5/2$. By 6.7.1, the first simple and free point on each of these branches is a base point. Since the coordinates (from 5.3.4) of these points are the roots of $z^2 - az + 1 = 0$ and the satellite points still have coordinates 0 and ∞, arguing as in the former example, a^2 is an analytic invariant of γ_a.

Notice that all germs γ_a are equisingular but, by 6.7.2, γ_a and γ_b are non-isomorphic if $a^2 \neq b^2$. The special cases $a^2 = 4$ may be included since they give rise to generic polars with a different equisingularity type. The reader may easily see, to complete the example, that γ_a and γ_{-a} are isomorphic.

6.8 Polar invariants of irreducible germs

Back to considering properties of germs of curve related to their equisingular-ity types, in this section we will introduce and compute the polar invariants of irreducible germs. Also named polar quotients, the polar invariants of an irreducible germ γ are equisingularity invariants of γ related to its polar germs. We will see that the multiplicity and the polar invariants of an irreducible germ determine its equisingularity class. Along the way we will also obtain some in-formation about the branches of the polar germs in terms of the equisingularity class of γ.

The case of non-irreducible germs is not so easy and results are not so com-plete (for instance, the multiplicity and polar invariants of a non-irreducible germ do not determine its equisingularity class, see 6.11.7), mainly due to the far more complicated combinatorics of the non-unibranched case. Because of this, we make a separate treatment of irreducible germs in this section and will devote sections 6.9 to 6.11 to studying branches of polars and polar invariants of many-branched germs.

Let γ be an irreducible and singular germ of curve at O and fix an analytic function x defining a smooth germ η at O. Take such a function as the first of a pair of local coordinates x, y, so that η is the (germ at O of the) y-axis.

Let s be a Puiseux series of γ relative to coordinates x, y. Since we know (5.6.2) that the characteristic exponents of s depend only on the character-istic exponents of γ and the intersection multiplicity of γ with the y-axis (η in our case) it is clear that the characteristic exponents of s do not depend on the second coordinate y. Let us denote such characteristic exponents by $m_1/n, \ldots, m_k/n$, n the polydromy order of s, and write $n^i = \gcd(n, m_1, \ldots, m_i)$, for $0 \le i \le k$ ($n_0 = n$). The next result, due to Merle ([59]), is the main one in this section:

Theorem 6.8.1 (Merle) *Let $\zeta = P_x(\gamma)$ be any x-polar of γ: for any branch ζ' of ζ, there exists i, $1 \le i \le k$, such that*

$$\frac{[\gamma.\zeta']}{[\eta.\zeta']} = \frac{(n - n^1)m_1 + \cdots + (n^{i-2} - n^{i-1})m_{i-1} + n^{i-1}m_i}{n}. \tag{6.5}$$

Conversely, for any such i there is at least one branch ζ' of ζ for which the former equality holds.

PROOF: Let $\zeta = P_x(f)$, f an equation of γ. First of all, since nothing in the claim depends on the second coordinate, it is not restrictive to assume that we choose it in such a way that the initial term of the Puiseux series s is its first characteristic term, say $ax^{m_1/n}$ (if γ is tangent to the y-axis, then $m_1/n < 1$ and the first characteristic term is always the initial term. Otherwise, as in the proof of 5.2.2, take an x-axis having maximal contact with γ). Then, since γ is irreducible, its Newton polygon has a single side Γ, with ends $(0, n), (m_1, 0)$. Furthermore, the equation associated with Γ has a single class of conjugate roots, and hence, up to a multiplicative constant, it has the form $F_\Gamma(z) =$

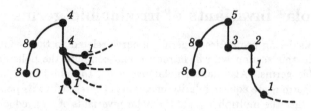

Figure 6.6: Enriques diagrams of generic polars of two irreducible germs with single characteristic exponent 22/9: that on the left is as described in 6.6.5 while the one on the right comes from the germ $y^9 - x^{22} = 0$. As the reader may easily check using the Noether formula, 6.8.1 (or 6.8.2) holds true for the branches of both polars, despite their different equisingularity types.

$(z^{n/n^1} - a^{n/n^1})^{n^1}$. It follows in particular that all monomials in the equation f of γ corresponding to points on Γ are non-zero. Then since the Newton diagram of $\partial f/\partial y$ comes from the Newton diagram of f by deleting the points on the first axis and moving all other points one step downwards, if $n^1 > 1$ the Newton polygon of ζ has a side Γ' from $(0, n-1)$ to $(m_1 - m_1/n^1, n/n^1 - 1)$. If $n^1 = 1$ such a side does not exist (or is reduced to a point), but in any case all further sides of $N(\zeta)$ are beyond $(m_1 - m_1/n^1, n/n^1 - 1)$ and have slope bigger than $-n/m_1$.

Assume first that the branch ζ' does not correspond to Γ', that is, that ζ' corresponds to a side beyond $(m_1 - m_1/n^1, n/n^1 - 1)$ or ζ' is the x-axis. Notice that there exists at least one such ζ' because $n/n^1 - 1 > 0$. If $\bar{s}$ is a Puiseux series of ζ', $\bar{n}$ denotes the polydromy order of $\bar{s}$ and $\bar{m}/\bar{n} = o_x\bar{s}$ ($\bar{s} = 0$, $\bar{n} = 1$ and $\bar{m} = \infty$ if ζ' is the x-axis), then $\bar{m}/\bar{n} > m_1/n$. Therefore $\bar{n}m_1 \leq \bar{n}\alpha + \bar{m}\beta$ for (α, β) in the Newton diagram of f and the equality holds if and only if $(\alpha, \beta) = (m_1, 0)$, the last end of $N(f)$. The initial term of $f(x, \bar{s}(x))$ has thus degree m_1 and so $[\gamma.\zeta'] = m_1\bar{n}$. Since $\bar{n} = [\eta.\zeta']$ we have the formula in the claim for $i = 1$.

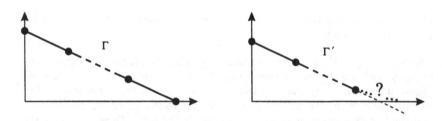

Figure 6.7: $N(\gamma)$ and the well determined side Γ' of $N(\zeta)$ in the proof of 6.8.1. Note that the slope of any further side of $N(\zeta)$ is bigger than the slope of Γ, which is a key fact for 6.8.1 to hold.

If $k = 1$ the proof is done, because in this case there is no side Γ'. We assume thus in the sequel that $k > 1$, and hence $n^1 > 1$, and that our branch ζ' corresponds to the side Γ'. Thus now $m_1/n = \bar{m}/\bar{n}$ if $\bar{s}$ is still a Puiseux series of ζ', $\bar{n}$ the polydromy order of $\bar{s}$ and $\bar{m}/\bar{n} = o_x\bar{s}$.

Denote by p the first point on γ after those depending on the characteristic exponent m_1/n. The equation associated with Γ' is, after an easy computation,

$$F_{\Gamma'}(z) = z^{1-n/n^1} \frac{d}{dz} F_\Gamma(z) = n(z^{n/n^1} - a)^{n^1 - 1}.$$

Since it has the same roots as $F_\Gamma(z)$, ζ' has a Puiseux series with leading term $ax^{m/n}$ and so it goes through the point p. Then we consider the strict transforms with origin at p of γ, the polar ζ and its branch ζ', denoted by $\tilde{\gamma}$, $\tilde{\zeta}$ and $\tilde{\zeta}'$, respectively. It is clear that $\tilde{\zeta}'$ is a branch of $\tilde{\zeta}$, which in turn, by 6.2.4, is a polar of $\tilde{\gamma}$ relative to an equation $\tilde{x}$ of the germ E of the exceptional divisor at p:

$$\tilde{\zeta} = P_{\tilde{x}}(\tilde{\gamma}).$$

Complete $\tilde{x}$ to a pair of local coordinates $\tilde{x}, \tilde{y}$ at p and let $\tilde{s}$ be a Puiseux series of $\tilde{\gamma}$ relative to such coordinates. By 5.6.3, the characteristic exponents of $\tilde{s}$ are

$$\frac{m_2 - m_1}{n^1}, \ldots, \frac{m_k - m_1}{n^1}.$$

Assume, using induction, that the claim is true for irreducible germs with $k - 1$ characteristic exponents. Then, for some $i = 2, \ldots, k$, the quotient $[\tilde{\gamma}.\tilde{\zeta}']/[E.\tilde{\zeta}']$ equals

$$\frac{(n^1 - n^2)(m_2 - m_1) + \cdots + (n^{i-2} - n^{i-1})(m_{i-1} - m_1) + n^{i-1}(m_i - m_1)}{n^1}$$

$$= \frac{(n^1 - n^2)m_2 + \cdots + (n^{i-2} - n^{i-1})m_{i-1} + n^{i-1}m_i - n^1 m_1}{n^1}. \tag{6.6}$$

Since an easy computation gives

$$\sum_{q<p} e_q(\gamma)e_q(\zeta') = m_1\bar{n} = \bar{m}n,$$

we have

$$[\gamma.\zeta'] = \bar{n}m_1 + [\tilde{\gamma}.\tilde{\zeta}']$$

and of course $\bar{n} = [\eta.\zeta']$. Put $\bar{n}^1 = \gcd(\bar{m}, \bar{n})$, then $\bar{n}^1 = [E.\tilde{\zeta}']$, $m_1/n^1 = \bar{m}/\bar{n}^1$ and $n/n^1 = \bar{n}/\bar{n}^1$. We put all this together to get

$$[\gamma.\zeta'] = \bar{n}m_1 + \bar{n}^1 \left[\frac{(n^1 - n^2)m_2 + \cdots + (n^{i-2} - n^{i-1})m_{i-1} + n^{i-1}m_i - n^1 m_1}{n^1} \right]$$

$$= \bar{n}m_1 + \bar{n} \left[\frac{(n^1 - n^2)m_2 + \cdots + (n^{i-2} - n^{i-1})m_{i-1} + n^{i-1}m_i - n^1 m_1}{n} \right]$$

$$= \bar{n} \left[\frac{(n - n^1)m_1 + (n^1 - n^2)m_2 + \cdots + (n^{i-2} - n^{i-1})m_{i-1} + n^{i-1}m_i}{n} \right]$$

which is just the formula in the claim.

If $i = 1$, we have already seen that there are branches ζ' for which the equality 6.5 in the claim holds true. For $i > 1$, by the induction hypothesis, there is some branch $\tilde{\zeta}'$ of $\tilde{\zeta}$ for which the equality 6.6 above holds true. Then, by the former computations, 6.5 holds true for the branch ζ' of ζ whose strict transform is $\tilde{\zeta}'$. ◇

Merle's original proof of 6.8.1 in [59] is interesting, as it makes a clever use of arithmetical properties of intersection numbers, see also [29]. The above proof has a more geometrical character and is closer to the one we will give for many-branched germs.

Though the non-transverse case is needed in our proof and has its own interest, the most interesting consequences of 6.8.1 are for transverse polars: assume that x is chosen so that η is not tangent to γ. Then the characteristic exponents m_i/n are the characteristic exponents of γ and furthermore, since by 6.1.7 no branch of ζ is tangent to η, $[\eta.\zeta'] = e_O(\zeta')$. The quotients

$$\frac{[\zeta'.\gamma]}{e_O(\zeta')},$$

for ζ' ranging on all branches of a transverse polar ζ of γ, are called the *polar quotients* or the *polar invariants* of γ, they were introduced by Teissier [80] for hypersurfaces. The next corollary is a direct consequence of 6.8.1. It presents an explicit formula for the polar invariants which in particular proves that the polar invariants do not depend on the transverse polar used in their definition:

Corollary 6.8.2 *Let γ be an irreducible germ of order n and characteristic exponents $m_1/n, \ldots, m_k/n$. Let ζ be any transverse polar of γ. For any branch ζ' of ζ there is an i, $1 \le i \le k$ such that*

$$\frac{[\gamma.\zeta']}{e_O(\zeta')} = \frac{(n - n^1)m_1 + \cdots + (n^{i-2} - n^{i-1})m_{i-1} + n^{i-1}m_i}{n},$$

where $n^i = \gcd(n, m_1, \ldots, m_i)$, for $0 \le i \le k$. Conversely, for each $i = 1, \ldots, k$ there is at least one branch ζ' for which the former equality holds.

Remark 6.8.3 The second term of the formula may be related to the generators of the semigroup of γ introduced in section 5.8, namely, by 5.8.1,

$$\frac{(n - n^1)m_1 + \cdots + (n^{i-2} - n^{i-1})m_{i-1} + n^{i-1}m_i}{n} = \frac{n^{i-1}\bar{m}_i}{n}.$$

A direct consequence of 6.8.2 is:

Corollary 6.8.4 *If γ is an irreducible germ with k characteristic exponents, all its transverse polars have at least k branches.*

The next corollary of 6.8.2 should be highlighted because of the fact, already explained in 6.5.1, that the equisingularity type of γ does not determine the equisingularity type of its transverse or generic polars, nor even their number of branches (see figure 6.6).

Corollary 6.8.5 *If γ is an irreducible germ, then the multiplicity and the polar invariants of γ determine and are in turn determined by the equisingularity class of γ.*

PROOF: That the polar invariants are equisingularity invariants of γ is clear from 6.8.2, as so are the characteristic exponents (5.5.4). Conversely, an easy computation shows that the formulas in 6.8.2 allow us to obtain the integers m_i, $i = 1, \ldots, k$, from n and the polar invariants. Then it is enough to use 5.5.4. ◇

To close this section let us show that the polar invariants may be also computed using a second polar of γ instead of γ itself. Let x, y be any system of local coordinates and assume, after reversing its order if necessary, that the y-axis is not tangent to γ. Then the polar invariants are the intersections of the branches ζ' of $P_x(f)$ with γ. Using once again the equality

$$\frac{d}{dt} f(x(t), y(t)) = \frac{\partial f}{\partial x}(x(t), y(t)) \frac{dx}{dt}$$

where $x(t), y(t)$ is a uniformizing map of a representative of ζ', we get

$$[\gamma.\zeta'] = [P_y(f).\zeta'] + [\eta.\zeta']$$

where η still denotes the y-axis. Since, by 6.1.7, ζ' cannot be tangent to the y-axis, $[\eta.\zeta'] = e_O(\zeta')$ and the last equality gives the polar invariant corresponding to ζ':

$$\frac{[\gamma.\zeta']}{e_O(\zeta')} = \frac{[P_y(f).\zeta']}{e_O(\zeta')} + 1.$$

Thus one may also determine the equisingularity type of γ from a pair of its polar germs, namely:

Corollary 6.8.6 *If γ is an irreducible germ, fix an equation f of γ and any pair of local coordinates x, y: then the equisingularity type of γ is determined by its polars $P_x(f), P_y(f)$.*

PROOF: First of all choose from $P_x(f), P_y(f)$ a polar germ with minimal multiplicity and call it ζ: by 6.1.7, ζ is a transverse polar and so $e_O(\gamma) = e_O(\zeta) + 1$, which determines the multiplicity of γ. Then the polar invariants may be computed from the intersections of the branches of ζ with the other polar as shown above, after which the claim follows from 6.8.5. ◇

6.9 More on infinitely near points

This section is devoted to introducing some further notions about infinitely near points. They are rather technical, but needed in the next sections 6.10 and 6.11.

Assume that either $p = O$ or p is a free point infinitely near to O. A satellite point q infinitely near to p has been said to be a *satellite of p* if there are no free points between p and q (see section 3.6). Now, we will deal with the free points that come just after the satellite points of p. We say that a point p' infinitely near to p is a *point next* p if and only if p' is free and there are no free points between p and p'. It is clear that for any given germ ξ and any p on ξ, there are finitely many points next p on ξ, just a single one if ξ is irreducible.

For a better handling of the non-transverse polars, we need to introduce modified versions of the notions of free and satellite point and related ones, relative to a smooth germ η through O. Assume that a smooth germ $\eta : g = 0$ with origin at O has been fixed. A point p infinitely near to O will be called a *g-free point* if and only if it is a free point and does not belong to η. Equivalently, p is *g*-free if and only if the total transform $\bar{\eta}_p$ is smooth. Otherwise p will be called a *g-satellite point*: *g*-satellite points are thus the points infinitely near to O that either belong to η or are satellite points. We easily get relative versions of related notions, just by substituting *g-free* for *free* and *g-satellite* for *satellite*. Indeed, assume that either $p = O$ or p is *g*-free, that p' is infinitely near to p and that there are no *g*-free points between p and p'. Then if p' is *g*-free, it will be called a *point g-next* p. Otherwise, if p' is *g*-satellite, we will say that it is a *g-satellite of* p.

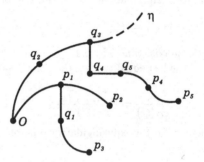

Figure 6.8: The points $p_1, \ldots, p_5$ are η-free and q_1, q_2, q_3 are η-satellite. p_1 and p_4 are next O, while p_2, p_3, p_5 are not.

Obviously, all such notions depend just on η and not on g. Thus we will also use the prefix η- instead of g- by saying η-free for g-free, etc.

Let ξ be any germ having not η as a component: then it is clear that there are finitely many g-satellite points on ξ. Furthermore, it follows from the definitions that if no branch of ξ is tangent to η, then a point p on ξ is η-free (resp. η-satellite) if and only if it is free (resp. satellite) so that, regarding the points on ξ, all notions relative to η agree with the corresponding absolute ones.

Assume that local coordinates x, y at O are fixed: the next lemma describes the x-satellite points of O on an irreducible germ γ different from the y-axis:

Lemma 6.9.1 *If γ is irreducible and has Puiseux series*

$$s = ax^{m/n} + \cdots, \quad a \neq 0,$$

then

(a) *if $m/n \geq 1$, there are no x-satellite points of O on γ, and therefore the point on γ in the first neighbourhood of O is the point x-next O on γ.*

(b) *if $m/n < 1$, all points depending on the characteristic exponent m/n are x-satellite points of O and the first free point after them is the point x-next O on γ.*

PROOF: Follows from the definitions and 5.5.1. ◇

Now we will show a fact about irreducible germs through a given point which will be quite useful later on:

Proposition 6.9.2 *Let q be any point infinitely near to O. For any two irreducible germs γ, γ' going through q and having a free point in its first neighbourhood*

$$\frac{e_q(\gamma)}{e_q(\gamma')} = \frac{e_p(\gamma)}{e_p(\gamma')}$$

for any point p, $O \leq p < q$.

PROOF: Since the point in the first neighbourhood of q on γ is free, no point p preceding q has a proximate point on γ after q and so the proximity equality for p reads

$$e_p(\gamma) = \sum_{\substack{p' \leq q \\ p' \text{ proximate to } p}} e_{p'}(\gamma).$$

Since the same arguments apply to γ', also

$$e_p(\gamma') = \sum_{\substack{p' \leq q \\ p' \text{ proximate to } p}} e_{p'}(\gamma').$$

Then the claim follows from these equalities using induction on the number of points between p and q. ◇

Let p and p' be free points infinitely near to O, p' equal or infinitely near to p. Denote by E and E' the germs of the exceptional divisors at p and p' respectively, choose any irreducible germ γ through p' and put

$$r(p, p') = \frac{[\tilde{\gamma}_p . E]}{[\tilde{\gamma}_{p'} . E']}.$$

Then we have:

Lemma 6.9.3 *$r(p, p')$ is an integer and depends only on p and p', not on the germ γ used for defining it.*

PROOF: Let q and q' be the points just before p and p', respectively: by 3.5.3,

$$r(p, p') = \frac{e_q(\gamma)}{e_{q'}(\gamma)}.$$

If a second irreducible germ γ' is used instead of γ, then we get

$$\frac{[\tilde{\gamma}'_p.E]}{[\tilde{\gamma}'_{p'}.E']} = \frac{e_q(\gamma')}{e_{q'}(\gamma')}$$

which equals $r(p,p')$ by 6.9.2. Lastly, it is enough to choose γ with $[\tilde{\gamma}_{p'}.E'] = 1$ (4.2.9) to see that $r(p,p') \in \mathbb{Z}$. $\diamond$

Remark 6.9.4 The integer $r(p,p')$ may be easily computed from an Enriques diagram representing the points up to p'. It is enough to take γ with $[\tilde{\gamma}_{p'}.E'] = 1$ and, as explained in section 3.9, compute backwards the multiplicities of γ at the preceding points till getting that of the point q just before p, as then $e_q(\gamma) = [\tilde{\gamma}_p.E] = r(p,p')$.

Remark 6.9.5 The proof of 6.9.3 shows that actually $r(p,p')$ does not depend on the point p' but only on the point q' whose first neighbourhood p' belongs to: $r(p,p')$ is the same for all free points p' in the first neighbourhood of q'.

We extend the definition of $r(p,p')$ to the case $p = O$ by taking

$$r(O,p') = \frac{e_O(\gamma)}{[\tilde{\gamma}_{p'}.E']}$$

for $p' \neq O$, γ being any irreducible germ through p', and

$$r(O,O) = 1.$$

The reader may easily extend lemma 6.9.3 and remarks 6.9.4 and 6.9.5 to such cases. In particular $r(p,p')$ still does not depend on the irreducible germ γ.

Assume as before that η is a smooth germ at O, and that g is an equation of η. We need a modified version of $r(p,p')$ relative to g (or to η): If p and p' are g-free and still p' is equal or infinitely near to p, take

$$r_g(p,p') = r(p,p'),$$

while we put

$$r_g(O,p') = \frac{[\gamma.\eta]}{[\tilde{\gamma}_{p'}.E']}$$

for any irreducible germ γ through p' and, of course,

$$r_g(O,O) = 1.$$

The reader may easily check that, as above, the definition does not depend on γ and $r_g(p,p') \in \mathbb{Z}$. The next remark easily follows from the definition:

Remark 6.9.6 If the point preceding p' in the first neighbourhood of O does not belong to η, then any irreducible germ γ through p' is not tangent to η, $[\gamma.\eta] = e_O(\gamma)$ and hence

$$r_g(p,p') = r(p,p')$$

for $p = O$ and, obviously, also for any g-free (i.e., free) point p preceding p'.

For future reference let us quote:

Lemma 6.9.7 *Let $p \le p' \le p''$ be points either equal to O or free and infinitely near to O. We have*

$$r(p,p'') = r(p,p')r(p',p'').$$

If there is given g defining a smooth germ through O and the points are assumed to be either equal to O or g-free, still

$$r_g(p,p'') = r_g(p,p')r_g(p',p'').$$

PROOF: Obvious from the definitions. ◇

Lemma 6.9.8 *We have $r(p,p') = 1$ (resp. $r_g(p,p') = 1$) if and only if there are no satellite (resp. g-satellite) points between p and p'. Therefore p' is next (resp. g-next) p and $r(p,p') = 1$ (resp. $r_g(p,p') = 1$) if and only if p' is in the first neighbourhood of p.*

PROOF: We make the proof in the relative case, as by 6.9.6 the absolute one follows from it by taking $\eta : g = 0$ missing the point in the first neighbourhood of O that precedes p'.

Obviously $r_g(p,p) = 1$, so for proving the *if* part we may assume $p \ne p'$. Let q be the point whose first neighbourhood p' belongs to. Write E for the germ at p of the exceptional divisor and take $E = \eta$ if $p = O$. Fix any irreducible germ γ going through p'. If there are no g-satellite points between p and p', then no point on γ infinitely near to p belongs to E and hence $[\tilde{\gamma}_p . E] = e_p(\gamma)$. Furthermore, since p' and all points between p and p' are free, we know from 3.5.8 or 5.5.1 that $e_p(\gamma) = e_q(\gamma)$, which together with the proximity equality $e_q(\gamma) = [\tilde{\gamma}_{p'} . E_{p'}]$ gives $r_g(p,p') = 1$.

For the converse, assume that there is some g-satellite point p_1 between p and p': we will prove $r_g(p,p') > 1$. By 6.9.7, it is not restrictive to assume that p_1 is in the first neighbourhood of p. Then, since p_1 belongs to E,

$$[\tilde{\gamma}_p . E] \ge e_p(\gamma) + e_{p_1}(\gamma) > e_p(\gamma) \ge e_q(\gamma) = [\tilde{\gamma}_{p'} . E_{p'}],$$

as claimed.

The second part of the claim directly follows from the first one and the definition of g-next. ◇

In some cases one may compute $r_g(O,p)$ from the Newton diagram of a germ through p:

Lemma 6.9.9 *Let x, y be local coordinates at O, ξ a germ of curve, Γ a side of the Newton polygon of ξ relative to x, y and p the point x-next O on a branch γ of ξ corresponding to Γ. Assume that Γ has slope $-n/m \le -1$, $n, m \in \mathbb{N}$, $\gcd(m,n) = 1$. Then $r_x(O,p) = n$.*

PROOF: We know from section 2.2 that a Puiseux series of γ has the form

$$s = ax^{m/n} + \cdots, \quad a \neq 0$$

and polydromy order equal to a multiple of n, say to dn. Denote by η the y-axis and by E the exceptional divisor at p.

If $m/n = 1$, that is, $n = 1$, then p is in the first neighbourhood of O and not on the y-axis. It is clear in such a case that $r_x(O, p) = 1 = n$, using for instance that $[\gamma.\eta] = e_O(\gamma) = [\tilde{\gamma}_p.E]$.

Assume thus $m/n < 1$. Denote by q the last point depending on the characteristic exponent m/n, so that p is in the first neighbourhood of q and we know from 5.2.7 or 5.5.1 that

$$[\tilde{\gamma}_p.E] = e_q(\gamma) = \gcd(md, nd) = d.$$

On the other hand, the intersection multiplicity of γ and the y-axis is the polydromy order of s, that is, nd, and so the claim follows. ◇

The reader may see as exercise that 6.9.9 is still true for $m/n > 1$ provided p is taken to be either the first point on γ after those depending on the characteristic exponent m/n if $n > 1$, or the point in the m-th neighbourhood of O if $n = 1$.

6.10 Components of the polar germs

In this section we will describe a decomposition of the polar germs of a many-branched germ ξ into non-necessarily irreducible components. Indeed, for each free point p on ξ we will compute the intersection multiplicity of $\eta : g = 0$ and the germ composed of all branches of a g-polar whose last point on ξ is either p or a satellite point of p. In the transverse case, this will give partial information about the number and the orders of such branches (6.10.4). The results in this section essentially come from [47], although they will be given here a more geometrical treatment. The interested reader may also see a nice early precedent in [76], 18. Since we will deal with both transverse and non-transverse polars, relative notions such as g-free and g-next will be used instead of the absolute ones.

First of all we need an auxiliary lemma. Assume that ξ is a germ of curve and that, once local coordinates x, y are fixed, Γ is a side of the Newton polygon $N(\xi)$ of ξ. Assume furthermore that Γ has slope $-n/m$, $\gcd(m, n) = 1$, and that its associated equation is $\Omega(z) = 0$. Then select a factor G of Ω so that $G \in \mathbb{C}[z^n]$ and thus it is invariant under conjugation $z \mapsto \varepsilon z$ by the n-th roots ε of unity. Assume furthermore that G has no common root with $G' = \Omega/G$. Define the germ ξ_G as composed of all branches of ξ that correspond to the side Γ and to a conjugacy class of roots of $G(z) = 0$, each with the same multiplicity it has in ξ. In other words the branches of ξ_G are the branches of ξ which have a Puiseux series (and hence, since $G \in \mathbb{C}[z^n]$, all Puiseux series) of the form

$$s = bx^{m/n} + \cdots, \quad G(b) = 0,$$

and they are taken with the same multiplicity they already have as components of ξ. After all such conventions we have:

Lemma 6.10.1 *The intersection multiplicity of ξ_G and the y-axis is just the degree of G.*

Remark 6.10.2 Lemma 6.10.1 applies in particular to the case $G = \Omega$ and then it says that the intersection multiplicity of the y-axis and the component of ξ corresponding to Γ is the height of Γ.

PROOF OF 6.10.1 For any non-zero series

$$F = \sum_{\alpha, \beta \geq 0} a_{\alpha, \beta} x^\alpha y^\beta \in \mathbb{C}[[x, y]]$$

define its (m, n)-twisted order as being

$$k(F) = \min\{n\alpha + m\beta \,|a_{\alpha, \beta} \neq 0\}$$

and its (m, n)-twisted initial form as

$$\hat{F} = \sum_{n\alpha + m\beta = k(F)} a_{\alpha, \beta} x^\alpha y^\beta.$$

As is clear, if $\mathbf{N}(F)$ has a side Γ of slope $-n/m$, then

$$\hat{F} = \sum_{(\alpha, \beta) \in \Gamma} a_{\alpha, \beta} x^\alpha y^\beta$$

and hence the equation associated with Γ is

$$\Omega_\Gamma(\zeta) = \hat{F}(1, z) z^{-\beta_1}$$

if β_1 is the ordinate of the last end of Γ.

If $\mathbf{N}(F)$ has no side with slope $-n/m$, then $\hat{F}$ has a single monomial, namely the one corresponding to the vertex of $\mathbf{N}(F)$ where $n\alpha + m\beta$ takes its minimal value.

On the other hand, for any two series F_1, F_2, clearly,

$$k(F_1 F_2) = k(F_1) + k(F_2), \quad \widehat{F_1 F_2} = \hat{F}_1 \hat{F}_2.$$

Back to considering our germs ξ and ξ_G, write an equation of ξ in the form $f = f_1 f_2$, f_1 being an equation of ξ_G and hence f_2 defining a third germ ξ_2 so that $\xi = \xi_G + \xi_2$. We know (1.1.4) that the intersection multiplicity of ξ_G and the y-axis equals the height of the Newton polygon of ξ_G. By the definition of ξ_G, its Newton polygon has a single side, this side has slope $-n/m$ and its associated equation is $\hat{f}_1(1, z) = 0$. Thus the intersection multiplicity of ξ_G and the y-axis equals the degree of $\hat{f}_1(1, z)$ and so, to reach the claim it is enough to see that $\widehat{f}_1(1,z)$ and G agree up to a non-zero multiplicative constant, which we will do next.

It follows from the definition of ξ_G on one hand and the Newton–Puiseux algorithm on the other that the polynomials G and $\widehat{f}_1(1,z)$ have the same roots, namely the coefficients of the leading terms of all Puiseux series of all branches of ξ_G.

By the definition of ξ_G, no branch of ξ_2 has a Puiseux series with initial term $bx^{m/n}$, $G(b) = 0$ and so no root of G may be a root of $\widehat{f}_2(1,z)$, which, as explained above, is either a power of z, or the product of a power of z and the first member of the equation associated with the side of $\mathbf{N}(f_2)$ of slope $-n/m$. Since we already know that no root of G is either zero or a root of $G' = \Omega/G$, in the equality

$$\widehat{f}_1(1,z)\widehat{f}_2(1,z) = \widehat{f}(1,z) = \Omega(z)z^{\bar{\beta}} = G(z)G'(z)z^{\bar{\beta}}$$

G and $\widehat{f}_1(1,z)$ have the same prime factors, from which no one divides either $\widehat{f}_2(1,z)$ or $G'z^{\bar{\beta}}$. It follows that the polynomials G and $\widehat{f}_1(1,z)$ divide each other and hence the claim. ⬦

Theorem 6.10.3 *Assume that ξ is a non-empty germ of curve with origin at O and that $g \in \mathcal{O}_O$ defines a smooth germ η which is not a component of ξ. Let $\zeta = P_g(\xi)$ be any g-polar of ξ and let p be O or any g-free point on ξ. Denote by ζ^p the germ composed of all branches of ζ that go through p and miss all points p' g-next p on ξ, each branch taken with the multiplicity it has in ζ. We have*

$$[\zeta^p.\eta] = r_g(O,p)\left(\sum_{p'} r_g(p,p') - 1\right),$$

the summation running on all points p' g-next p on ξ.

PROOF: We will use induction on the order of the neighbourhood the point p is belonging to and hence we will deal with the case $p = O$ first.

Choose local coordinates x,y at O such that $x = g$ and the x-axis, $y = 0$, is not tangent to ξ. In particular η is now the y-axis. Let

$$f = \sum_{\alpha,\beta \geq 0} a_{\alpha,\beta}x^\alpha y^\beta$$

be the equation of ξ for which $\zeta = P_x(f)$. Assume that the sides of the Newton polygon of ξ are $\Gamma_1, \ldots, \Gamma_k$, and also that each Γ_i has ends $(\alpha_{i-1}, \beta_{i-1})$, (α_i, β_i) and slope $-n_i/m_i$, $n_i, m_i > 0$, $\gcd(n_i, m_i) = 1$, $i = 1, \ldots, k$. Since no one of the axes is a component of ξ, we have $\alpha_0 = \beta_k = 0$.

Write $d_i n_i$ and $d_i m_i$ for the height and the width of Γ_i, that is, $d_i n_i = \beta_{i-1} - \beta_i$ and $d_i m_i = \alpha_i - \alpha_{i-1}$. We will denote by $\Omega_i(z) = 0$ the equation associated with Γ_i,

$$\Omega_i(z) = \sum_{(\alpha,\beta)\in\Gamma_i} a_{\alpha,\beta}z^{\beta-\beta_i}.$$

Using that $\Omega_i(z) \in \mathbb{C}[z^{n_i}]$ (1.3.2) we may also write

$$\Omega_i(z) = a_{\alpha_{i-1},\beta_{i-1}} \prod_{j=1}^{\ell_i} (z^{n_i} - b_{i,j}^{n_i})^{\rho_{i,j}} \tag{6.7}$$

where $b_{i,j} \neq 0$ and $b_{i,j}^{n_i} \neq b_{i,j'}^{n_i}$ if $j \neq j'$.

Each branch of ξ has a Puiseux series of the form

$$s = b_{i,j} x^{m_i/n_i} + \cdots$$

and since ξ has no branches tangent to the x-axis, $m_i/n_i \leq 1$ for all $i = 1, \ldots, k$. Then one may use lemma 6.9.1 on all branches of ξ to show that the points x-next O on ξ are, for each $i = 1, \ldots, k$ and each $j = 1, \ldots, \ell_i$, the first point $p_{i,j}$ on any branch with Puiseux series

$$s = b_{i,j} x^{m_i/n_i} + \cdots$$

after the points depending on the characteristic exponent m_i/n_i (or just after O if $m_i/n_i = 1$). Notice that $p_{i,j}$ is the point of coordinate $b_{i,j}^{n_i}$ if one uses the absolute coordinate of 5.3.4. We have seen in 6.9.9 that $r_x(O, p_{i,j}) = n_i$.

Now we will deal with the polar ζ: it has equation $\partial f/\partial y$ and so it is clear that its Newton diagram $\Delta(\partial f/\partial y)$ comes from that of f by deleting the points on the α-axis and moving all the other points one step downwards. It follows that the first $k-1$ sides of the Newton polygon of ζ, $\Gamma_1', \ldots, \Gamma_{k-1}'$, are the translations of $\Gamma_1, \ldots, \Gamma_{k-1}$ by the vector $(0, -1)$. Hence Γ_i' still has slope $-n_i/m_i$ and its ends are $(\alpha_{i-1}, \beta_{i-1} - 1), (\alpha_i, \beta_i - 1)$, for $i = 1, \ldots, k-1$.

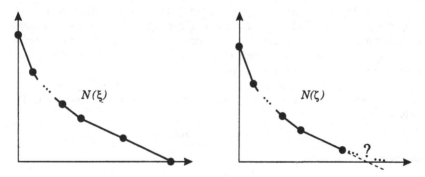

Figure 6.9: $N(\xi)$ and the well determined part of $N(\zeta)$ in the proof of 6.10.3. As for irreducible germs (figure 6.7) all further sides of $N(\zeta)$ have slope bigger than the slope of any side of $N(\xi)$.

Unlike the other sides of $N(f)$, the whole of Γ_k does not give rise to a side of $N(\partial f/\partial y)$ because its last end lies on the α-axis and is thus deleted by derivation. Denote by $(\bar{\alpha}, \bar{\beta})$ the last point of the Newton diagram of f on Γ_k

before its end $(\alpha_k, 0)$. We have $\alpha_{k-1} \leq \bar{\alpha} < \alpha_k$, $\beta_{k-1} \geq \bar{\beta} > 0$ and $a_{\bar{\alpha},\bar{\beta}} \neq 0$. If $(\bar{\alpha}, \bar{\beta}) \neq (\alpha_{k-1}, \beta_{k-1})$ we have a new side of $N(\partial f / \partial y)$ that goes from the end of Γ'_{k-1} to the point $(\bar{\alpha}, \bar{\beta} - 1)$. We will denote this side by Γ'_k: notice that it still has slope $-n_k/m_k$. In the case $(\bar{\alpha}, \bar{\beta}) = (\alpha_{k-1}, \beta_{k-1})$ the side Γ'_k does not exist.

There may be many other sides of $N(\partial f / \partial y)$ beyond $(\bar{\alpha}, \bar{\beta})$, all we need to know about them is that all have slope strictly bigger than $-n_k/m_k$: this is clear because, by the definition of $(\bar{\alpha}, \bar{\beta})$, all points $(\alpha, \beta) \in \Delta(f)$ with $\beta > 0$ and $\alpha > \bar{\alpha}$ are on the open half-plane above Γ_k and hence all points $(\alpha, \beta) \in \Delta(\partial f / \partial y)$ with $\alpha > \bar{\alpha}$ are above the line through $(\bar{\alpha}, \bar{\beta} - 1)$ of slope $-n_k/m_k$.

Let γ be a branch of ζ and assume first that it corresponds to a side Γ'_i, $i < k$. Then the leading term of (any of) its Puiseux series has the form bx^{m_i/n_i} where b is a root of the equation $\Omega'_i(z) = 0$ associated with the side Γ'_i. An easy computation shows that

$$\Omega'_i(z) z^{\beta_i - 1} = \frac{d}{dz}(\Omega_i(z) z^{\beta_i}),$$

so that, using equality 6.7 above, we easily get

$$\Omega'_i(z) = a_{\alpha_{i-1}, \beta_{i-1}} \prod_{j=1}^{\ell_i} (z^{n_i} - b_{i,j}^{n_i})^{\rho_{i,j} - 1} G(z)$$

where G is a polynomial of degree $n_i \ell_i$, invariant under conjugation by the n_i-th roots of unity and with no root $b_{i,j}^{n_i}$, $j = 1, \dots, \ell_i$. Since the leading term of its Puiseux series has degree m_i/n_i, it is clear that γ goes through no point $p_{i',j}$ with $i' \neq i$. Furthermore, by 5.3.4, γ goes through $p_{i,j}$ if and only if $b^{n_i} = b_{i,j}^{n_i}$, so that γ goes through no point $p_{i,j}$ g-next O on ξ if and only if $G(b) = 0$. Denote by ζ_i the germ composed of all branches of ζ corresponding to the side Γ'_i and going through no point x-next O on ξ, each taken with the same multiplicity as it has in ζ. Then it is enough to use 6.10.1 to see that the intersection multiplicity of ζ_i and the y-axis is just $n_i \ell_i$.

A similar argument applies to the branches of ζ corresponding to the side Γ'_k in the case when it does exist. Then the equation $\Omega'_k(z)$ associated with Γ'_k satisfies

$$\Omega'_k(z) z^{\bar{\beta} - 1} = \frac{d}{dz}(\Omega_i(z))$$

and hence has the form

$$\Omega'_k(z) = a_{\alpha_{k-1}, \beta_{k-1}} \prod_{j=1}^{\ell_k} (z^{n_k} - b_{k,j}^{n_k})^{\rho_{k,j} - 1} G(z)$$

where G still is invariant under conjugation and has no root b with $b^{n_k} = b_{k,j}^{n_k}$, its degree being now $n_k \ell_k - \bar{\beta}$. Denote by ζ_k the germ composed of all branches of ζ corresponding to the side Γ'_k and going through no point g-next O on ξ, each taken with the same multiplicity as it has in ζ: it follows again from 6.10.1 that the intersection multiplicity of ζ_k and the y-axis is $n_k \ell_k - \bar{\beta}$.

If Γ'_k does not exist, then necessarily $a_{\alpha,\beta} = 0$ if (α, β) is in the interior of Γ_k and so $\Omega_k(z) = a_{\alpha_{k-1},\beta_{k-1}} z^{\beta_{k-1}} + a_{\alpha_k,0}$. This implies $\ell_k = d_k$ and since $\bar{\beta} = \beta_{k-1} = d_k n_k$, $n_k \ell_k - \bar{\beta} = 0$. Thus, if Γ'_k does not exist, we take $\zeta_k = \emptyset$ and still its intersection multiplicity with the y-axis is $n_k \ell_k - \bar{\beta}$.

Lastly, the branches of ζ not yet considered are those corresponding to the sides of $\mathbf{N}(\zeta)$ beyond $(\bar{\alpha}, \bar{\beta} - 1)$: all these sides having slope bigger than $-n_k/m_k$, their corresponding branches go through no point $p_{i,j}$ x-next O. Therefore we add up all these branches, counted according to the multiplicities they have in ζ, to get a germ ζ_{k+1}. The intersection multiplicity of ζ_{k+1} and the y-axis is, by 6.10.2, the sum of the heights of the sides of $\mathbf{N}(\zeta)$ beyond $(\bar{\alpha}, \bar{\beta} - 1)$, that is, $\bar{\beta} - 1$.

By construction we have $\zeta^O = \zeta_1 + \cdots + \zeta_k + \zeta_{k+1}$, so it is enough to add up the intersection multiplicities we have already computed to find that the intersection multiplicity of ζ^O and the y-axis is $n_1 \ell_1 + \cdots + n_k \ell_k - 1$ which equals the intersection number in the claim by 6.9.9.

Once we have dealt with the case $p = O$, assume that p is infinitely near to O: then p is equal or infinitely near to just one of the points x-next O on ξ. Assume that this point is $p_{i,j}$ and let us write in the sequel, in order to simplify the notations, $p_{i,j} = q$ and $n_i = n, m_i = m$. Note that next and x-next are equivalent notions if applied to points infinitely near to q.

We know from 6.2.4 that the strict transform of ζ with origin at q still is a polar germ, namely

$$\tilde{\zeta}_q = P_{\tilde{x}}(\tilde{\xi}_q)$$

where $\tilde{x}$ is an equation of germ E_q of the exceptional divisor at q. The reader may easily verify that since q does not belong to the y-axis, a point p' is x-next p if and only if it is next p and in turn this occurs if and only if p' is $\tilde{x}$-next p (p and p' viewed as points infinitely near to q in the latter case). It is then clear that the strict transform of ζ^p with origin at q is composed of the branches of $\tilde{\zeta}_q$ that go through p and miss all points p' $\tilde{x}$-next p on $\tilde{\xi}_q$, that is,

$$(\widetilde{\zeta^p})_q = (\tilde{\zeta}_q)^p.$$

The induction hypothesis gives thus

$$[(\widetilde{\zeta^p})_q . E_q] = r_{\tilde{x}}(q, p)(\sum_{p'} r_{\tilde{x}}(p, p') - 1),$$

the summation running on the points p' $\tilde{x}$-next p on $\tilde{\xi}_q$. Note that these are just the points x-next p on ξ.

Since the y-axis does not contain p or q, $r_{\tilde{x}}(q, p) = r(q, p) = r_x(q, p)$ and, similarly, $r_{\tilde{x}}(p, p') = r_x(p, p')$. Then the former equality may be written

$$[(\widetilde{\zeta^p})_q . E_q] = r_x(q, p)(\sum_{p'} r_x(p, p') - 1).$$

On the other hand, any branch of ζ^p may be used in the definition of $r_x(O, q)$: by adding up the corresponding equalities for all branches of ζ^p we get

$$[\zeta^p.\eta] = [(\widetilde{\zeta^p})_q.E_q] r_x(O, q),$$

η being the y-axis. Then the claim follows from the last two displayed equalities using the multiplicativity of r_x (6.9.7). $\diamond$

Of course the most interesting consequence of 6.10.3 is for transverse polars:

Corollary 6.10.4 *Assume that ξ is a non-empty germ of curve with origin at O and ζ a transverse polar of ξ. For either $p = O$ or p a free point on ξ define the germ ζ^p as being composed of all branches of ζ going through p and missing all points next p on ξ, each branch counted according to the multiplicity it has in ζ. Then the multiplicity of ζ^p is*

$$e_O(\zeta^p) = r(O, p)(\sum_{p'} r(p, p') - 1),$$

with the summation extended to all points p' next p on ξ.

PROOF: Since ζ is a transverse polar, then all its branches, in particular all branches of ζ^p, are not tangent to the y-axis, by 6.1.7. Then $e_O(\zeta^p) = [\zeta^p.\eta]$ and the claim follows from 6.10.3. $\diamond$

One may use 6.10.4 to determine the cases in which $\zeta^p = \emptyset$, i.e., all branches of a transverse polar through p are going through a point next p: this is equivalent to

$$\sum_{p'} r(p, p') - 1 = 0$$

which in turn is true if and only if there is a single point p' next p on ξ and furthermore such p' belongs to the first neighbourhood of p (6.9.8). A similar claim may be obtained from 6.10.3 for the non-transverse case.

The reader may check as an exercise that if ξ is reduced, 6.10.4 gives

$$\sum_p e_O(\zeta^p) = e_O(\xi) - 1,$$

which on the other hand is clear as in such a case $\sum_p \zeta^p = \zeta$.

We close this section by just stating a rather technical result we have already established while proving 6.10.3, as it will be useful in the forthcoming chapter 8.

Corollary 6.10.5 (of the proof of 6.10.3) *Assume that ξ is a non-empty germ of curve and x, y are local coordinates at the origin of ξ taken so that the y-axis is not a component of ξ and the x-axis is not tangent to ξ. Assume furthermore that the sides $\Gamma_1, \ldots, \Gamma_k$ of the Newton polygon of ξ, relative to x, y, have slopes $-n^1/m^1, \ldots, -n^k/m^k$ and that they have been ordered so that*

$m_{i-1}/n_{i-1} < m_i/n_i$, $i = 2, \ldots, k$. If a branch of an x-polar of ξ has Puiseux series

$$s = ax^{m/n} + \cdots, \quad a \neq 0,$$

then either $m/n = m_i/n_i$ for some $i = 1, \ldots, k$, or $m/n > m_k/n_k$.

6.11 Polar invariants: the general case

Assume that ξ is a non-empty reduced germ with origin at O. In this section we will introduce the polar invariants of ξ and compute them from an Enriques diagram of ξ. This will in particular prove that they are equisingularity invariants of ξ. Combinatorics of the many-branched case being more complicated, the determination of the polar invariants is less explicit than in the unibranched case. Polar invariants are related to the dual graph of ξ and given an interesting topological meaning in [51].

As in preceding sections we will deal with all polar germs, the results for transverse polars will appear as corollaries of the general ones. We begin by proving an easy auxiliary result we shall need later on:

Lemma 6.11.1 Let $\xi : f = 0$ be a germ of curve. Assume that its Newton polygon, relative to local coordinates x, y, has sides $\Gamma_1, \ldots, \Gamma_k$, the last end of the last side Γ_k being $(\bar{\alpha}, 0)$. Write $-n_i/m_i$ for the slope of Γ_i and $\Omega_i(z) = 0$ for its associated equation. If γ is any branch with Puiseux series

$$s(x) = ax^{m/n} + \cdots,$$

$a \neq 0$, and η is the germ of the y-axis, then

(a) $o_x f(x, s(x)) = [\xi.\gamma]/[\eta.\gamma]$.

(b) $[\xi.\gamma]/[\eta.\gamma] \geq \min\{\alpha + \beta m/n\}$, where (α, β) ranges over the vertices of $\mathbf{N}(\xi)$.

(c) The above inequality is strict if and only if, for some i, $m/n = m_i/n_i$ and $\Omega_i(a) = 0$.

(d) If either $m/n > m_k/n_k$, or $m/n = m_k/n_k$ and $\Omega_k(a) \neq 0$, then $[\xi.\gamma]/[\eta.\gamma] = \bar{\alpha}$.

PROOF: If ν is the polydromy order of s, then, by 2.6.6, $\nu = [\eta.\gamma]$ and $[\xi.\gamma] = \nu o_x f(x, s(x))$ which proves (a). Direct computation of the initial term of $f(x, s(x))$ as in section 1.3 gives (b), (c) and (d). ◇

Let ξ be a non-empty reduced germ with origin at O and assume that $g \in \mathcal{O}_O$ defines a smooth germ η at O which is not a component of ξ. We define a subset $\mathcal{R}_g(\xi)$ of the set of points on ξ in the following way: a point q on ξ belongs to $\mathcal{R}_g(\xi)$ if and only if either

(a) q is a g-satellite point and there is at least a g-free point on ξ in its first neighbourhood, or

(b) q is either O or a g-free point and there are at least two g-free points on ξ in its first neighbourhood.

It is clear that $\mathcal{R}_g(\xi)$ is a finite set, and also that it is empty if and only if ξ is smooth and non-tangent to $g = 0$. For each $q \in \mathcal{R}_g(\xi)$ we choose γ^q to be an irreducible germ through q whose point in the first neighbourhood of q is g-free and does not belong to ξ. To fix the ideas, the reader may add the condition of such point to be non-singular on γ^q, even if this is not necessary at all. In any case the existence of γ^q is guaranteed by 4.2.9 and the multiplicities of γ^q at the points preceding q are well determined by $e_q(\gamma^q)$, as by the definition of γ^q, no point after q is proximate to one preceding q and so, for $p < q$

$$e_p(\gamma) = \sum_{\substack{p' \leq q \\ p' \text{ proximate to } p}} e_{p'}(\gamma). \tag{6.8}$$

Define rational numbers

$$I_g(q) = \frac{[\xi \cdot \gamma^q]}{[\eta \cdot \gamma^q]}$$

for each $q \in \mathcal{R}_g(\xi)$. It is not difficult to see that these numbers are well defined:

Lemma 6.11.2 *The number $I_g(q)$ does not depend on the branch γ^q we choose to define it.*

PROOF: Compute both the numerator and the denominator of the formula defining $I_g(q)$ using the Noether formula and then apply 6.9.2 ◇

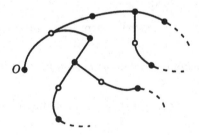

Figure 6.10: Enriques diagram of a germ ξ: the points in $\mathcal{R}(\xi)$ are the white ones.

Remark 6.11.3 The rational numbers $I_g(q)$ may be directly computed from an Enriques diagram of the germ $\xi + \eta$. Indeed, one may take γ^q with $e_q(\gamma^q) = 1$ and compute the $e_p(\gamma^q)$, $p < q$ using the recursive relations 6.8 above. Then both the numerator and denominator of the formula defining $I_g(q)$ may be computed using the Noether formula, as in the proof of 6.11.2.

Remark 6.11.4 It is clear from the definitions that if no branches of ξ are tangent to η, then neither $\mathcal{R}_g(\xi)$ nor the $I_g(q)$, $q \in \mathcal{R}_g(\xi)$, depend on g. In such a case we will just write $\mathcal{R}(\xi)$ and $I(q)$ instead,

$$I(q) = \frac{[\xi \cdot \gamma^q]}{e_O(\gamma^q)}$$

and the set $\{I(q)|q \in \mathcal{R}(\xi)\}$ depends only on the equisingularity type of ξ as it may be computed, as in 6.11.3, from just the Enriques diagram of ξ.

The next theorem is the main result in this section:

Theorem 6.11.5 *Assume that ξ is a non-empty reduced germ of curve with origin at O and that $g \in \mathcal{O}_O$ defines a smooth germ η not a component of ξ. If $\gamma_1, \ldots, \gamma_\ell$ are the branches of the polar $\zeta = P_g(\xi)$, then*

$$\left\{ \frac{[\xi \cdot \gamma_i]}{[\eta \cdot \gamma_i]} \right\}_{i=1,\ldots,\ell} = \{I_g(q)\}_{q \in \mathcal{R}_g(\xi)}.$$

Of course 6.11.5 holds true for ξ irreducible and hence 6.8.1 may be derived from it. Indeed, assume that ξ is irreducible and g has been taken as the first local coordinate. By its own definition, $q \in \mathcal{R}_g(\xi)$ if and only if it is the last point depending on one of the characteristic exponents of the Puiseux series of ξ, after which direct computation of the $I_g(q)$ gives 6.8.1.

Before proving 6.11.5, it is worth making a separate statement for the particular and most important case of transverse polars:

Corollary 6.11.6 *If $\zeta = P_g(\xi)$ is any transverse polar of ξ and it has branches $\gamma_1, \ldots, \gamma_\ell$, then*

$$\left\{ \frac{[\xi \cdot \gamma_i]}{e_O(\gamma_i)} \right\}_{i=1,\ldots,\ell} = \{I(q)\}_{q \in \mathcal{R}(\xi)}$$

and in particular this set depends only on the equisingularity type of ξ.

As for ξ irreducible, the rational numbers

$$\frac{[\xi \cdot \gamma_i]}{e_O(\gamma_i)}, \quad i = 1, \ldots, \ell$$

are called the *polar invariants* or the *polar quotients* of ξ. By corollary 6.11.6 they do not depend on the transverse polar used in their definition and are equisingularity invariants of ξ. Clearly, 6.11.6 and 6.11.3 allow a direct computation of the polar invariants of ξ from its Enriques diagram.

Even if 6.11.6 says that for each branch γ_i of ζ there exists an irreducible germ γ^q such that

$$\frac{[\xi \cdot \gamma_i]}{e_O(\gamma_i)} = \frac{[\xi \cdot \gamma^q]}{e_O(\gamma^q)}$$

(and conversely), the reader may notice that the polar germ ζ is in general not composed of branches γ^q, no matter how they are chosen. To see an easy but representative example, assume that ξ is irreducible and has a single characteristic exponent: then $\mathcal{R}(\xi) = \{q\}$, q being the last satellite point on ξ. By 5.3.4, any branch γ^q has the same first characteristic exponent as ξ. This implies $e_O(\gamma^q) > e_O(\xi) - 1$ and hence, by 6.1.7, that γ^q cannot be a component of the polar germ.

As a final remark before proving 6.11.5, let us point out that the equisingularity type of a many-branched germ is in general not determined by its multiplicity and polar invariants (see 6.8.5). Next we give an easy example:

Example 6.11.7 Let ξ_1 be a germ with two non-tangent branches, both with single characteristic exponent 5/3, and ξ_2 a germ with six smooth branches giving three double points in the first neighbourhood and no multiple point in the successive neighbourhoods. One may take, for instance,

$$\xi_1 : (y^5 - x^3)(y^3 - x^5) = 0$$
$$\xi_2 : (x^4 - y^2)(x^2 - y^4)(x^4 - (x-y)^2) = 0.$$

Clearly $e_O(\xi_1) = e_O(\xi_2) = 6$ and an easy computation using 6.11.6, shows that both germs have polar invariants $\{6,8\}$. Nevertheless, they are not equisingular (see figure 6.11).

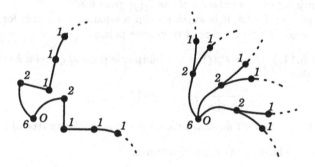

Figure 6.11: The Enriques diagrams of ξ_1 and ξ_2 in 6.11.7.

PROOF OF 6.11.5: Take all conventions and notations as in the proof of 6.10.3, in particular $g = x$ is the first local coordinate.

We will deal first with the branches of the polar going through no point $p_{i,j}$ x-next O on ξ. For $i = 1, \ldots, k$, denote by q_i the point whose first neighbourhood the points $p_{i,1}, \ldots, p_{i,\ell_i}$ are belonging to. We will prove that if a branch γ of ζ misses all points x-next O, then there is an i such that $q_i \in \mathcal{R}_x(\xi)$ and

$$\frac{[\xi \cdot \gamma]}{[\eta \cdot \gamma]} = I_x(q_i),$$

and also that for each i with $q_i \in \mathcal{R}_x(\xi)$ there is a branch γ of ζ missing all points $p_{i,j}$ for which the above equality holds.

Assume first that γ corresponds to a side Γ_i', $1 \le i < k$, of $\mathbf{N}(\zeta)$. Then, since $m_i/n_i < 1$, necessarily q_i is an x-satellite point and, by definition, $q_i \in \mathcal{R}_x(\xi)$. Furthermore, since γ corresponds to Γ_i', it goes through q_i while by hypothesis it does not go through any $p_{i,j}$: thus one may take $\gamma^{q_i} = \gamma$ and then, obviously

$$\frac{[\xi \cdot \gamma]}{[\eta \cdot \gamma]} = I_x(q_i)$$

as wanted. Conversely it is clear that for $i < k$ there exists at least a branch γ of ζ corresponding to Γ_i and going through no point $p_{i,j}$ because it has been shown in the proof of 6.10.3 that the factor G of the equation associated with Γ_i' has degree $n_i \ell_i > 0$. Hence, the former equality holds for at least one branch γ of ζ for each $i < k$.

We have seen that $q_i \in \mathcal{R}_x(\xi)$ for all $i < k$. This is not always the case if $i = k$: it is clear that $q_k \in \mathcal{R}_x(\xi)$ if it is an x-satellite point, because there is at least $p_{k,1}$ in its first neighbourhood. Since q_k has been defined as the last point before the $p_{k,j}$ and these are x-next O, the only possibility for q_k being not an x-satellite point is $q_k = O$. This clearly occurs if and only if $m_k/n_k = 1$, that is, if and only if $n_k = 1$, because it is assumed that $m_k/n_k \le 1$. In this case, by the definition of $\mathcal{R}_x(\xi)$, $O = q_k \in \mathcal{R}_x(\xi)$ if and only if $\ell_k > 1$. Then we conclude that $q_k \notin \mathcal{R}_x(\xi)$ if and only if Γ_k has slope -1 and its associated equation has a single root, i.e., $n_k = \ell_k = 1$. In this case an easy computation shows $\bar{\beta} = 1$, still using the notations of the proof of 6.10.3.

Assume that there is some branch γ of ζ going through no point $p_{i,j}$ and corresponding to the side Γ_k' of $\mathbf{N}(\zeta)$: as seen in the proof of 6.10.3, this implies $n_k \ell_k - \bar{\beta} > 0$ and then, since by its definition $\bar{\beta} \in (n_k)$, $\ell_k > 1$ and $q_k \in \mathcal{R}_x(\xi)$. Thus, as in the former case, one may take $\gamma^{q_k} = \gamma$ and then

$$\frac{[\xi \cdot \gamma]}{[\eta \cdot \gamma]} = I_x(q_k).$$

Assume now that γ, still going through no $p_{i,j}$, corresponds to a side of $\mathbf{N}(\zeta)$ beyond $(\bar{\alpha}, \bar{\beta} - 1)$: this obviously implies that $\bar{\beta} - 1 > 0$ and thus, that $q_k \in \mathcal{R}_x(\xi)$. Then choose any γ^{q_k}: both γ and γ^{q_k} satisfy the hypothesis of 6.11.1.(d) and so again

$$\frac{[\xi \cdot \gamma]}{[\eta \cdot \gamma]} = I_x(q_k).$$

Conversely, if $q_k \in \mathcal{R}_x(\xi)$, $n_k \ell_k > 1$ so that either $n_k \ell_k - \bar{\beta} \ge 1$ or $\bar{\beta} - 1 > 0$. In the first case there exists the side Γ_k' and the factor G of its associated equation has positive degree, so there exists a branch γ of ζ corresponding to Γ_k' and going through no one of the $p_{i,j}$. In the second case there is at least a side of $\mathbf{N}(\zeta)$ beyond $(\bar{\alpha}, \bar{\beta} - 1)$ and one branch γ corresponding to this side. In both cases we have seen that

$$\frac{[\xi \cdot \gamma]}{[\eta \cdot \gamma]} = I_x(q_k)$$

so that $I_x(q_k)$ occurs from a branch γ of ζ going through no point $p_{i,j}$ x-next O.

Now, to complete the proof we need to consider the branches of ζ going through some $p_{i,j}$ and, on the other side, the points in $\mathcal{R}_x(\xi)$ other than $q_1, \ldots, q_k$. We will use induction on the number of multiple points on ξ, the case of ξ smooth being obvious.

Assume first that γ is a branch of ζ going thorough $p_{i,j}$, write $p = p_{i,j}$ and keep all other notations as in the proof of 6.10.3. The strict transform of γ with origin at p is a branch of

$$\tilde{\zeta}_p = P_{\tilde{x}}(\tilde{\xi}_p).$$

By the induction hypothesis there is $q \in \mathcal{R}_{\tilde{x}}(\tilde{\xi}_p)$ and a branch $\tilde{\gamma}^q$ with origin at p, going through q and with an $\tilde{x}$-free point not on $\tilde{\xi}_p$ in the first neighbourhood of q, such that

$$\frac{[\tilde{\xi}_p \cdot \tilde{\gamma}_p]}{[E_p \cdot \tilde{\gamma}_p]} = I_{\tilde{x}}(q) = \frac{[\tilde{\xi}_p \cdot \tilde{\gamma}^q]}{[E_p \cdot \tilde{\gamma}^q]},$$

the second equality being the definition of $I_{\tilde{x}}(q)$.

Obviously $q \in \mathcal{R}_x(\xi)$ and blowing down $\tilde{\gamma}^q$ we get a branch γ^q with origin at O whose strict transform is $\tilde{\gamma}^q$. Therefore it may be used for defining $I_x(q)$.

By multiplying the former equality by $r_x(O, p)$, computed twice using both γ and γ^q, we get

$$\frac{[\tilde{\xi}_p \cdot \tilde{\gamma}_p]}{[\eta \cdot \gamma]} = \frac{[\tilde{\xi}_p \cdot \tilde{\gamma}^q]}{[\eta \cdot \gamma^q]}. \tag{6.9}$$

On the other hand, using 6.9.2 and the Noether formula on the denominators it is easy to see that for any $p' < p$,

$$\frac{e_{p'}(\gamma)}{[\eta \cdot \gamma]} = \frac{e_{p'}(\gamma^q)}{[\eta \cdot \gamma^q]},$$

from which it follows that

$$\frac{1}{[\eta \cdot \gamma]} \sum_{p' < p} e_{p'}(\xi) e_{p'}(\gamma) = \frac{1}{[\eta \cdot \gamma^q]} \sum_{p' < p} e_{p'}(\xi) e_{p'}(\gamma^q).$$

Adding the last equality and 6.9 above and using then the Noether formula we obtain

$$\frac{[\xi \cdot \gamma]}{[\eta \cdot \gamma]} = \frac{[\xi \cdot \gamma^q]}{[\eta \cdot \gamma^q]} = I_x(q)$$

as wanted, as obviously $q \neq q_i$, $i = 1, \ldots, k$.

To see the converse, let $q \in \mathcal{R}_x(\xi)$, $q \neq q_i$, $i = 1, \ldots, k$. Then, since the points $p_{i,j}$, $i = 1, \ldots, k$, $j = 1, \ldots, \ell_i$, are the points x-next O on ξ, q needs to be equal or infinitely near to some of them, say to $p_{i,j} = p$. It is clear that $q \in \mathcal{R}_{\tilde{x}}(\tilde{\xi}_p)$ so that by the induction hypothesis there is a branch $\tilde{\gamma}$ of

$$P_{\tilde{x}}(\tilde{\xi}_p) = \tilde{\zeta}_p$$

such that

$$\frac{[\tilde{\xi}_p.\tilde{\gamma}]}{[E_p.\tilde{\gamma}]} = I_{\tilde{x}}(q).$$

The branch $\tilde{\gamma}$ is the strict transform of a branch γ of ξ that goes through p, and it is enough to repeat the above argument to show that

$$\frac{[\xi.\gamma]}{[\eta.\gamma]} = I_x(q)$$

and hence that $I_x(q)$ occurs from a branch γ of ζ, which ends the proof. ◇

In fact, as the reader may easily check, our proof of 6.11.5 allows a slightly more precise claim we state in the next corollary for future use. Still take all hypotheses and notations as in 6.11.5. For $p = O$ or p a g-free point on ξ, denote by $\mathcal{R}_g^p(\xi)$ the set of points $q \in \mathcal{R}_g(\xi)$ which are either equal to p or g-satellite points of p.

Corollary 6.11.8 (of the proof of 6.11.5) *If p is either O or any g-free point on ξ, then the set of all polar quotients $[\xi.\gamma]/[\eta.\gamma]$, for γ a branch of ζ going through p and missing all points g-next p on ξ, is just $\{I_g(q) | q \in \mathcal{R}_g^p(\xi)\}$.*

6.12 Exercises

6.1 Let f and g be non-constant analytic functions defined on a (connected) surface S. Prove that there is a uniquely determined subset J of S (the *Jacobian locus of f, g*) such that for each open subset U of S where analytic coordinates x, y are defined, $J \cap U = \{p \,|\, \partial(f, g)/\partial(x, y)(p) = 0\}$. Prove that a point p belongs to J if and only if the curves $f - f(p) = 0$ and $g - g(p) = 0$ are not transverse at p.

6.2 Give an example of a non-empty germ of curve ξ at O one of whose polars ζ has $e_O(\zeta) > e_O(\xi)$. Prove that ζ cannot be a transverse polar of ξ.

6.3 Let f, g be non-invertible germs of analytic functions at O. Assume that the germ of curve $\eta : g = 0$ is smooth and is not a branch of $\xi : f = 0$.

(a) Show that, for a fixed e, all germs $f + \lambda g^e = 0$, $\lambda \in \mathbb{C}$, share a g-polar and prove, using 6.3.4, that all but finitely many of them are reduced.

(b) Prove that for any given $e, e' \in \mathbb{N}$ there is a reduced germ of curve of multiplicity e one of whose polars has multiplicity non-less than e'.

6.4 Prove that the polar germs $P_x(f)$, $P_y(f)$ have a common branch γ if and only if γ is a multiple branch of $\xi : f = 0$ (see 6.4.2). (*Hint:* compute $df(x(t), y(t))/dt$, $(x(t), y(t))$ a uniformizing map of a representative of γ, to prove that $\gamma \subset \xi$ and then apply 6.3.4).

6.5 Recall that the discriminant of a polynomial in y, $f = \prod_{i=1}^n (y - a_i)$, is defined as $D(f) = \prod_{i \neq j}(a_i - a_j)$. Assume that f is the Weierstrass equation of a non-empty germ ξ relative to local coordinates x, y and prove that $d_x(\xi) = o_x(D(f))$, where $d_x(\xi) = [\xi.P_x(\xi)]$ is the discriminantal index of ξ relative to x, hence its name.

6.6 *On obsolete local invariants.* Let ξ be a reduced germ of curve and take all notations as in sections 6.3 and 6.4. Following Smith ([76]), define the *cuspidal index* of ξ as $\kappa(\xi) = e_O(\xi) - r$, and the *nodal index* of ξ as $\delta'(\xi) = \frac{1}{2}(d(\xi) - 3\kappa(\xi))$.

(a) Prove that if ξ is a node, then

$$\kappa(\xi) = 0, \quad \delta'(\xi) = 1, \quad \delta(\xi) = 1, \quad d(\xi) = 2, \quad \mu(\xi) = 1,$$

while if ξ is an ordinary cusp,

$$\kappa(\xi) = 1, \quad \delta'(\xi) = 0, \quad \delta(\xi) = 1, \quad d(\xi) = 3, \quad \mu(\xi) = 2.$$

(b) Prove that for any ξ,

$$\delta(\xi) = \delta'(\xi) + \kappa(\xi)$$
$$d(\xi) = 2\delta'(\xi) + 3\kappa(\xi),$$

and so regarding δ and d, ξ is, in Smith's words, *equivalent in a certain manner to $\delta'(\xi)$ common nodes and $d(\xi)$ common cusps* ([76], p. 160).

(c) Show that, unfortunately, the above equivalence does not hold regarding other invariants, as in general

$$\mu(\xi) \neq \delta'(\xi) + 2\kappa(\xi).$$

6.7 Let ξ be a germ of curve, $P_g(\xi)$ a polar of ξ and η the germ $g = 0$. Prove that if $[\xi.P_g(\xi)] = [\xi.\eta] - 1$ holds, then ξ is smooth. Prove also that if ξ is smooth, then the former equality holds for any polar $P_g(\xi)$ of ξ.

6.8 *Genus formula for algebraic plane curves.* Let $C : F(X_0, X_1, X_2) = 0$ be an irreducible algebraic plane curve of degree d in $\mathbb{P}_2$ and $\pi : \tilde{C} \longrightarrow C$ its non-singular model, the points of $\tilde{C}$ being in one to one correspondence with the branches of C (see exercise 2.19). Assume that the projective coordinates X_0, X_1, X_2 have being chosen so that the point $P = [0, 0, 1]$ does not belong to C.

(a) If D is the curve $\partial F/\partial X_2 = 0$ and $O \in C$, then the germ of D at O is a g-polar germ of the germ of C, g being a local equation at O of the line OP.

(b) Let $\varphi : \tilde{C} \longrightarrow \mathbb{P}_1$ be the rational map defined by the rational function X_1/X_0 (or by the projection from P). Prove that for any $q \in \tilde{C}$ the ramification index ε_q of φ at q is

$$\varepsilon_q = [\gamma.\ell]$$

where γ is the branch of C corresponding to q and ℓ the line joining its origin to P.

(c) Use Plücker's formula 6.3.2 to prove that, for any $O \in C$,

$$\sum_{\pi(q)=O} (\varepsilon_q - 1) = [C.D]_O - 2\delta_O(C).$$

(d) Add up the above equalities and use the Bezout and Hurwitz theorems ([37], V.3 and VIII.6 or [43], I.7.8 and IV.2, for instance) to get the classical formula for the (geometric) genus of C (or $\tilde{C}$, it is the same), namely

$$g(C) = \frac{(d-1)(d-2)}{2} - \sum_p \frac{e_p(C)(e_p(C) - 1)}{2},$$

the summation running on all ordinary or infinitely near points on C (see [74], 114 or [35], IV.II.16).

6.9 *First Plücker formula.* Let $C : F = 0$ be an irreducible algebraic curve in $\mathbb{P}_2$ of degree $d > 1$. Fix a point $P = [a_0, a_1, a_2] \in \mathbb{P}_2 - C$ and let

$$D : a_0 \frac{\partial F}{\partial X_0} + a_1 \frac{\partial F}{\partial X_1} + a_2 \frac{\partial F}{\partial X_2} = 0$$

be the (projective) polar of C relative to P.

(a) Prove that for any O in C the germ of D at O is a g-polar of the germ of C, g being a local equation of the line OP.

(b) Prove that the sum of the classes (see exercise 5.2) of the branches of C at O that have principal tangent OP equals $[C.D]_O - 2\delta_O(C) - e_O(C) + r_O(C)$, $r_O(C)$ being the number of branches of C at O.

(c) Prove that the sum of the classes of the branches of C whose principal tangent goes through P is

$$m = d(d-1) - \sum_{O \in C} (2\delta_O(C) + e_O(C) - r_O(C)).$$

The integer m above is called the *class* of the projective curve C. Clearly C has finitely many principal tangents at its multiple points (2.3.9) and it is a well known fact (see [43] IV, ex. 2.3) that both the tangent lines to C at its inflexion points and the tangent lines that have two or more contact points (multiple tangents) are finitely many. Prove that if P does not belong to either a principal tangent at a multiple point, or to an inflexional or multiple tangent (i.e., if P is a in general position), then m equals the number of lines through P that are tangent to C at simple points.

6.10 Assume that local coordinates at O are fixed. A germ of curve ξ with origin at O is said to be *non-degenerate* with respect to its Newton polygon $N(\xi)$ if and only if $N(\xi)$ has both ends on the axis and no one of the equations associated with its sides has a multiple root (see exercise 2.17).

Show by an example that the above condition depends on the coordinates and characterize it in terms of the equisingularity class of the germ and the relative position of its branches and the coordinate axes.

Prove that if ξ is a non-empty germ, non-degenerate with respect to its Newton polygon, then

$$\mu(\xi) = 2a - w - h + 1,$$

where a is the area limited by $N(\xi)$ and the coordinate axes, and w and h are the width and height of $N(\xi)$ (see [46]). (*Hint:* use induction on the number of sides of $N(\xi)$)

6.11 Prove that if ξ is a reduced germ of curve with origin at O, then

$$\mu(\xi) = e_O(\xi)(e_O(\xi) - 1) - \ell + 1 + \sum_{i=1}^{\ell} \mu(\tilde{\xi}_{p_i}),$$

where $p_1, \ldots, p_\ell$ are the points on ξ in the first neighbourhood of O.

6.12 *Topological meaning of the Milnor number.* Let O be the origin in $\mathbb{C}^2$ and ξ be a curve reduced at O and defined in a open set $|x| < \varepsilon_0$, $\varepsilon_0 \in \mathbb{R}^+$, by a Weierstrass polynomial f of degree n. Denote by ζ its polar curve $\zeta : \partial f/\partial y = 0$.

(a) Prove that there is an $\varepsilon_1 \in \mathbb{R}^+$, $\varepsilon_1 < \varepsilon_0$ so that no curve $\xi_\lambda : f - \lambda = 0$, $\lambda \in \mathbb{C}$, $\lambda \neq 0$, has a singular point in $U_{\varepsilon_1} = \{(x,y)| |x| < \varepsilon_1\}$. (*Hint:* use 2.3.3 and 6.4.2 to avoid points $p \neq O$ with $(\partial f/\partial x)(p) = (\partial f/\partial y)(p) = 0$.

(b) Prove that there are a δ and an ε in $\mathbb{R}^+$, $\varepsilon < \varepsilon_1$, so that, for $0 < |\lambda| < \delta$, $\sum_{p \in U_\varepsilon} [\xi_\lambda . \zeta]_p$ equals $r = n - 1 + \mu_O(\xi)$. (*Hint:* Apply exercise 2.15 to the branches of ζ, as it may be non-reduced.)

(c) Fix any λ, $0 < |\lambda| < \delta$, take $X_\lambda = \xi_\lambda \cap U_\varepsilon$, $T = X_\lambda \cap \zeta$, let $\pi : \mathbb{C}^2 \longrightarrow \mathbb{C}$ be the projection on the second factor and D_ε the open disk $\{x \in \mathbb{C}| |x| < \varepsilon\}$. Prove that X_λ is a topological surface and that $X_\lambda - T$ is a (topological, unramified) covering of $D_\varepsilon - \pi(T)$ via π.

(d) Let Q be a (finite) polygonal decomposition of a slightly smaller closed disk $\bar{D} \subset D_\varepsilon$, $\bar{D}$ and Q being chosen in such a way that all points in $\pi(T)$ are interior points of $\bar{D}$ and vertices of Q. If k_0, k_1, k_2 are the numbers of vertices, edges and faces of Q, respectively, prove that Q lifts to a polyhedral surface Q' in X_λ with $nk_0 - r$ vertices, nk_1 edges and nk_2 faces. (*Hint:* Use 6.7.)

(e) Prove that the Euler characteristic of X_λ is $1 - \mu_O(\xi)$.

Note: X_λ is called a *Milnor fibre* of ξ at O. In fact $\mu_O(\xi)$ equals its middle Betti number, as the Milnor fibres have the homotopy type of a bouquet of circles, see [60], 6.5.

6.13 Prove that the last polar invariant of an irreducible germ γ with origin at O and last characteristic exponent m_k/n, $n = e_O(\gamma)$, is

$$I_k = \frac{1}{n}(\mu(\gamma) + m_k - 1) + 1.$$

6.14 Let γ be the irreducible germ $(x^4 - y^5)^3 - x^{14}y = 0$ (see exercise 5.3). Compute the x-polar quotients of γ directly from its definition and check the result using 6.8.1. (Answer: 12, 74/5).

6.15 Use the results of section 6.6 to compute, directly from the branches of a transverse polar, the polar invariants of an irreducible germ with a single characteristic exponent and generic in its equisingularity class. Check the result using 6.8.2.

6.16 (*Ephraim*, [36]). Let γ be an irreducible germ of curve, ζ a g-polar of γ and η the germ $g = 0$. Prove that for any branch ζ' of ζ,

$$[\gamma . \zeta'] \neq [\eta . \zeta'][\eta . \gamma].$$

Prove that if ζ is a transverse polar

$$[\gamma . \zeta'] > [\eta . \zeta'][\eta . \gamma],$$

while the reversed inequality may hold true if ζ is non-transverse (take $\gamma : x^5 - y^7 + x^4 y^2 = 0$, $g = x$).

6.17 Let γ be an irreducible germ and p a free point on γ whose first neighbouring point on γ is also free. Prove that all transverse polars of γ have effective multiplicity $e_p(\gamma) - 1$ at p.

6.18 (*Merle*, [59], 3.1). Let γ be an irreducible germ of curve and $\zeta = P_g(\gamma)$ a polar of γ. Assume that x, y are local coordinates with $g = x$ and that a Puiseux series of γ has characteristic exponents $m_1/n, \ldots, m_k/n$. Put $n^i = \gcd(n, m_1, \ldots, m_i)$ and denote by ζ_i the germ composed by all branches of ζ that go through the last free point depending on m_i/n but not through its next free point on γ. Denote by η the germ of the y-axis and prove that:

(a) The quotients $[\gamma.\zeta']/[\eta.\zeta']$ are the same for all branches ζ' of ζ_i.

(b) For any $i = 1, \ldots, k$,

$$[\eta.\zeta_i] = \frac{n}{n^i} - \frac{n}{n^{i-1}}.$$

Rewrite both claims assuming that ζ is a transverse polar.

6.19 All notations and hypotheses being as in 6.10.3, assume that there are h different g-satellite points of p on ξ that have a point g-next p on ξ in its first neighbourhood. Prove that ζ^p has at least h branches.

6.20 Let $\xi : f = 0$ be a reduced germ of curve with origin at O and x, y local coordinates at O with the y-axis non-tangent to ξ. Let ζ be the x-polar of ξ which has equation $\partial f/\partial y = 0$. Denote by Ψ the analytic map from a neighbourhood of O to $\mathbb{C}^2$ defined by the rule $\Psi(x, y) = (x, f(x, y))$. Use exercises 2.13 and 2.14 to prove that the polar invariants of ξ are the opposites of the inverses of the slopes of the sides of the Newton polygon of $\Psi_*(\zeta)$ (the *jacobian Newton polygon*, see [81]).

Chapter 7

Linear families of germs

This chapter is devoted to studying pencils and linear systems of germs of curve. We introduce their clusters of base points and prove that all but finitely many germs in a pencil and generic germs in a linear system go sharply through the corresponding clusters of base points. This allows us to prove restricted versions (for pencils and linear systems) of the μ-constant theorem. Linear systems are then used for determining which monomials in the equation of a reduced germ ξ are irrelevant to either the equisingularity type (E-sufficiency) or the analytic type (A-sufficiency) of the germ.

7.1 Linear series on a projective line

In this section we recall some elementary facts about linear series on a complex projective line. They will be applied to the linear series cut out by pencils and linear systems on exceptional divisors.

Let $\mathbb{P}_1$ be a one-dimensional complex projective space, and assume that X_0, X_1 are projective coordinates on $\mathbb{P}_1$. An *effective divisor* or *group of points* of $\mathbb{P}_1$ is a formal sum of finitely many points in $\mathbb{P}_1$ or, equivalently, an almost zero map from $\mathbb{P}_1$ into the set of the non-negative integers. A group of points D is thus uniquely written

$$D = \sum_{p \in \mathbb{P}_1} n_p p$$

where the coefficients n_p are non-negative integers, all but finitely many equal to zero. The integer n_p is usually called the *multiplicity* of p in D. Points of positive multiplicity are said to belong to D. Points p with $n_p = 1$ (resp. $n_p > 1$) are called *simple* (resp. *multiple*) *points of* D and $\sum_p n_p$ is the *degree* or the *number of points* of D. Addition of groups of points is defined in the obvious way, by adding up the multiplicities of the points.

If $F \in \mathbb{C}[X_0, X_1]$ is homogeneous of degree d, it has an essentially unique factorization

$$F = \prod_{i=1}^{r} (b_i X_0 - a_i X_1)^{d_i}$$

where the (a_i, b_i) are determined up to a multiplicative constant and we assume

$$\begin{vmatrix} a_i & b_i \\ a_j & b_j \end{vmatrix} \neq 0,$$

for $i \neq j$. Then a point $p = [a, b] \in \mathbb{P}_1$ gives $F(p) = F(a, b) = 0$ if and only if $p = p_i = [a_i, b_i]$ (i.e., $(a, b) = \rho(a_i, b_i)$, $\rho \in \mathbb{C} - \{0\}$) for some $i = 1, \ldots, r$. The group of points

$$\sum_{i=1}^{r} d_i p_i$$

is called the *group* (or *divisor*) *of zeros* of F, we will often denote it as the group $F = 0$. Conversely, any group of points D is clearly the group of zeros of a homogeneous polynomial which is determined by D up to a multiplicative constant: such a polynomial will be called an *equation* of the group. Notice that its degree equals the degree of D.

The easy proof of the next lemma is left to the reader.

Lemma 7.1.1 *A point p is a multiple point of a group $F = 0$ if and only if $(\partial F/\partial X_0)(p) = 0$ and $(\partial F/\partial X_1)(p) = 0$.*

A *linear series of degree d* on $\mathbb{P}_1$ is any set of groups of points of $\mathbb{P}_1$ of the form $\ell = \{F = 0 | F \in L - \{0\}\}$ for L a linear subspace of $\mathbb{C}[X_0, X_1]_d$, the vector space of the homogeneous polynomials of degree d. Notice that the linear space L is in turn determined by ℓ itself as $L - \{0\}$ is the set of all equations of all groups in ℓ. Notice also that ℓ has a natural structure of projective space induced by L. In particular the dimension of ℓ is defined as $\dim \ell = \dim L - 1$. Obviously $\dim \ell \leq d$. The groups in a linear series of degree d and dimension k have thus the form

$$\lambda_0 F_0 + \lambda_1 F_1 + \cdots + \lambda_k F_k = 0,$$

where the F_i are linearly independent homogeneous polynomials of degree d and the λ_i are complex numbers, not all of them equal to zero.

Of course zero-dimensional linear series have a single group of points and any group of points may be considered as a zero-dimensional series. In particular one may consider the zero-dimensional linear series whose group of points is empty: it is the only linear series of degree zero, defined by the subspace of the forms of degree zero, $\mathbb{C} \subset \mathbb{C}[X_0, X_1]$.

Let ℓ be a linear series of degree d and dimension k as above. The *fixed part* of ℓ is the (maybe empty) group of points $G = 0$, where G is the greatest common divisor of all polynomials in L (or just of the elements of a basis of L). The points in $G = 0$ are called the *fixed points* of ℓ. More precisely, a point of multiplicity m in $G = 0$ is called a fixed point of ℓ of multiplicity m. Notice that a fixed point of multiplicity m belongs with multiplicity $\geq m$ to all groups of the linear series. By taking away the group of fixed points from all groups in ℓ one obtains a linear series ℓ' of degree $d - \deg G$ and dimension k, defined by the subspace $G^{-1}L$. The series ℓ' has no fixed points and is called the *variable part* of ℓ. In fact the groups of ℓ may be written in the form

$$G(\lambda_0 F'_0 + \lambda_1 F'_1 + \cdots + \lambda_k F'_k) = 0,$$

the $F'_i = F_i/G$ being a basis of $G^{-1}L$.

The reader may easily check that if p is any point, the groups of a linear series ℓ that contain p describe a linear series of dimension $\geq \dim \ell - 1$ and the equality holds if and only if p is not a fixed point of ℓ. If in particular ℓ has dimension one and no fixed points and p is any point, there is one and only one group in the series containing p. The next lemma determines when the point p is multiple in the group of the series that contains it.

Lemma 7.1.2 *If $\ell : \lambda F + \mu G = 0$ is a one-dimensional linear series with no fixed part and degree d, then the jacobian determinant $\partial(F, G)/\partial(X_0, X_1)$ is a form of degree $2d - 2$. A point p is multiple in the only group of ℓ that contains it if and only if p belongs to the group*

$$J(\ell) : \frac{\partial(F, G)}{\partial(X_0, X_1)} = 0.$$

More precisely, the multiplicity of p in $J(\ell)$ is one less than its multiplicity in the group of ℓ that contains it.

PROOF: A straightforward computation shows that the jacobian determinants of the forms F, G relative to two different systems of projective coordinates differ in a non-zero multiplicative constant (the jacobian determinant of the linear substitution of coordinates). Thus, after fixing a point $p \in \mathbb{P}_1$, it is not restrictive to make the proof using projective coordinates for which $p = [0, 1]$. Since there are no fixed points, after reversing the roles of F and G if needed, it is not restrictive to assume that $G(0, 1) \neq 0$ and hence the group in the series containing p is $F + \mu G = 0$ for a suitable μ. Assume that p has multiplicity n in this group, that is, $F + \mu G = X_0^n H$, $H(0, 1) \neq 0$. Then one may substitute $F + \mu G$ for F in the jacobian determinant in order to get, after an easy computation,

$$\frac{\partial(F, G)}{\partial(X_0, X_1)} = n X_0^{n-1} H \frac{\partial G}{\partial X_1} + X_0^n H'$$

for some polynomial H'. By its definition H has no factor X_0, and neither has $\partial G/\partial X_1$, because G itself has no factor X_0 and $\deg G = d \geq \dim \ell = 1 > 0$. Hence the last equality proves that the jacobian determinant has the factor X_0 with multiplicity just $n - 1$, as wanted. This in particular proves that $\partial(F, G)/\partial(X_0, X_1)$ does not identically vanish, after which, clearly, it is homogeneous of degree $2d - 2$. ◇

Corollary 7.1.3 *If ℓ is a one-dimensional linear series on $\mathbb{P}_1$ with no fixed points and degree d, then finitely many groups of ℓ have a multiple point. Furthermore, if $d > 1$, there are at least two different groups of ℓ having a multiple point.*

PROOF: The first part of the claim is obvious, since by 7.1.2 there are finitely many multiple points in the groups of ℓ and each point determines the group of ℓ that contains it. If all multiple points belong to the same group of ℓ, then the sum of their multiplicities in the group is at most the degree d, and so the sum of their multiplicities in $J(\ell)$ is at most $d - 1$, which in turn is strictly less than $2d - 2$, the degree of $J(\ell)$, if $d > 1$. $\diamond$

7.2 Base points of pencils and linear systems

This section is devoted to introducing the weighted clusters of base points of pencils and linear systems with no fixed part, and to proving their main properties. We fix a point O on a smooth surface S and write $\mathcal{O} = \mathcal{O}_{S,O}$, $\mathcal{M} = \mathcal{M}_{S,O}$. Assume that $\mathcal{L}$ is the linear system defined by an ideal $I \neq 0$ of $\mathcal{O}$, that is $\mathcal{L} = \{\xi : f = 0 | f \in I - \{0\}\}$. Recall that $\mathcal{L}$ is called irrelevant if and only if $I = (1)$ (see section 2.7).

Lemma 7.2.1 *Let ζ be a fixed germ at O and write $k = \min\{[\zeta.\zeta'] | \zeta' \in \mathcal{L}\}$. Then the set U of all elements $g \in I$ for which $\xi : g = 0$ has $[\zeta.\xi] = k$ is a non-empty Zariski-open subset of I.*

PROOF: If ζ is irreducible, it follows from 2.5.4 that 0 and the non-zero elements $g \in I$ for which $\xi : g = 0$ has $[\zeta.\xi] > k$ are the elements of an ideal J contained in I. Clearly $J \neq I$ and furthermore J is $\mathcal{M}$-primary as it contains a suitable power of $\mathcal{M}$. J is thus Zariski-closed, by 2.7.5, and so in this case the claim is proved.

The claim is obviously true if $\zeta = \emptyset$. Otherwise decompose ζ into its branches, say $\zeta = n_1\gamma_1 + \cdots + n_r\gamma_r$ and write $k_i = \min\{[\gamma_i.\zeta'] | \zeta' \in \mathcal{L}\}$. Since the claim has been proved for irreducible germs, we use it for the branches of ζ: we call U_i the non-empty Zariski-open set which corresponds to γ_i according to the claim. Then the intersection $U_1 \cap \cdots \cap U_r$ is a non-empty Zariski-open subset by 2.7.7. Furthermore it is clear that for any germ $\xi : f = 0$ in $\mathcal{L}$, one has $[\zeta.\xi] = \sum n_i k_i$ if $f \in U_1 \cap \cdots \cap U_r$ and $[\zeta.\xi] > \sum n_i k_i$ otherwise. It follows that $k = \sum n_i k_i$ and $U = U_1 \cap \cdots \cap U_r$, which proves the claim. $\diamond$

Set $e = e_O(\mathcal{L}) = \min\{e_O(\xi) | \xi \in \mathcal{L}\}$. In other words, I is contained in $\mathcal{M}^e$ but not in $\mathcal{M}^{e+1}$. Obviously, $e_O(\mathcal{L}) = 0$ if and only if $\mathcal{L}$ is irrelevant. All germs with equation in $I - \mathcal{M}^{e+1}$ have multiplicity e at O and so, in particular, a generic germ in $\mathcal{L}$ has multiplicity e. We will call $e = e_O(\mathcal{L})$ the *multiplicity* of $\mathcal{L}$ at O. As is clear, if $\mathcal{G}$ is a set of germs generating $\mathcal{L}$, then also $e_O(\mathcal{L}) = \min\{e_O(\xi) | \xi \in \mathcal{G}\}$.

If p is any point in the first neighbourhood of O, denote by $\mathcal{O}_p$ its local ring, by $\varphi_p : \mathcal{O} \longrightarrow \mathcal{O}_p$ the morphism induced by blowing up, and by z an equation at p of the exceptional divisor. The ideal generated by $z^{-e}\varphi_p(I)$ obviously does not depend on the particular equation z we are using and defines a linear system $\mathcal{L}_p$ of germs at p that is generated by the virtual transforms of any set of germs generating $\mathcal{L}$, O being taken with virtual multiplicity e. The linear system $\mathcal{L}_p$ will be called the *transform of $\mathcal{L}$ with origin at p*. We extend this definition to all points in the successive neighbourhoods by using induction on the order of the neighbourhood: if p is infinitely near to q and q is in the first neighbourhood of O, take $\mathcal{L}_p = (\mathcal{L}_q)_p$ and still call it the *transform of $\mathcal{L}$ with origin at p*. If needed we will take $\mathcal{L}_O = \mathcal{L}$. We write $e_p(\mathcal{L}) = e_p(\mathcal{L}_p)$ and call this number the *multiplicity* of $\mathcal{L}$ at p.

Our first aim is to prove that all but finitely many transforms $\mathcal{L}_p$ of any given linear system $\mathcal{L}$ with no fixed part are irrelevant, or, equivalently, that $e_p(\mathcal{L}) = 0$ for all but finitely many points p. To this end we introduce a control function as follows: define, for an arbitrary linear system $\mathcal{L}$, $\kappa = \kappa(\mathcal{L}) = \min\{[\zeta.\zeta']|\zeta,\zeta' \in \mathcal{L}\}$. By 2.7.1, $\kappa(\mathcal{L}) < \infty$ if and only if $\mathcal{L}$ has no fixed part and, obviously, $\kappa(\mathcal{L}) = 0$ if and only if $\mathcal{L}$ is irrelevant. Behaviour of κ under blowing up is described next.

Lemma 7.2.2 *If $\mathcal{L}$ has no fixed part and p is in the first neighbourhood of O, then $\kappa(\mathcal{L}) \geq e_O(\mathcal{L})^2 + \kappa(\mathcal{L}_p)$ and in particular $\mathcal{L}_p$ has no fixed part either. Furthermore, $\mathcal{L}_p$ is irrelevant for all but finitely many p in the first neighbourhood of O.*

PROOF: Let us show first that one may choose $\zeta,\zeta' \in \mathcal{L}$ so that $e_O(\zeta) = e_O(\zeta') = e = e_O(\mathcal{L})$ and $\kappa(\mathcal{L}) = [\zeta.\zeta']$. For this take any two germs $\xi,\xi' \in \mathcal{L}$ so that $\kappa = [\xi.\xi']$. The equations of the germs ζ such that $e_O(\zeta) = e$ describe a non-empty Zariski-open subset of I and so, by 7.2.1 and the irreducibility of I (2.7.7), one may find a germ ζ such that $e_O(\zeta) = e$ and still $\kappa = [\zeta.\xi']$. The same argument used on ξ' gives rise to another germ ζ' with $e_O(\zeta') = e$ and $\kappa = [\zeta.\zeta']$ as wanted.

Since both the strict transforms $\tilde{\zeta}_p$ and $\tilde{\zeta}'_p$ of ζ and ζ' agree with the virtual ones, they belong to $\mathcal{L}_p$ and the claimed inequality follows using Noether's formula 3.3.4:

$$k(\mathcal{L}) = [\zeta.\zeta'] \geq e_O(\mathcal{L})^2 + [\tilde{\zeta}_p.\tilde{\zeta}'_p] \geq e_O(\mathcal{L})^2 + \kappa(\mathcal{L}_p).$$

To close, since $\tilde{\zeta}_p \in \mathcal{L}_p$, $\mathcal{L}_p$ is irrelevant for all points p but those lying on ζ. ◇

Proposition 7.2.3 *If $\mathcal{L}$ is a linear system with no fixed part, for all but finitely many points p infinitely near to O its transform $\mathcal{L}_p$ is irrelevant.*

As is clear from the definitions (and follows also from 7.2.2), all transforms of an irrelevant linear system are irrelevant. Thus if $\kappa(\mathcal{L}) = 0$, then $\mathcal{L}$ is irrelevant and there is nothing to prove.

Otherwise, $\mathcal{L}$ is not irrelevant and so $e_O(\mathcal{L}) > 0$. By 7.2.2 there are points p_i, $i = 1, \ldots, r$, $r \geq 0$, in the first neighbourhood of O so that if p lies in the first neighbourhood of O and $p \neq p_i$, $i = 1, \ldots, r$, then $\mathcal{L}_p$ is irrelevant, and hence so are all its transforms $\mathcal{L}_q$, $q > p$. Thus it is enough to prove that each $\mathcal{L}_{p_i}$, $i = 1, \ldots, r$, has at most finitely many non-irrelevant transforms, and this is true by induction, because $\mathcal{L}_{p_i}$ has no fixed part and $\kappa(\mathcal{L}_{p_i}) < \kappa(\mathcal{L})$, again by 7.2.2. $\diamond$

Now let $\mathcal{L}$ be a non-irrelevant linear system with no fixed part. We define the (weighted) *cluster of base points* of $\mathcal{L}$, $BP(\mathcal{L})$ by taking the points p equal or infinitely near to O for which $\mathcal{L}_p$ is not irrelevant, each p taken with virtual multiplicity $e_p(\mathcal{L})$. $BP(\mathcal{L})$ is actually a weighted cluster. Indeed, if K denotes its set of points, it is non-empty because $\mathcal{L}$ is non-irrelevant and is finite by 7.2.3; furthermore, if $p \in K$ then also $q \in K$ for all $q < p$ because, as already noticed, all transforms of an irrelevant linear system are irrelevant too.

The points in $BP(\mathcal{L})$ will be called the *base points* of $\mathcal{L}$. Note that p is a base point of $\mathcal{L}$ if and only if $e_p(\mathcal{L}) > 0$. In such a case $e_p(\mathcal{L})$, the virtual multiplicity of p in $BP(\mathcal{L})$, will be called the *multiplicity of the base point p*.

If I is the ideal defining $\mathcal{L}$, we will often write $BP(I)$ for $BP(\mathcal{L})$ and call it the *cluster of base points of I*. Its points will be called *base points* of I and we will write $e_p(I)$ for $e_p(\mathcal{L})$ as well.

Remark 7.2.4 It follows from the definition that all germs in $\mathcal{L}$ go through $BP(\mathcal{L})$. It is also clear from the definition that if a point p effectively belongs to all germs in $\mathcal{L}$, then it is a base point. Even more, since the virtual transforms always contain the strict ones, if $e_p(\xi) \geq e > 0$ for all $\xi \in \mathcal{L}$, then the virtual multiplicity of p in $BP(\mathcal{L})$ is non-less than e. The reader may notice that the converse is far from true and there may be germs in $\mathcal{L}$ failing to effectively go through some base points.

Remark 7.2.5 If q lies in the first neighbourhood of some base point of $\mathcal{L}$, then it is clear, still using induction, that $\mathcal{L}_q$ is generated by the virtual transforms, relative to the virtual multiplicities of $BP(\mathcal{L})$, of any set of germs generating $\mathcal{L}$.

Clusters of base points have been defined only for non-irrelevant systems with no fixed part. In the sequel these linear systems will be called *neat*. Equivalently, (see section 2.7) a linear system is neat if and only if the ideal defining it is $\mathcal{M}$-primary. The next lemma will be useful later on. Assume that $\mathcal{L}$ is neat and that all virtual transforms are taken with respect to the virtual multiplicities in $BP(\mathcal{L})$:

Lemma 7.2.6 *Let $\mathcal{G}$ be a set of germs that generate $\mathcal{L}$. If p is a base point, the virtual transform with origin at p of some $\xi \in \mathcal{G}$ has multiplicity $e_p(\mathcal{L})$ at p. If q is not a base point and lies in the first neighbourhood of a base point p, then the virtual transform with origin at q of some germ in $\mathcal{G}$ is empty.*

PROOF: By 7.2.5 the virtual transforms with origin at p, $\check{\xi}_p$, of the germs $\xi \in \mathcal{G}$ generate $\mathcal{L}_p$, thus $e_p(\mathcal{L}) = e_p(\mathcal{L}_p) = \min_{\xi \in \mathcal{G}} \{e_p(\check{\xi}_p)\}$, which proves the first claim. For the second one, assume that the virtual transforms with origin at q of the elements of $\mathcal{G}$ are all non-empty. Then so are all germs in $\mathcal{L}_q$, again by 7.2.5, and q is a base point of $\mathcal{L}$ against the hypothesis. $\diamond$

Assume now that $\mathcal{P}$ is a pencil with no fixed part, $\mathcal{P} = \{\alpha f + \beta g = 0 | (\alpha, \beta) \in \mathbb{C}^2 - \{(0,0)\}\}$, and let us denote by $\bar{\mathcal{P}}$ the linear system generated by $\mathcal{P}$, which clearly is neat and corresponds to the ideal (f, g). We define the *cluster of base points* of $\mathcal{P}$ as being that of $\bar{\mathcal{P}}$: $BP(\mathcal{P}) = BP(\bar{\mathcal{P}})$. Its points will be called the *base points* of $\mathcal{P}$. We will write $e_p(\mathcal{P}) = e_p(\bar{\mathcal{P}})$ and, if $p \in BP(\mathcal{P})$, still call this number the *multiplicity of the base point*. We have:

Proposition 7.2.7 *Virtual transforms being relative to the virtual multiplicities of $BP(\mathcal{P})$, for each $p \in BP(\mathcal{P})$ the virtual transforms with origin at p of the germs in $\mathcal{P}$ describe a pencil $\mathcal{P}_p$. By mapping each germ to its virtual transform one gets a projectivity $\mathcal{P} \longrightarrow \mathcal{P}_p$. Furthermore, all but at most one of the germs in $\mathcal{P}_p$ have multiplicity $e_p(\mathcal{P})$.*

PROOF: The first part of the claim obviously follows from 4.1.1. As for any pencil, all germs in $\mathcal{P}_p$ have the same multiplicity m, but for at most one, whose multiplicity may be bigger than m (2.7.2). By 7.2.5, the germs in $\mathcal{P}_p$ generate $\bar{\mathcal{P}}_p$ so that $m = e_p(\bar{\mathcal{P}}_p) = e_p(\bar{\mathcal{P}}) = e_p(\mathcal{P})$ as wanted. $\diamond$

Let $\xi : f = 0$ be any germ with origin at O. If E is the exceptional divisor of blowing up O, we will denote by $\tilde{\xi}.E$ the group of points of E

$$\sum_{q \in E} [\tilde{\xi}_q.E]_q q$$

and call it the group of points *cut out* by ξ on the first neighbourhood of O. Note that points are counted in $\tilde{\xi}.E$ according to intersection multiplicities. Identify E and the pencil of lines tangent to the surface S at O through the projectivity τ of 3.2.2. Then the initial form of f may be considered as a form of degree $e_O(\xi)$ on E and, by 3.2.2, its group of zeros is just $\tilde{\xi}.E$.

In order to deal with linear systems and pencils simultaneously, assume for a while that $\mathcal{F}$ is a linear family of germs at $\mathcal{O}$, defined (see section 2.7) by a linear subspace $F \subset \mathcal{O}$. Let e be the minimal multiplicity of the germs in $\mathcal{F}$. We have $F \subset \mathcal{M}^e$ and by composing this inclusion with the natural morphism onto the quotient, we get a linear morphism $\psi : F \longrightarrow \mathcal{M}^e/\mathcal{M}^{e+1}$. Let $f \in F$: if the germ $\xi : f = 0$ has multiplicity e, then $\psi(f)$ is the initial form of f, otherwise $\psi(f) = 0$. Assume that $\psi(f) \neq 0$: then $\psi(f)$ is a form of degree e on E whose group of zeros is $\tilde{\xi}.E$, as noticed above. If ξ describes the set of all germs in $\mathcal{F}$ of multiplicity e, the group $\tilde{\xi}.E$ describes the linear series on E defined by $\psi(F)$: we shall call it *the linear series cut out on E by $\mathcal{F}$*.

Assume now that $\mathcal{F}$ is either a pencil or a neat linear system, let p be a base point of $\mathcal{F}$ and call E_p the exceptional divisor of blowing up p: the *linear series that $\mathcal{F}$ cuts out on the first neighbourhood of p (or on E_p)* is, by definition, the linear series cut out on E_p by $\mathcal{F}_p$. We have:

Lemma 7.2.8 *The points of the fixed part of the linear series cut out by $\mathcal{F}$ on E_p are just the base points of $\mathcal{F}$ in the first neighbourhood of p.*

PROOF: It is clearly enough to deal with the case $p = O$ which in turn easily follows from the definition of base points. ⋄

Remark 7.2.9 Once again it is enough to check the case $p = O$ to see that there are two possibilities for the linear series cut out by a pencil $\mathcal{P}$ on the first neighbourhood of a base point p: either the germs in $\mathcal{P}_p$ have variable tangent cone (2.7.2), and then the linear series is one-dimensional and the map sending each germ to the group it cuts out is a projectivity, or these germs have constant tangent cone and then the linear series cut out by the pencil is zero-dimensional, i.e., a single group.

The main properties of base points of pencils and linear systems are stated in the next theorems:

Theorem 7.2.10 *If $\mathcal{P}$ is a pencil of germs at O with no fixed part, then*

(a) All germs in $\mathcal{P}$ go through $BP(\mathcal{P})$.

(b) All but finitely many germs in $\mathcal{P}$ go sharply through $BP(\mathcal{P})$.

(c) No point besides the base ones lies on two different germs in $\mathcal{P}$ going sharply through $BP(\mathcal{P})$.

(d) All but finitely many germs in $\mathcal{P}$ are reduced and have the same equisingularity type.

PROOF: Part (a) obviously follows from the definitions, as noticed in 7.2.4. Proposition 7.2.7 applied to all base points shows that all but finitely many germs go through $BP(\mathcal{P})$ with effective multiplicities equal to the virtual ones. To see non-singularity outside of the base points for all but finitely many of these germs, let us fix a base point p: the strict transform $\tilde{\xi}_p$ of any germ ξ going through $BP(\mathcal{P})$ with effective multiplicities equal to the virtual ones, cuts out on the first neighbourhood of p a group of the linear series ℓ cut out by $\mathcal{P}$. The fixed points of ℓ are base points by 7.2.8. If $\dim \ell = 0$, then all points of ξ in the first neighbourhood of p are base points and we have nothing to prove there. If $\dim \ell = 1$, then finitely many of the groups of the variable part of ℓ either contain a satellite point or have a multiple point (by 7.1.2), which means that only finitely many germs can have a singular point other than the base ones in the first neighbourhood of p. Since there are finitely many base points p, claim (b) has been proved. Claim (c) follows from 7.2.6 as any two different germs in $\mathcal{P}$ generate $\bar{\mathcal{P}}$ and (d) is a direct consequence of (b) and 4.2.6. ⋄

Remark 7.2.11 Theorem 7.2.10 characterizes $BP(\mathcal{P})$: its points are those belonging to all but finitely many germs in the pencil, and the virtual multiplicity of a point p is the effective multiplicity at p of all but finitely many germs in $\mathcal{P}$.

Remark 7.2.12 The reader may notice that 7.2.10 encloses a sort of local Bertini theorem for pencils, as it shows that the germs in a pencil do not have variable multiple infinitely near points.

We have for linear systems a result similar to 7.2.10, namely,

Theorem 7.2.13 *Let $\mathcal{L}$ be a neat linear system of germs of curve at O and T a finite set of points infinitely near to O, no one a base point of $\mathcal{L}$. All germs in $\mathcal{L}$ go through $BP(\mathcal{L})$. Generic germs in $\mathcal{L}$ go sharply through $BP(\mathcal{L})$ and miss all points in T and, in particular, they are reduced and have the same equisingularity type.*

PROOF: That all germs in $\mathcal{L}$ go through $BP(\mathcal{L})$ has been already noticed (7.2.4). Let I be the ideal defining $\mathcal{L}$. For each base point p let $\mathcal{K}(p)$ be the weighted cluster that consists of all points preceding p in $BP(\mathcal{L})$ with the same virtual multiplicities and the point p itself with its virtual multiplicity in $BP(\mathcal{L})$ increased by one. Then, by 7.2.6, $I - H_{\mathcal{K}(p)}$ is a non-empty Zariski-open set and therefore so is $I - \bigcup_p H_{\mathcal{K}(p)}$. This proves that generic germs in $\mathcal{L}$ go through $BP(\mathcal{L})$ with effective multiplicities equal to the virtual ones.

Now we will prove that generic germs miss all points in T. Clearly we need only consider the case in which each point in T is in the first neighbourhood of a base point. Let q be a point in T and denote by $\mathcal{K}(q)$ the weighted cluster that contains all points in $BP(\mathcal{L})$ with the same virtual multiplicities and furthermore the point q with virtual multiplicity one. Then, again by 7.2.6, $I - H_{\mathcal{K}(q)}$ is a non-empty Zariski-open set and the claim follows as above.

Now we can see that generic germs in $\mathcal{L}$ actually go sharply through $BP(\mathcal{L})$. For this, using what we have proved so far, we choose first a germ $\xi \in \mathcal{L}$ going through $BP(\mathcal{L})$ with effective multiplicities equal to the virtual ones. Then we choose a second germ $\zeta \in \mathcal{L}$ still going through all points in $BP(\mathcal{L})$ with effective multiplicities equal to the virtual ones and missing all non-base points on ξ which are in the first neighbourhood of some base point. We choose equations for ξ, ζ and fix a pencil $\mathcal{P}$ containing both germs: of course $\mathcal{P} \subset \mathcal{L}$. The reader should find no difficulty in proving that $\mathcal{P}$ has no fixed component and $BP(\mathcal{P}) = BP(\mathcal{L})$, after which it follows from 7.2.9 that there is a germ in $\mathcal{P}$, and hence in $\mathcal{L}$, going sharply through $BP(\mathcal{P}) = BP(\mathcal{L})$. By 4.2.7, the equations of the germs going sharply through $BP(\mathcal{L})$ describe a non-empty Zariski-open set V in $H_{BP(\mathcal{L})}$: we have just seen that $V \cap I$ is non-empty and hence the claim follows.

To close, that all germs going sharply through $BP(\mathcal{L})$ have the same equisingularity type follows from 4.2.6. ◊

Corollary 7.2.14 *All clusters of base points of pencils and neat linear systems are consistent.*

PROOF: Follows from 7.2.10, 7.2.13 and 4.2.2. ◊

Remark 7.2.15 Theorem 7.2.13 characterizes the points of $BP(\mathcal{L})$ as those lying on generic germs in $\mathcal{L}$, and their virtual multiplicities as the effective multiplicities of generic germs in $\mathcal{L}$.

Corollary 7.2.16 *A neat linear system $\mathcal{L}$ may be generated by finitely many germs going sharply through $BP(\mathcal{L})$, any two of them sharing no point other than the base ones.*

PROOF: Our argument follows the usual one for proving that ideals of noetherian rings are finitely generated. Let I be the ideal of $\mathcal{L}$. Assume we have determined finitely many $f_i \in I$, $i = 1, \ldots, r$ such that all germs $f_i = 0$ go sharply through $BP(\mathcal{L})$ and any two of them share base points only. For $r = 2$ this is easily achieved using 7.2.13. Then, in particular, the f_i, $i = 1, \ldots, r$, have no common factor and hence the ideal $(f_1, \ldots, f_r)$ is $\mathcal{M}$-primary (1.8.10). If $I = (f_1, \ldots, f_r)$, then we have a system of generators as claimed. If not, $V = I - (f_1, \ldots, f_r)$ is a non-empty Zariski-open subset of I. Let T be the set of points which are not base points, belong to some germ $f_i = 0$ and lie in the first neighbourhood of a base point: using 7.2.13 for such a T we get f_{r+1} in V so that the germ $f_{r+1} = 0$ goes sharply through $BP(\mathcal{L})$ and shares with $f_i = 0$ no points other than the base ones, for $i = 1, \ldots, r$. Then the f_i, $i = 1, \ldots, r+1$ define germs that go sharply through $BP(\mathcal{L})$ and from which no two share a non-base point. Furthermore the ideal $(f_1, \ldots, f_{r+1})$ is strictly bigger than $(f_1, \ldots, f_r)$. Since an infinite increasing sequence of ideals is not allowed because $\mathcal{O}$ is noetherian, the above procedure actually stops giving a set of generators of I as claimed. ◇

Now, we may compute the integer $\kappa(\mathcal{L})$ used for defining the cluster of base points:

Corollary 7.2.17 *If $\mathcal{L}$ is a neat linear system, then*

$$\kappa(\mathcal{L}) = \min\{[\zeta.\zeta']|\zeta, \zeta' \in \mathcal{L}\} = \sum_{p \in BP(\mathcal{L})} (e_p(\mathcal{L}))^2.$$

PROOF: Assume that $\kappa(\mathcal{L}) = [\zeta.\zeta']$. Using 7.2.1 and 7.2.13 one gets a germ $\xi \in \mathcal{L}$ going sharply through $BP\mathcal{L}$ and such that $\kappa(\mathcal{L}) = [\xi.\zeta']$. After this one gets in the same way another germ $\xi' \in \mathcal{L}$ still going sharply through $BP(\mathcal{L})$, sharing with ξ base points only, and such that $\kappa(\mathcal{L}) = [\xi.\xi']$. Then the Noether formula gives the claim. ◇

Corollary 7.2.18 *If $\mathcal{L}$ is a neat linear system, defined by an ideal I, then*

$$\sum_{p \in BP(\mathcal{L})} (e_p(\mathcal{L}))^2 \geq dim_{\mathbb{C}} \mathcal{O}/I \geq \sum_{p \in BP(\mathcal{L})} \frac{e_p(\mathcal{L})(e_p(\mathcal{L}) + 1)}{2}.$$

PROOF: If $f, g \in I$ define germs whose intersection multiplicity takes the minimal value $\kappa(\mathcal{L})$, then the claim follows from 7.2.17 and the inclusions $(f, g) \subset I \subset H_{BP(\mathcal{L})}$, using 3.11.10 and 4.7.1. ◇

The number $dim_{\mathbb{C}} \mathcal{O}/I$, which is defined for any linear system without fixed part, will be called the *codimension* of the linear system. The codimension of $\mathcal{L}$ may be understood as the number of (linearly independent) conditions that the equation of a germ ξ needs to satisfy in order to have $\xi \in \mathcal{L}$.

We leave the proof of the next lemma to the reader, who may argue as in the first part of the proof of 7.2.13.

Lemma 7.2.19 *If the germ* $\xi : f = 0$ *goes through a weighted cluster* $\mathcal{K}$ *with effective multiplicities equal to the virtual ones and a second germ* $\zeta : g = 0$, $\zeta \neq \xi$, *just goes through* $\mathcal{K}$, *then all but finitely many of the germs in the pencil* $\{\lambda f + \mu g = 0\}$ *go through* $\mathcal{K}$ *with effective multiplicities equal to the virtual ones.*

7.3 μ-constant equisingularity criteria

Let $\xi_t : f(x,y,t) = 0$, $|t| < \varepsilon$, $f \in \mathbb{C}\{x,y,t\}$ be an analytic family of germs of curve. A very deep theorem proved by Lê and Ramanujam ([52], see also [55],[49]), often named the *μ-constant theorem*, asserts that for $t \neq 0$ small enough, all germs ξ_t are equisingular, $\mu(\xi_t) \leq \mu(\xi_0)$ and the equality holds if and only if ξ_0 and ξ_t are equisingular. Unfortunately, proving this theorem is far beyond the scope of this book. Our main goal in this section is to prove a restricted version of it that applies to pencils and linear systems. It is worth noting that in this restricted version, constancy of the Milnor number implies that the particular germ goes sharply through the base points, which is stronger than being equisingular to generic germs.

To begin with, let us assume that $\mathcal{K} = (K, \nu)$ is a weighted cluster and denote its excesses by ρ_p, $p \in K$. We define the *Milnor number*, $\mu(\mathcal{K})$, of $\mathcal{K}$ by the formula

$$\mu(\mathcal{K}) = \sum_{p \in K} \nu_p(\nu_p - 1) - \sum_{p \in K} \rho_p + 1.$$

Naming $\mu(\mathcal{K})$ the Milnor number of $\mathcal{K}$ is justified by the next proposition:

Proposition 7.3.1 *If* $\mathcal{K}$ *is a consistent cluster and* ξ *is a non-empty germ of curve going sharply through* $\mathcal{K}$, *then* $\mu(\xi) = \mu(\mathcal{K})$.

PROOF: Immediate from 4.2.6 and the definition of $\mu(\xi)$ in section 6.4. ◇

Behaviour of the Milnor number by a tame unloading step (see section 4.7) is described next:

Lemma 7.3.2 *If* $\mathcal{K} = (K, \nu)$ *is a non-consistent weighted cluster with origin at* O, *and* $\mathcal{K}' = (K, \nu')$ *is the one it gives rise to by tame unloading on a point* $p \in K$, *then*

$$\mu(\mathcal{K}') - \mu(\mathcal{K}) = \ell + \ell' - 3 + r$$

where ℓ *is the number of points of* $\mathcal{K}$ *proximate to* p, ℓ' *is the number of points of* $\mathcal{K}$ *in the first neighbourhood of* p *and* r *is the number of satellite points in the first neighbourhood of* p *that do not belong to* $\mathcal{K}$.

Note that $\ell' + r$ is the number of points in the first neighbourhood of p that either are satellite or belong to $\mathcal{K}$.

PROOF OF 7.3.2: After the definitions, one may clearly write

$$\mu(\mathcal{K}) = 2c(\mathcal{K}) - 2\sum_{q\in K}\nu_q - \sum_{q\in K}\rho_q + 1$$

and similarly for $\mathcal{K}'$. Since the unloading is tame, $c(\mathcal{K}) = c(\mathcal{K}')$ (4.7.2) and a single unit of multiplicity is unloaded ($n = 1$). Then the definition of unloading directly gives $2\sum_q \nu_q - 2\sum_q \nu'_q = 2(\ell - 1)$. Lastly, again from the definitions, the variation of excesses by unloading is as follows:

$$\rho'_q - \rho_q = \begin{cases} \ell + 1 & \text{if } q = p \\ -1 & \text{if } q \text{ is maximal among the points proximate to } p \text{ in } K \\ -1 & \text{if } p \text{ is maximal among the points proximate to } q \text{ in } K \\ 0 & \text{otherwise.} \end{cases}$$

Since the points proximate to p describe totally ordered sequences from the first neighbourhood of p onwards, the number of maximal points among those proximate to p is ℓ'. On the other hand, p is the last point in K proximate to just r points q. After this the claim follows by adding up the differences computed above. $\diamond$

Keep all hypotheses and notations of 7.3.2 in the next two lemmas.

Lemma 7.3.3 *Assume that p is a satellite point and K contains either a free point in the first neighbourhood of p or at least two points proximate to p. Then*

$$\mu(\mathcal{K}') - \mu(\mathcal{K}) > 0.$$

PROOF: In the first case $\ell \geq 1$ and $\ell' + r \geq 3$ while in the second one $\ell \geq 2$ and $\ell' + r \geq 2$. So in both cases the claim follows from 7.3.2. $\diamond$

Lemma 7.3.4 *If p is neither the origin nor a maximal point of K and $\mu(\mathcal{K}') - \mu(\mathcal{K}) < 0$, then p is a free point, the satellite point in its first neighbourhood is its only proximate point in K and $\mu(\mathcal{K}') - \mu(\mathcal{K}) = -1$.*

PROOF: The point p being not maximal, $\ell \geq \ell' \geq 1$. Thus, by 7.3.2, $\ell = \ell' = 1$ and $r = 0$, from which the claim. $\diamond$

In the sequel we will call *strictly consistent* the consistent clusters that have no points of virtual multiplicity zero. Strictly consistent clusters will play an important role in the forthcoming sections 8.3 and 8.4. Next are a couple of remarks about them.

Remark 7.3.5 Since all points in a consistent cluster have non-negative multiplicities, as already seen in section 4.2, a cluster is strictly consistent if and only if all its virtual multiplicities are positive.

Remark 7.3.6 Let $\mathcal{K}' = (K', \nu)$ be a consistent cluster. Consistency clearly implies that if $\nu_p > 0$, then also $\nu_q > 0$ for all q preceding p. Thus one may drop from $\mathcal{K}'$ all 0-fold points in order to get a strictly consistent cluster $\mathcal{K}$ obviously equivalent to $\mathcal{K}'$.

The main result in this section reads as follows:

Theorem 7.3.7 (μ-constant for clusters) *Let $\mathcal{K} = (K, \nu)$ and $\mathcal{T} = (T, \tau)$ be strictly consistent clusters and assume that $H_{\mathcal{K}} \supset H_{\mathcal{T}}$. Then $\mu(\mathcal{K}) \leq \mu(\mathcal{T})$ and the equality holds if and only if $\mathcal{T}$ comes from $\mathcal{K}$ by adding free points of virtual multiplicity one, that is, $K \subset T$, $\nu_p = \tau_p$ if $p \in K$, and p is free and $\tau_p = 1$ if $p \in T - K$.*

PROOF: The *if* part of the second half of the claim being trivial, we will just prove the inequality $\mu(\mathcal{K}) \leq \mu(\mathcal{T})$ and that the equality $\mu(\mathcal{K}) = \mu(\mathcal{T})$ implies that $\mathcal{T}$ comes from $\mathcal{K}$ by adding free points virtually counted once.

In the case $H_{\mathcal{K}} = H_{\mathcal{T}}$ we have $\mathcal{K} = \mathcal{T}$ by 4.6.4 and then the claim is obviously satisfied. Thus we may proceed by induction on $\dim H_{\mathcal{K}}/H_{\mathcal{T}}$.

Case (a): Assume that $K \subset T$ and $\nu_p = \tau_p$ for all $p \in K$. Take $q \in T - K$ in the first neighbourhood of some point in K. Add to $\mathcal{K}$ the point q with assigned virtual multiplicity τ_q thus getting a new weighted cluster $\mathcal{K}_1$. Clearly $\mathcal{K}_1$ is strictly consistent (because $\mathcal{T}$ is), $H_{\mathcal{K}} \supset H_{\mathcal{K}_1} \supset H_{\mathcal{T}}$ and, by 4.2.2, $H_{\mathcal{K}} \neq H_{\mathcal{K}_1}$. Direct computation from the definition gives $\mu(\mathcal{K}_1) - \mu(\mathcal{K}) = \tau_q(\tau_q - 1)$ if q is a free point and $\mu(\mathcal{K}_1) - \mu(\mathcal{K}) = \tau_q^2$ if q is satellite, so that $\mu(\mathcal{K}_1) - \mu(\mathcal{K}) \geq 0$ and the equality holds if and only if q is free and $\tau_q = 1$. Then the proof ends by induction.

Case (b): Assume now that there is a $q \in K$ so that either $q \notin T$ or $\tau_q \neq \nu_q$. We will prove that in this case $\mu(\mathcal{K}) < \mu(\mathcal{T})$. It is not restrictive to assume that q is minimal among the points satisfying one of the above conditions, and hence that all points p preceding q belong to T and have $\tau_p = \nu_p$. Then necessarily $q \in T$, otherwise a germ going sharply through $\mathcal{T}$ and missing q (4.2.8) would not go through $\mathcal{K}$ against $H_{\mathcal{K}} \supset H_{\mathcal{T}}$. Similarly, if $\tau_q < \nu_q$ a germ going sharply through $\mathcal{T}$ would not go through $\mathcal{K}$. Thus we have $\tau_q > \nu_q$. Choose a free point q' in the first neighbourhood of q not already in K or T and define $\mathcal{K}_1$ by adding to $\mathcal{K}$ the point q' virtually counted once. The conditions for going through $\mathcal{K}$ or $\mathcal{T}$ at points preceding q being identical, the inequality $\tau_q > \nu_q$ ensures that all germs through $\mathcal{T}$ are going through $\mathcal{K}_1$ too, and so $H_{\mathcal{K}_1} \supset H_{\mathcal{T}}$. On the other hand it is clear that $H_{\mathcal{K}} \supset H_{\mathcal{K}_1}$ and, again by 4.2.8, $H_{\mathcal{K}} \neq H_{\mathcal{K}_1}$, at which point we let the proof fork into two subcases:

Subcase (b1): The excess ρ_q of $\mathcal{K}$ at q is positive. Then $\mathcal{K}_1$ is strictly consistent and by direct computation $\mu(\mathcal{K}) = \mu(\mathcal{K}_1)$. On the other hand, since $q' \notin T$, the induction hypothesis gives $\mu(\mathcal{K}_1) < \mu(\mathcal{T})$ and hence the claim.

Subcase (b2): $\rho_q = 0$. Now, its excess at q being -1, $\mathcal{K}_1$ is not consistent. On one hand we have $\dim \mathcal{O}/H_{\mathcal{K}_1} > \dim \mathcal{O}/H_{\mathcal{K}}$. On the other, by 4.7.3, direct computation and the codimension formula 4.7.1,

$$\dim \mathcal{O}/H_{\mathcal{K}_1} \leq c(\mathcal{K}_1) = c(\mathcal{K}) + 1 = \dim \mathcal{O}/H_{\mathcal{K}} + 1.$$

It follows thus that $\dim \mathcal{O}/H_{\mathcal{K}_1} = c(\mathcal{K}_1)$ and so, by 4.7.3, all unloadings performed from $\mathcal{K}_1$ are tame.

Make a first unloading on q (there is no other choice) to get a new cluster $\mathcal{K}'_1$ and let us show that it increases the Milnor number. Indeed, there is at least one point of $\mathcal{K}$ in the first neighbourhood of q, otherwise $\rho_q > 0$. Thus $\mathcal{K}_1$ has at least two points in the first neighbourhood of q, which gives $\ell + \ell' \geq 4$ in 7.3.2 and hence $\mu(\mathcal{K}'_1) - \mu(\mathcal{K}_1) > 0$.

Now we will build a finite sequence of weighted clusters,

$$\mathcal{K}'_1, \mathcal{K}'_2, \ldots, \mathcal{K}'_s,$$

by the modified unloading procedure of 4.6.3, namely $\mathcal{K}'_s$ is consistent and equivalent to $\mathcal{K}'_1$, for any j, $1 < j \leq s$, $\mathcal{K}'_j$ comes from $\mathcal{K}_j$ by an unloading step, and $\mathcal{K}_j$ is in turn obtained from $\mathcal{K}'_{j-1}$ by dropping all 0-fold points that are either maximal or followed by 0-fold points only. Since we have already seen that $\dim \mathcal{O}/H_{\mathcal{K}'_1} = \dim \mathcal{O}/H_{\mathcal{K}_1} = c(\mathcal{K}_1) = c(\mathcal{K}'_1)$ and, clearly, $c(\mathcal{K}'_{j-1}) = c(\mathcal{K}_j)$, arguing as in the proof of 4.7.3 we see that all the above unloading steps are tame, no matter which points we choose to unload on. In particular just a single unit of multiplicity is unloaded at each step.

It is worth noting that in no case is an unloading on a point preceding q performed. Indeed, assume that the first unloading on a point preceding q is performed on p. In the resulting cluster $\mathcal{K}'_i$ the point p has virtual multiplicity $\nu_p + 1 = \tau_p + 1$, while its preceding points p' still keep virtual multiplicities $\nu_{p'} = \tau_{p'}$. Then a germ going sharply through $\mathcal{T}$ would not go through $\mathcal{K}'_i$, against the fact that all germs through $\mathcal{T}$ go through $\mathcal{K}_1$ and $\mathcal{K}_1 \sim \mathcal{K}'_i$.

We need to make a special choice for certain of the above unloading steps in order to keep all virtual multiplicities non-negative. Inductively assume that $\mathcal{K}_j$ has non-negative virtual multiplicities, which is obviously true for $j = 2$. If $\mathcal{K}_j$ has no 0-fold point, we make no special choice, as unloading a single unit of multiplicity will not produce any negative multiplicity. Otherwise we choose to unload on a point p which is maximal among the 0-fold points of $\mathcal{K}_j$. Note that p is not maximal in $\mathcal{K}_j$ and all points after it have positive virtual multiplicity. This assures us not only that the excess at p is negative (and so unloading on p is allowed), but also that, as in the former case, no negative virtual multiplicity will appear after unloading.

Once it is guaranteed that no negative virtual multiplicity does appear in the whole procedure, it is clear that no unloading is performed on a maximal point. We will see next that no unloading is performed on the origin O either. If $O < q$ this has been already seen, so assume that $O = q$. Take $\xi = \xi_1 + \xi_2$ where ξ_1 goes sharply through $\mathcal{K}$ and ξ_2 is smooth. It is clear that ξ goes through $\mathcal{K}_1$ and $e_\xi = \nu_O + 1$. Then ξ goes through $\mathcal{K}'_1$ and has effective multiplicity at O equal to the virtual one. This proves that no cluster equivalent to $\mathcal{K}'_1$ may have a bigger virtual multiplicity at O and so no further unloading on O may be performed.

Clearly $\mu(\mathcal{K}'_i) = \mu(\mathcal{K}_{i+1})$. Thus, in order to see that $\mu(\mathcal{K}'_1) \leq \mu(\mathcal{K}'_s)$, we only need to care about the differences $\mu(\mathcal{K}'_i) - \mu(\mathcal{K}_i)$. Since no unloading is made on either O or a maximal point, the only possibility for having $\mu(\mathcal{K}'_i) - \mu(\mathcal{K}_i) < 0$ is that already described in 7.3.4. Assume thus that there is a free point $p_1 \in \mathcal{K}$ whose only proximate point in $\mathcal{K}$ is the satellite point in its first neighbourhood, we call it p_2. Call p_0 the point immediately preceding p_1. For $i = 3, \ldots, k$ and as far as it is the only point proximate to p_{i-1} in K, call p_i the point proximate to p_0 in the $(i-1)$-th neighbourhood of p_1: thus $k \geq 2$ and, unless p_k is maximal, there are in K either two points proximate to p_k, or a free point in its first neighbourhood. Assume that some of the unloadings giving rise to the clusters $\mathcal{K}'_i$, $i > 1$, are performed on p_1: we will see next that for each such unloading, at least one unloading has been already performed on p_k, which is thus, in particular, not maximal. Then 7.3.3 and 7.3.4 prove that the Milnor number does not decrease by the combined effects of both unloadings.

Let us write ν^i_j for the virtual multiplicity of p_j in $\mathcal{K}_i$ (take $\mathcal{K} = \mathcal{K}_0$ and $\nu^i_j = 0$ if p_j has been already dropped). Put $u_i = \max\{\nu^i_2 - \nu^i_1, \ldots, \nu^i_k - \nu^i_1\}$. If $i > 1$ there is no point in $\mathcal{K}_i$ proximate to p_1 other than p_2 and therefore the unloading giving rise to $\mathcal{K}'_i$ is not performed on p_1 as far as $u_i \leq 0$. Note first that, $\mathcal{K}$ being consistent, $u_0 \leq 0$. Since we assume that an unloading on p_1 is performed, either $q < p_1$ or $q = p_1$: in any case it easy to see that after unloading on q, still $u_1 \leq 0$. Since u_i is not modified by unloading on either p_0 or any point $p' \neq p_j$, $j = 0, \ldots, k$ and it does not increase by unloading on p_j $j = 2, \ldots, k-1$, it is clear that at least one unloading on p_k needs to be performed before unloading on p_1. More precisely, since u_i drops by at least one unit by unloading on p_1 and is increased by at most one unit by unloading on p_k, it is clear that at least one unloading on p_k is performed before each unloading on p_1, as already claimed.

Points p_k associated with different points p_1 being different, the argument may be repeated to prove that $\mu(\mathcal{K}'_1) \leq \mu(\mathcal{K}'_s)$. All together we have constructed a consistent cluster $\mathcal{K}'_s$ such that $H_\mathcal{K} \supset H_{\mathcal{K}'_s} \supset H_\mathcal{T}$, $H_\mathcal{K} \neq H_{\mathcal{K}'_s}$ and $\mu(\mathcal{K}'_s) > \mu(\mathcal{K})$. Dropping from $\mathcal{K}'_s$ all points of virtual multiplicity zero we get an equivalent strictly consistent cluster $\mathcal{K}_{s+1}$ (see 7.3.6) with the same Milnor number. Thus still $H_\mathcal{K} \supset H_{\mathcal{K}_{s+1}} \supset H_\mathcal{T}$, $H_\mathcal{K} \neq H_{\mathcal{K}_{s+1}}$ and $\mu(\mathcal{K}_{s+1}) > \mu(\mathcal{K})$, after which the proof ends by induction as in former cases. ◇

Corollary 7.3.8 (μ-constant for pencils) *If $\mathcal{P}$ is a pencil without fixed part and $\xi \in \mathcal{P}$, then $\mu(BP(\mathcal{P})) \leq \mu(\xi)$ and the equality holds if and only if ξ goes sharply through $BP(\mathcal{P})$.*

PROOF: We assume that ξ is reduced, as otherwise $\mu(\xi) = \infty$ and the claim is obvious. Put $\mathcal{K} = (K, \nu) = BP(\mathcal{P})$. Let T be the set of points on ξ that either belong to $\mathcal{K}$ or are singular points of ξ. Define the weighted cluster $\mathcal{T}$ by taking $\tau_p = e_p(\xi)$ as the virtual multiplicity of each $p \in T$. Clearly ξ goes sharply through $\mathcal{T}$ and $\mathcal{T}$ is strictly consistent. Next we check that $H_\mathcal{K} \supset H_\mathcal{T}$: in fact, by 7.2.16, it is enough to see that any germ going sharply through $\mathcal{T}$ goes also through $\mathcal{K}$. If ζ is such a germ, $e_p(\zeta) \geq e_p(\xi)$ for all $p \in K$ as one has equality if $p \in T$ and $e_p(\xi) = 0$ otherwise. Since ξ actually goes through $\mathcal{K}$, these inequalities imply that also ζ goes through $\mathcal{K}$, as wanted.

Now 7.3.7 applies to $\mathcal{K}$ and $\mathcal{T}$ giving $\mu(\mathcal{K}) \leq \mu(\mathcal{T}) = \mu(\xi)$ as wanted. If the equality holds, still by 7.3.7, $\mathcal{T}$ comes from $\mathcal{K}$ just by adding free points of virtual multiplicity one, and then ξ goes sharply through $\mathcal{K}$ just because it is going sharply through $\mathcal{T}$. The converse is obvious from 7.3.1. $\diamond$

The next corollary is a direct consequence of 7.3.8. The reader may notice that it is a bit weaker, as for a couple of germs, going sharply through the same weighted cluster is stronger that just being equisingular.

Corollary 7.3.9 *Let $f, g \in \mathcal{O}$ be non-invertible germs of analytic function with no common factor and consider the family of germs at O $\{\xi_t : f + tg = 0 | t \in \mathbb{C}\}$. There is an $\varepsilon \in \mathbb{R}$, $\varepsilon > 0$ so that all germs ξ_t, $0 < |t| < \varepsilon$, are reduced, have the same equisingularity type and (hence) the same Milnor number μ. Furthermore, $\mu \leq \mu(\xi_0)$ and the equality holds if and only if ξ_0 is reduced and equisingular to the germs ξ_t, $0 < |t| < \varepsilon$.*

PROOF: The germs $\lambda f + \mu g = 0$, $(\lambda, \mu) \in \mathbb{C}^2 - \{(0,0)\}$, describe a pencil $\mathcal{P}$ without fixed part. Then the first half of the claim follows from 7.2.10, $\mu = \mu(BP(\mathcal{P}))$ by 7.3.1 and 7.2.10 again, after which 7.3.8 completes the proof. $\diamond$

A similar result may be stated for linear systems:

Corollary 7.3.10 (μ-constant for linear systems) *If $\mathcal{L}$ is a neat linear system and $\xi \in \mathcal{L}$, then $\mu(\xi) \geq \mu(BP(\mathcal{L}))$ and the equality holds if and only if ξ goes sharply through $BP(\mathcal{L})$*

PROOF: Let f be an equation of ξ. Since $\min\{[\zeta.\xi] | \zeta \in \mathcal{L}\}$ is finite by 2.7.1 and reached by generic germs in $\mathcal{L}$ by 7.2.1, generic germs in $\mathcal{L}$ share no component with ξ. Therefore, by 7.2.13 we may choose $\zeta : g = 0$ in $\mathcal{L}$ going sharply through $BP(\mathcal{L})$ and sharing no branch with ξ. Then the pencil $\mathcal{P} : \lambda f + \lambda' g = 0$ has no fixed part, all its germs go through $BP(\mathcal{L})$, and it contains a germ going sharply through $BP(\mathcal{L})$. Again by 7.2.13, all but finitely many germs in $\mathcal{P}$ go sharply through $BP(\mathcal{L})$ and so, by 7.2.10, there is a germ going sharply through both $BP(\mathcal{L})$ and $BP(\mathcal{P})$. This clearly implies that the points belonging to both clusters have in them equal virtual multiplicities, while the points belonging to just one of the clusters are free and have virtual multiplicity one. It follows that $\mu(BP(\mathcal{L})) = \mu(BP(\mathcal{P}))$ and hence, by 7.3.8, $\mu(\xi) \geq \mu(BP(\mathcal{P})) = \mu(BP(\mathcal{L}))$, as claimed.

Furthermore, since the clusters $BP(\mathcal{L})$ and $BP(\mathcal{P})$ differ only in free points of virtual multiplicity one and ξ goes through both of them, it is easy to check that ξ goes sharply through one of them if and only if it goes sharply through the other, after which the remaining part of the claim follows from 7.3.8. $\diamond$

Let $\mathcal{L}$ be either a pencil without fixed part or a neat linear system. A germ of curve in $\mathcal{L}$ is called *special* if and only if it does not go sharply through $BP(\mathcal{L})$. Generic germs in a neat linear system and all but finitely many germs in a pencil without fixed part are non-special by 7.2.13 and 7.2.10. The next corollary shows that special germs are not equisingular to the non-special ones.

Corollary 7.3.11 *Let* $\mathcal{L}$ *be, as above, either a pencil without fixed part or a neat linear system, and let* $\xi \in \mathcal{L}$ *be a non-special germ. A germ* $\zeta \in \mathcal{L}$ *is special if and only if it is not equisingular to* ξ.

PROOF: The *if* part obviously follows from 4.2.6. Conversely, if ζ is special, then $\mu(\zeta) > \mu(BP(\mathcal{L})) = \mu(\xi)$ by either 7.3.8 or 7.3.10 and 7.3.1. Therefore ξ and ζ are not equisingular. ◇

7.4 Special germs and jacobian determinants

As in preceding sections, fix a point O on a surface S and write $\mathcal{O} = \mathcal{O}_{S,O}$. Let $\mathcal{P} = \{\alpha f + \beta g = 0\}$ be a pencil of germs of curve at O with no fixed part, and consider the jacobian determinant $j = j(f,g) = \partial(f,g)/\partial(x,y) \in \mathcal{O}$, x, y local coordinates at O. In this section we will see that $\theta : j = 0$ is a germ of curve at O whose branches are closely related to the special germs of $\mathcal{P}$.

The next formula, due to Delgado ([28]), is the key to the results in this section. Assume that local coordinates x, y at O are fixed:

Proposition 7.4.1 *Let* $\mathcal{P} = \{\alpha f + \beta g = 0\}$, $(\alpha, \beta) \in \mathbb{C}^2 - \{(0,0)\}$, *be a pencil at* O *with no fixed part and* $\xi \in \mathcal{P}$. *For any branch* γ *of* ξ,

$$o_\gamma j(f,g) = [\gamma.P_h(\xi)] - [\gamma.\eta] + [\gamma.\zeta],$$

where $j(f,g) = \partial(f,g)/\partial(x,y)$, ζ *is any germ in* $\mathcal{P}$ *different from* ξ *and* $h \in \mathcal{O}$ *defines a smooth germ* $\eta : h = 0$, $\eta \neq \gamma$.

PROOF: Using different coordinates just multiplies $j(f,g)$ by an invertible factor, hence $o_\gamma j(f,g)$ does not depend on the local coordinates and we may take them so that $h = x$. Furthermore, since $j(f,g)$ obviously remains unmodified if independent $\mathbb{C}$-linear combinations of f, g are taken instead of f, g themselves, we may also assume that ξ and ζ are $f = 0$ and $g = 0$, respectively. Let $t \mapsto (x(t), y(t))$ be a uniformizing map of a representative of γ. Along the proof we will write $\bar{z} = z(x(t), y(t)) \in \mathbb{C}\{t\}$ for any $z \in \mathcal{O}$, and also $\dot{z} = d\bar{z}/dt$. Note that, by hypothesis, $\gamma \neq \eta$, $\bar{x} \neq 0$ and hence also $\dot{x} \neq 0$, because $\bar{x}(0) = 0$. From $\bar{f} = 0$ we get

$$0 = \overline{\frac{\partial f}{\partial x}}\dot{x} + \overline{\frac{\partial f}{\partial y}}\dot{y}$$

which gives

$$\overline{j(f,g)} = \overline{\frac{\partial f}{\partial x}}\,\overline{\frac{\partial g}{\partial y}} - \overline{\frac{\partial f}{\partial y}}\,\overline{\frac{\partial g}{\partial x}}$$

$$= -\frac{\dot{y}}{\dot{x}}\overline{\frac{\partial f}{\partial y}}\,\overline{\frac{\partial g}{\partial y}} - \overline{\frac{\partial f}{\partial y}}\,\overline{\frac{\partial g}{\partial x}}$$

$$= -\frac{1}{\dot{x}}\overline{\frac{\partial f}{\partial y}}\left(\overline{\frac{\partial g}{\partial y}}\dot{y} + \overline{\frac{\partial g}{\partial x}}\dot{x}\right).$$

On the other hand, since g is not invertible, $o_t\bar{g} > 0$ and so

$$[\gamma.\zeta] - 1 = o_t\bar{g} - 1 = o_t\dot{g} = o_t\left(\overline{\frac{\partial g}{\partial x}}\dot{x} + \overline{\frac{\partial g}{\partial y}}\dot{y}\right).$$

Both equalities together give

$$o_\gamma j(f.g) = o_t\overline{j(f,g)} = [\gamma.P_x(\xi)] - [\gamma.\eta] + [\gamma.\zeta]$$

as claimed. ◇

Remark 7.4.2 The difference $[\gamma.P_h(\xi)] - [\gamma.\eta]$ does not depend on the choice of h provided $\eta \neq \gamma$, as, by 6.3.1,

$$[\gamma.P_h(\xi)] - [\gamma.\eta] = \sum_{p\in\gamma} e_p(\gamma)(e_p(\xi) - 1) - 1.$$

Also $[\gamma.\zeta]$ is independent of the choice of ζ, $\zeta \neq \xi$, because $[\gamma.\zeta] < \infty = [\gamma.\xi]$ and then 2.7.3 applies.

One may choose ξ reduced (by 7.2.10), in which case $o_\gamma j(f,g) < \infty$. Hence, in particular, $j(f,g) \neq 0$ and therefore defines a germ of curve $\theta : j(f,g) = 0$ that will be called a *jacobian germ* of the pencil $\mathcal{P}$. Clearly, the jacobian germ θ does not depend on the choice of the coordinates, but the reader may notice that it depends on the choice of the equations f, g: equations $uf, ug, u \in \mathcal{O}$ invertible, give rise to the same pencil and, in general, to a different jacobian germ. A jacobian germ may be empty: it follows directly from its definition that this occurs if and only if the germs in the pencil are smooth and have variable tangent.

In the next corollary we just rewrite 7.4.1 using 7.4.2 and a jacobian germ.

Corollary 7.4.3 *If $\mathcal{P}$ is a pencil without fixed part, $\xi \in \mathcal{P}$, γ is a branch of ξ and θ is a jacobian germ of $\mathcal{P}$, then*

$$[\gamma.\theta] = \sum_{p\in\gamma} e_p(\gamma)(e_p(\xi) - 1) + [\gamma.\zeta] - 1,$$

where ζ is any germ in $\mathcal{P}$ different from ξ.

Since by definition polar germs are in particular jacobian germs, 7.4.3 generalizes 6.3.1. As seen in 2.7.4, the intersection multiplicity $[\xi.\zeta]$, $\xi \neq \zeta$, depends only on $\mathcal{P}$ and not on ξ or ζ, therefore we write it $[\xi.\zeta] = \kappa(\mathcal{P})$. A nicer equality easily comes out from 7.4.3, namely:

Corollary 7.4.4 *If $\mathcal{P}$ is a pencil with no fixed part and θ a jacobian germ of $\mathcal{P}$, for any $\xi \in \mathcal{P}$,*

$$[\xi.\theta] = \mu(\xi) + \kappa(\mathcal{P}) - 1.$$

PROOF: Just add up the equalities of 7.4.3 for the different branches of ξ using the definition of $\mu(\xi)$ given in section 6.4. ◇

Now it is enough to use 7.3.8 to get a characterization of the special germs in $\mathcal{P}$:

Corollary 7.4.5 *A germ $\xi \in \mathcal{P}$ is special if and only if $[\xi.\theta]$ is not minimal among the intersection multiplicities $[\zeta.\theta]$, $\zeta \in \mathcal{P}$.*

We may be a bit more precise. We know from 2.7.3 that for each branch v of θ, all germs in $\mathcal{P}$ have the same intersection multiplicity with v but for one germ whose intersection multiplicity is higher. So if we associate with each branch v of θ the only germ $\xi_v \in \mathcal{P}$ having higher intersection multiplicity with v, we see that the germs ξ_v have higher intersection with θ and therefore are special, while the remaining ones have not, and hence are non-special. Summarizing, notations and hypothesis being as above:

Corollary 7.4.6 *By mapping each branch of θ to the only germ in $\mathcal{P}$ that has higher intersection multiplicity with it, we get a map from the set of branches of θ onto the set of special germs in $\mathcal{P}$. In particular the number of branches of θ is an upper bound for the number of special germs in $\mathcal{P}$.*

We close this section with an example. Consider the pencil $\alpha f + \beta x^n = 0$ for a fixed $n > 1$, $f \notin (x)$. Its jacobian $\theta : x^{n-1}\partial f/\partial y = 0$ is composed of the y-axis $\eta : x = 0$ and the polar $P_x(\xi) : \partial f/\partial y = 0$ of $\xi : f = 0$. The germ $x^n = 0$ is special and corresponds to η as a branch of θ, while it corresponds also to the branches of $P_x(\xi)$, and therefore there is no other special germ, if and only if $n[\gamma.\eta] > [\gamma.\xi]$ for all branches γ of $P_x(\xi)$. If ξ is reduced and the y-axis is not tangent to ξ, this is a particular case of the forthcoming theorem 7.6.1. The reader may see in particular that the map of 7.4.6 need not be injective.

7.5 E-sufficiency degree

Still denote by $\mathcal{O} = \mathcal{O}_{S,O}$ the local ring of the surface S at O and by $\mathcal{M} = \mathcal{M}_{S,O}$ its maximal ideal. Let $\xi : f = 0$ be a non-empty and reduced germ of curve at O. A positive integer n is said to be E-*sufficient* for ξ (where E stands for equisingularity) if and only all $h \in f + \mathcal{M}^n$ are non-zero and define a reduced germ equisingular to ξ. The definition clearly does not depend on the particular equation f of ξ, as for any invertible $u \in \mathcal{O}$, the equations in $uf + \mathcal{M}^n = u(f + \mathcal{M}^n)$ clearly define the same germs as those in $f + \mathcal{M}^n$. The reader may notice that to say that n is E-sufficient for ξ means that the equisingularity class of ξ is determined by the class mod $\mathcal{M}^n$ of its equation f or, if $f = \sum_{i,j=1}^{\infty} a_{i,j}x^i y^j$, by the partial sum $f = \sum_{i,j=1}^{n-1} a_{i,j}x^i y^j$. It is also clear from the definition that if n is E-sufficient for ξ, then any $n' > n$ is E-sufficient for ξ too.

Since equisingularity may be understood as a topological equivalence of singularities, in current literature C^0-*sufficient* is often said for E-sufficient, C^0 standing for continuous or topological.

If there exist positive integers that are E-sufficient for ξ, then their minimum is called the E-*sufficiency degree* (or C^0 sufficiency degree) of ξ. In the present section we prove the existence of E-sufficient integers for any reduced germ and determine the E-sufficiency degree of an arbitrary reduced germ from its equisingularity class. Our argument is based on virtual multiplicities and pencils, it has been translated into terms of exceptional divisors and dual graphs in [54].

In fact, it is easy to prove that E-sufficient integers actually exist for any reduced germ. Indeed, if ξ is reduced the whole cluster $\mathcal{S}(\xi)$ (see section 3.8) may be computed using finitely many terms of an equation f of ξ, either by performing successive blowing-ups or by using the Newton–Puiseux algorithm. Since $\mathcal{S}(\xi)$ determines the equisingularity class of ξ, it follows that the equisingularity class of ξ does not depend on the monomials of f from a certain degree onwards, hence such a degree is E-sufficient for ξ. Anyway we do not need to pursue this argument here, as the existence of E-sufficient integers will follow from the determination of the degree of E-sufficiency, see 7.5.2 below.

Clearly, the E-sufficiency degree of a smooth germ is two. To avoid trivialities, for the remaining part of this section and throughout the next one we will assume that the germ ξ is singular.

Let K be the cluster of all singular points on the reduced germ $\xi : f = 0$: thus $p \in K$ if and only if p itself or a point infinitely near to p on ξ is either multiple or satellite. For any $p \in K$ we write $e_p = e_p(\xi)$ for the multiplicity of p on ξ and take these effective multiplicities as virtual multiplicities for the points of K, thus defining a consistent cluster $\mathcal{K} = (K, e)$ we call the *weighted cluster of singular points* of ξ. We will use vectors and matrices indexed by K, as in section 4.5. In particular we will write $e = (e_p)_{p \in K}$ for the column vector of effective multiplicities of ξ and still denote by $P = P_K$ the proximity matrix of K, as introduced in section 4.5. If $p \in K$, we will write 1_p for the K-indexed column vector whose q-th component is 1 if $q = p$ and 0 otherwise.

Some points in K need to be distinguished because they will play a different role in the sequel. Thus, let $T \subset K$ be defined by the following property: $p \in T$ if and only if p is free, ξ has a single free point p' in the first neighbourhood of p and furthermore p' is non-singular for ξ. Of course one may have $T = \emptyset$. Our main result in this section characterizes the E-sufficient integers in terms of the weighted cluster $\mathcal{K}$ of singular points of ξ.

Theorem 7.5.1 *Let $u = (u_p)$ be the vector $u = P^{-1}(n1_O - e)$. Then the following conditions are equivalent:*

(i) $u_p \geq 0$ for $p \in T$ and $u_p > 0$ for $p \in K - T$.

(ii) *For any $g \in \mathcal{M}^n$, $f + g \neq 0$ and the germ of curve $\zeta : f + g = 0$ goes sharply through $\mathcal{K}$.*

(iii) n *is E-sufficient for ξ.*

The reader may notice that, by 4.5.2, the condition (i) just says that any germ τ with multiplicity non-less than n goes through $\mathcal{K}$ and its virtual transforms have multiplicities strictly greater than the virtual ones at all points in

$K - T$. Notice also that condition (ii) says that all germs $f + g = 0$, $g \in \mathcal{M}^n$ have the same weighted cluster of singular points.

PROOF OF 7.5.3: That (ii) implies (iii) is clear, as if (ii) is satisfied, then both ξ and any $\zeta : f + g = 0$, $g \in \mathcal{M}^n$, go sharply through $\mathcal{K}$ and therefore are equisingular by 4.2.6.

Next we will prove that (i) implies (ii). Since $O \notin T$, $u_O > 0$ which forces (by 4.5.8) $n > e_O$ and consequently $f + g \neq 0$ for any $g \in \mathcal{M}^n$. The case $g = 0$ being obvious, we assume in the sequel $g \neq 0$ and call τ the germ $g = 0$.

First of all, let us see that τ goes through $\mathcal{K}$. By the hypothesis $g \in \mathcal{M}^n$ one has $e_K(\tau) \geq n1_O$. Since, by 4.5.8, all entries of P^{-1} are non-negative, $P^{-1}e_K(\tau) \geq P^{-1}n1_O$ and hence

$$u_K(\tau) = P^{-1}(e_K(\tau) - e) \geq P^{-1}(n1_O - e) = u. \tag{7.1}$$

All components of u being non-negative by hypothesis, τ actually goes through $\mathcal{K}$ by 4.5.2.

Fix $p \in K$ and assume, by induction, that for all q preceding p we have $e_q(\zeta) = e_q$ and furthermore all points on ζ in the first neighbourhood of q and not in K are non-singular for ζ. In the sequel all virtual transforms will be taken with respect to the system of virtual multiplicities $\{e_p\}_{p \in K}$. Denote by $\tilde{\eta}_p$ and $\breve{\eta}_p$, respectively, the strict and virtual transforms with origin at p of any germ η going through $\mathcal{K}$. Since the virtual multiplicities are the effective multiplicities of ξ, clearly $\tilde{\xi}_p = \breve{\xi}_p$ and by the induction hypothesis also $\tilde{\zeta}_p = \breve{\zeta}_p$. Then, by 4.1.1, one may take an equation of $\breve{\zeta}_p$ of the form $\breve{f}_p + \breve{g}_p$ where $\breve{f}_p$ and $\breve{g}_p$ are equations of $\tilde{\xi}_p = \breve{\xi}_p$ and $\breve{\tau}_p$ respectively. Furthermore we know from 4.5.2 and inequality 7.1 above that

$$e_p(\breve{\tau}_p) - e_p = u_p(\tau) \geq u_p. \tag{7.2}$$

Assume first that $p \in K - T$: then $u_p > 0$ and hence $e_p(\breve{\tau}_p) > e_p = e_p(\tilde{\xi}_p)$. Then, by 2.7.2, all germs $\breve{f}_p + \lambda\breve{g}_p = 0$, and in particular $\tilde{\xi}_p$ and $\tilde{\zeta}_p$, have the same multiplicity and tangent cone at p. This shows that $e_p(\zeta) = e_p$. Furthermore, if E_p is the exceptional divisor of blowing up p, since both $\tilde{\xi}_p$ and $\tilde{\zeta}_p$ have the same tangent cone, their strict transforms cut E_p at the same points and with the same multiplicities (by 3.2.2). In particular, for the points not in K this intersection multiplicity is one and the points themselves are free because ξ has no singularities outside of K (4.2.5), which shows that ζ has no singular points infinitely near to p outside of K.

Assume now that $p \in T$ and that $e_p(\tau) > 0$. Then $e_K(\tau) - n1_O \geq 0$ and furthermore, since $p \neq O$, its p-th component is strictly positive. This is still true for $P^{-1}(e_K(\tau) - n1_O)$, as all entries of P^{-1} are non-negative and those on the diagonal are equal to one (4.5.8). Hence, the p-th component of the vectorial inequality 7.1 above is a strict scalar inequality, namely $u_p(\tau) > u_p$. Since by hypothesis $u_p \geq 0$, we get $u_p(\tau) > 0$ after which the claim is reached arguing just as in the former case $p \in K - T$.

Lastly assume that still $p \in T$ but this time $e_p(\tau) = 0$. Since τ goes through K, one has $e_p(\check{\tau}_p) \geq e_p$. If $e_p(\check{\tau}_p) > e_p$ we conclude just as in the preceding cases. We assume thus $e_p(\check{\tau}_p) = e_p$. Since p is a free point, let us take local coordinates x, y at p so that x is a local equation of the exceptional divisor and the only free point on ξ in the first neighbourhood of p is on $y = 0$. Then, by 3.2.2, the equation $\check{f}_p$ of $\tilde{\xi}_p$ has initial form $ax^{e_p-1}y$, $a \in \mathbb{C} - \{0\}$. Let p_1 be the only satellite point in the first neighbourhood of p: it corresponds to the direction of $x = 0$. Notice that $e_p > 1$, otherwise, by the hypothesis $p \in T$, p would be a non-singular point of ξ against the definition of K: then p_1 lies on ξ and hence belongs to K, as it is a satellite point. On the other hand, since τ does not (effectively) go through p, the germ $\check{\tau}_p$ has the germ of the exceptional divisor as its only irreducible component, and so the initial form of $\check{g}_p$ is bx^{e_p} for a suitable $b \in \mathbb{C} - \{0\}$. It follows that the tangent cone to $\tilde{\zeta}_p = \check{\zeta}_p$ has equation $ax^{e_p-1}y + bx^{e_p} = 0$. Thus, clearly $e_p(\zeta) = e_p$ as wanted. Furthermore, since, as seen above, $p_1 \in K$, ζ has a single point besides those in K in the first neighbourhood of p, namely the point corresponding to the principal tangent $ay + bx = 0$ to $\tilde{\zeta}_p$: since this tangent has multiplicity one, such a point is non-singular for ζ, by 3.2.2 and 3.8.1.

It remains to prove that (iii) implies (i): to this end, assume that there is a point $p \in K$ such that either $u_p \leq 0$ if $p \notin T$ or $u_p < 0$ if $p \in T$. It is clearly non-restrictive to assume that no point preceding p satisfies the same condition, otherwise such a point would be taken instead of p. Thus we assume that for all q preceding p, either $u_q > 0$ if $q \in K - T$, or $u_q \geq 0$ if $q \in T$. Use all notations as before and fix any non-zero $g \in \mathcal{M}^n - \mathcal{M}^{n+1}$ such that the germ $\tau : g = 0$ misses all points on ξ in the first neighbourhood of O: then $e_K(\tau) = n1_O$. In the sequel we will prove that the pencil $\mathcal{P} = \{\alpha f + \beta g = 0\}$ contains a special germ other than τ, say $\zeta^\rho : f + \rho g = 0$. After this the claim follows, as then, by 7.3.11, either ξ itself is special and therefore non-equisingular to $\zeta^\lambda : f + \lambda g = 0$ for all but finitely many λ, or ξ is non-special and hence non-equisingular to the special germ ζ^ρ.

First of all notice that since we are assuming that $u_q \geq 0$ for all q preceding p and even $u_q > 0$ if $q \in K - T$, the same argument used above applies to λg to show that $e_q(\zeta^\lambda) = e_q$ for any λ and all q preceding p. Thus the virtual and strict transforms of ζ^λ with origin at p agree, $\check{\zeta}_p^\lambda = \tilde{\zeta}_p^\lambda$, and of course still $\check{\xi}_p = \tilde{\xi}_p$. Hence $\check{\zeta}_p^\lambda$ belongs to the pencil $\mathcal{P}_p$ and one may choose equations $\check{f}$ of $\tilde{\xi}_p$ and $\check{g}_p$ of $\check{\tau}_p$ so that $\check{\zeta}_p^\lambda$ has equation $\check{f}_p + \lambda\check{g}_p$. Furthermore, as $e_K(\tau) = n1_O$, this time we have equality in 7.1,

$$u_K(\tau) = P^{-1}(e_K(\tau) - e) = P^{-1}(n1_O - e) = u$$

and so, by 4.3.3, $e_p(\check{\tau}_p) = e_p + u_p$.

Assume first that $u_p < 0$. Then $e_p(\check{\tau}_p) < e_p(\tilde{\xi}_p)$ and thus for any $\lambda \neq 0$, $e_p(\zeta^\lambda) = e_p(\tilde{\zeta}_p^\lambda) = e_p(\check{\tau}_p) < e_p(\tilde{\xi}_p) = e_p(\xi)$, which shows that ξ is special in $\mathcal{P}$.

Assume now that $u_p = 0$, which in particular implies that $p \notin T$. We have $e_p(\check{\tau}_p) = e_p$ and so both the initial forms of $\check{f}_p$ and $\check{g}_p$ have degree e_p. Call F and G the initial forms of $\check{f}_p$ and $\check{g}_p$, respectively, and put $H = \gcd(F, G)$,

$F = F'H$, $G = G'H$, $d = \deg(F') = \deg(G')$. We will distinguish four cases, namely:

Case 1, $d = 0$: F and G are proportional and so, for a suitable λ, F and G cancel in $\check{f}_p + \lambda \check{g}_p$. Then $e_p(\zeta^\lambda) = e_p(\tilde{\zeta}_p^\lambda) > e_p(\tilde{\xi}_p) = e_p(\xi)$ and so either ξ or ζ^λ is special.

Otherwise the pencil $\mathcal{P} : \mu \check{f}_p + \lambda \check{g}_p = 0$ cuts out on the first neighbourhood of p a linear series with fixed part $H = 0$ and variable part, of dimension 1 and degree d, $\ell_d^1 : \mu F' + \lambda G' = 0$. Notice that different germs in $\mathcal{P}$ give rise to different groups of ℓ_d^1, by 7.2.9, and that, as for any one-dimensional linear series without fixed points, different groups of ℓ_d^1 are disjoint.

Case 2, $d > 1$: By 7.1.3, at least two different groups of ℓ_d^1 have a multiple point. Non-special germs having d different free points in the first neighbourhood of p besides the base points, these groups are cut by two different special germs, and of course at least one of them is different from τ.

In the remaining cases we have $d = 1$, which forces $p \neq O$. Indeed, we are assuming $e_p(\check{\tau}_p) = e_p$, which for $p = O$ is $n = e_O$. By the way the germ τ has been chosen, the pencil $\mathcal{P}$ has e_O variable points in the first neighbourhood of O, that is, $d = e_O$ and by hypothesis $e_O > 1$. Thus, in the sequel the point p is either satellite or free.

Case 3, $d = 1$ and p is a satellite point: The variable part of the series cut out by $\mathcal{P}$ on the first neighbourhood of p is now an ℓ_1^1. Since p is satellite, there are two satellite points in its first neighbourhood that constitute different groups of the ℓ_1^1 and therefore are cut out by two different special germs. Again one of them is not τ.

The last case cannot occur, as it is not allowed by our assumption $p \notin T$:

Case 4, $d = 1$ and p is a free point: As τ does not (effectively) go through p, the only irreducible component of $\check{\tau}_p$ is the germ of the exceptional divisor at p (which is irreducible because p is a free point) and so the full group cut out by $\check{\tau}_p$ on the first neighbourhood of p is the (only) satellite point counted e_p times. Then the fixed part $H = 0$ is the satellite point counted $e_p - 1$ times and so $\tilde{\xi}_p$ needs to cut on the first neighbourhood of p a group composed of the satellite point counted $e_p - 1$ times (the fixed part) plus a necessarily free point p'. If $e_p > 1$, then the satellite point belongs to ξ and hence also to K, so, in any case, ξ has p' as its only point in the first neighbourhood of p and not in K. Since, furthermore, the strict transform $\tilde{\xi}_{p'}$ and the exceptional divisor meet transversally at p', p' is a non-singular point of ξ (3.8.1). This means that $p \in T$, which has been already excluded. $\diamond$

Corollary 7.5.2 *For any reduced germ of curve ξ, there exist integers that are E-sufficient for ξ.*

PROOF: Since by 4.5.8 all entries in the first column of P^{-1} are positive, condition (i) in 7.5.1 is obviously satisfied if n is big enough. $\diamond$

The E-sufficiency degree of a germ ξ may be computed from its weighted cluster of singular points using 3.7.3. Next we show an example.

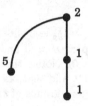

Figure 7.1: The weighted cluster $\mathcal{K}$ of the singular points of the germ $\xi : x^2 y^3 + xy^4 - x^6 = 0$ in 7.5.3.

Example 7.5.3 Let ξ be the germ $x^2 y^3 + xy^4 - x^6 = 0$, its cluster of singular points is shown in figure 7.1. An easy computation shows that in this case

$$P^{-1} = \begin{pmatrix} 1 & 0 & 0 & 0 \\ 1 & 1 & 0 & 0 \\ 2 & 1 & 1 & 0 \\ 3 & 1 & 1 & 1 \end{pmatrix} \quad \text{and} \quad e = \begin{pmatrix} 5 \\ 2 \\ 1 \\ 1 \end{pmatrix}.$$

Therefore,

$$u = \begin{pmatrix} n - 5 \\ n - 7 \\ 2n - 13 \\ 3n - 19 \end{pmatrix}$$

and the degree of E-sufficiency of ξ is 8.

7.6 E-sufficiency degree and polar invariants

In this section we will determine the E-sufficiency degree in terms of the polar invariants. We will add a fourth equivalent condition to those in 7.5.1 by proving:

Theorem 7.6.1 If ξ is a reduced and singular germ of curve and $\{I(q)|q \in \mathcal{R}(\xi)\}$ are the polar invariants of ξ, then a positive integer n is E-sufficient for ξ if and only if $n > \max\{I(q)|q \in \mathcal{R}(\xi)\}$.

Theorem 7.6.1 was first obtained by Teissier [80] for isolated singularities of hypersurfaces and, independently, by Kuo and Lu [47] for plane curves. Our approach here will be different, as we will just reformulate condition (i) in 7.5.1 in terms of polar invariants.

Before proving 7.6.1, we need to establish some auxiliary results. Keep all notations as in the preceding section and for each $p \in K$, fix an irreducible germ γ^p having multiplicity one at p and such that all its points after p are free (hence non-singular) and do not belong to ξ (4.2.9). First of all, we have:

Lemma 7.6.2 *For any $p \in K$, the p-th components of $P^{-1}1_O$ and $P^{-1}e$ are $e_O(\gamma^p)$ and $[\gamma^p.\xi]$ respectively.*

PROOF: Since all points after p on γ^p are free, no one of them may be proximate to a point preceding p. Thus if q precedes p, of course $q \in K$ and furthermore all points proximate to q on γ^p still belong to K. Thus the proximity equality at q may be written

$$e_q(\gamma^p) - \sum_{\substack{q' \in K \\ q' \text{ prox. to } q}} e_{q'}(\gamma^p) = 0.$$

The above equality becoming obviously true if q does not belong to γ^p, we have it for all $q \in K$, $q \neq p$. Again because all points on γ^p after p are free, the only point proximate to p on γ^p is its first neighbouring point, which is simple on γ^p. Then no point in K belongs to γ^p and is proximate to p, and so the proximity equality at p, actually $1=1$, may be written

$$e_p(\gamma^p) - \sum_{\substack{q' \in K \\ q' \text{ prox. to } p}} e_{q'}(\gamma^p) = 1.$$

The above equalities together give the vectorial one

$$e_K(\gamma^p)^t P = 1_p^t,$$

from which the p-th component of $P^{-1}1_O$ may be computed as

$$1_p^t P^{-1}1_O = e_K(\gamma^p)^t P P^{-1}1_O = e_K(\gamma^p)^t 1_O = e_O(\gamma^p),$$

as wanted.

Since all points shared by ξ and γ^p are in K, the Noether formula (3.3.1) may be written as

$$[\gamma^p.\xi] = e_K(\gamma^p)^t e = e_K(\gamma^p)^t P P^{-1} e = 1_p^t P^{-1} e,$$

just the second claimed equality. ◇

Remark 7.6.3 We have seen in the above proof that the p-th row of P^{-1} is $e_K(\gamma^p)^t$.

Let us associate with each point $p \in K$ a rational number, just as was done for the points $q \in \mathcal{R}(\xi)$ in section 6.11, when defining the polar invariants of ξ. Define, for $p \in K$,

$$J(p) = \frac{[\gamma^p.\xi]}{e_O(\gamma^p)}.$$

It is clear from 7.6.2 (or else from 6.9.2, as the proof of 6.11.2 still works) that $J(p)$ does not depend on the particular germ γ^p we are using. Of course, for $p \in \mathcal{R}(\xi)$ we have the polar invariants of ξ: $J(p) = I(p)$ if the notations are as in section 6.11.

Recall that $\mathcal{R}(\xi)$ was defined in section 6.11 as the set of points on ξ that either are free and have at least two free points on ξ in its first neighbourhood, or are satellite and have some free point on ξ in its first neighbourhood. In the sequel we will write $R = \mathcal{R}(\xi)$. Notice that, by its own definition, it is clear that $R \subset K$. We need a little bit more, namely,

Lemma 7.6.4 R *is a cofinal subset of K with respect to the natural ordering of infinitely near points.*

PROOF: If $p \in K$, p itself or some point after p is either a satellite point or a multiple point of ξ. Assume first that there is a satellite point p' on ξ equal or infinitely near to p: then, since there are finitely many satellite points on ξ, either p' itself or one of the satellite points on ξ after it has a free point on ξ in its first neighbourhood and hence belongs to R. Otherwise, p and all points infinitely near to p on ξ are free, and so p needs to be multiple on ξ. Let us assume that p' is maximal among the multiple points on ξ that are equal or infinitely near to p: then all points on ξ in the first neighbourhood of p' are simple. Furthermore, all points on ξ after p being free, so are the points proximate to p' on ξ and hence they all belong to the first neighbourhood of p'. Thus, by the proximity equality (3.5.3), there are $e_{p'}(\xi)$ different free points on ξ in the first neighbourhood of p', and since we are assuming $e_{p'}(\xi) > 1$, $p' \in R$ as wanted. ◇

Next we will establish some properties of the quotients $J(p)$. They may be used for comparing different polar invariants of the same germ.

Proposition 7.6.5 *If $p' \in K$ is a free point in the first neighbourhood of p, then $J(p) < J(p')$.*

PROOF: By the definition of $\gamma^{p'}$, $e_{p'}(\gamma^{p'}) = 1$ and all points after p' on $\gamma^{p'}$ are free. Thus p' is the only point on $\gamma^{p'}$ proximate to p and so, by the proximity equality, $e_p(\gamma^{p'}) = e_{p'}(\gamma^{p'}) = 1 = e_p(\gamma^p)$. Since both γ^p and $\gamma^{p'}$ have a free point in the first neighbourhood of p, 6.9.2 applies to both and the equality of multiplicities at p gives $e_q(\gamma^{p'}) = e_q(\gamma^p)$ for all points q preceding p. Thus, in particular we have $e_O(\gamma^{p'}) = e_O(\gamma^p)$. Furthermore, using the Noether formula we get

$$[\gamma^{p'}.\xi] = \sum_{O \leq q \leq p} e_q(\gamma^{p'})e_q(\xi) + e_{p'}(\xi) = \sum_{O \leq q \leq p} e_q(\gamma^p)e_q(\xi) + e_{p'}(\xi) = [\gamma^p.\xi] + e_{p'}(\xi),$$

which gives the wanted inequality, as clearly $e_{p'}(\xi) > 0$. ◇

Lemma 7.6.6 *Let p be a free point on ξ and denote by E the germ of the exceptional divisor at p. There is a rational number u so that for any branch γ through p*

$$\frac{[\gamma.\xi]}{e_O(\gamma)} = u + \frac{1}{r(O,p)}\frac{[\tilde{\gamma}_p.\tilde{\xi}_p]}{[\tilde{\gamma}_p.E]}.$$

PROOF: By the Noether formula 3.3.1,

$$[\gamma.\xi] = \sum_{q<p} e_q(\gamma)e_q(\xi) + [\tilde{\gamma}_p.\tilde{\xi}_p].$$

Then, since according to its definition in section 6.9,

$$r(O,p) = \frac{e_O(\gamma)}{[\tilde{\gamma}_p.E]},$$

it is enough to take

$$u = \frac{1}{e_O(\gamma)}\sum_{q<p} e_q(\gamma)e_q(\xi),$$

as it is independent of γ by 6.9.2. ◇

Remark 7.6.7 Just note, for further reference, that in the above proof we have seen that

$$u = \frac{1}{e_O(\gamma)}\sum_{q<p} e_q(\gamma)e_q(\xi).$$

Recall from section 3.6 that p' is called a satellite point of p if and only if p' is a satellite point and p is the last free point preceding it. Since all points in the first neighbourhood of O are free, any satellite point is satellite of some free point.

Proposition 7.6.8 *Let p be a free point in K.*

(a) If $p' \in K$ is a satellite point of p, then $J(p) \geq J(p')$.

(b) If there are no free points on ξ in the first neighbourhood of p, then there is a satellite point p' of p such that ξ has a free point in its first neighbourhood and $J(p) = J(p')$.

(c) If ξ has a single free point in the first neighbourhood of p and such a point is non-singular, then $J(p) > J(p')$ for all satellite points p' of p on ξ, and there is one of them whose first neighbourhood contains some free point on ξ and such that

$$J(p) = \frac{b}{c} > J(p') > \frac{b-1}{c}$$

for some positive integers b,c.

PROOF: Choose local coordinates at p so that the y-axis is the exceptional divisor and the x-axis is not tangent to $\tilde{\xi}_p$. Then assume that the Newton polygon N of $\tilde{\xi}_p$ relative to these coordinates has sides Γ_i with slopes $-n_i/m_i$ and associated equations $\Omega_i(z) = 0$, for $i = 1, \ldots, k$. Since the x-axis is not tangent to ξ_p, $m_i/n_i \leq 1$ for all i. Furthermore the last side Γ_k ends on the first axis, at the point $(e, 0)$, $e = e_p(\xi)$ (see 2.2.7). The strict transform of γ^p with origin at p being by definition smooth and non-tangent to $\tilde{\xi}_p$, $J(p) = u + e/r(O, p)$ by 7.6.6.

Let $p' \in K$ be any satellite of p. Write γ for the strict transform with origin at p of $\gamma^{p'}$ and assume that it has a Puiseux series of the form $s = ax^{m/n} + \cdots$. First of all let us recall that since γ goes through no free point on $\tilde{\xi}_p$ other than p itself, then there is no branch of $\tilde{\xi}_p$ with Puiseux series $s' = ax^{m/n} + \cdots$ (5.3.4), which in turn, by 1.3.1, implies that there is no side Γ_i of N for which $m/n = m_i/n_i$ and $\Omega_i(a) = 0$. Then 6.11.1 gives $[\gamma.\tilde{\xi}_p]/[\gamma.E] = \alpha + m\beta/n$, (α, β) a vertex of N giving minimal $n\alpha + m\beta$. Since, clearly, $\alpha + m\beta/n \leq e$, claim (a) follows by using 7.6.6.

If there are no free points on ξ in the first neighbourhood of p, then all branches of $\tilde{\xi}_p$ contain satellite points of p. Choosing p' to be the last satellite of p on one (in fact on all) of the branches corresponding to Γ_k forces $m/n = m_k/n_k$ and hence it is allowed to take $(\alpha, \beta) = (e, 0)$. It follows $[\gamma.\tilde{\xi}_p]/[\gamma.E] = e$ and then, by 7.6.6, $J(p) = J(p')$ as stated in claim (b).

To close, the hypothesis of (c) forces the last side Γ_k of N to have height 1 and slope -1. Then, since $m/n < 1$ because p' is a satellite of p, it is easy to see that $\alpha + m\beta/n < e$, which in turn forces $J(p') < J(p)$ by 7.7.4. Furthermore, Γ_k is not the only side of N, as otherwise p would be a non-singular point of ξ. Thus side Ω_{k-1} does exist, it has $m_{k-1}/n_{k-1} < 1$ and so the branches of $\tilde{\xi}_p$ corresponding to it contain satellite points of p. By taking p' to be the last satellite of p on one of these branches, $m/n = m_{k-1}/n_{k-1}$, $(\alpha, \beta) = (e - 1, 1)$ and so $e - 1 < [\gamma.\tilde{\xi}_p]/[\gamma.E] < e$. Then, as in former cases, 7.6.6 ends the proof. ◇

Now we are able to prove Theorem 7.6.1 stated at the beginning of this section.

PROOF OF 7.6.1: We know from 7.5.1 and 7.6.2 that the E-sufficiency degree is the smallest integer n such that $n \geq J(p)$ if $p \in T$ and $n > J(p)$ if $p \in K - T$. Since the polar invariants are $I(p) = J(p)$ for $p \in R$ and it is clear from their definitions that $T \cap R = \emptyset$, we need only see that if $n > J(q)$ for any $q \in R$, then $n > J(p)$ for any $p \in K - T$ and $n \geq J(p)$ for any $p \in T$, the converse being obvious. Since our argument will fork many times, we will enumerate cases and subcases.

We will deal with the free points first. Thus we assume (case a) that $p \in K$ is a free point and also, using induction and working backwards, that for any free point $p_1 \in K$ infinitely near to p, $n \geq J(p_1)$ and even $n > J(p_1)$ if $p_1 \notin T$. If $p \in K$ is maximal (case $a.1$), then $p \in R$ by 7.6.4 and there is nothing to prove. If the first neighbourhood of p contains some free point $p_1 \in K$ (case *a.2*), then the claim is obvious from the induction hypothesis $n \geq J(p_1)$ and 7.6.5.

There remains case *a.3*, in which all points in K in the first neighbourhood of p are satellite points. In such a case ξ may still have some free point in the first neighbourhood of p, but such a point needs to be non-singular as it does not belong to K.

Assume first (case *a.3.1*) that all points on ξ in the first neighbourhood of p are satellite points. Then 7.6.8(b) applies: there is p' satellite of p so that $J(p) = J(p')$ and ξ has at least a free point p_1 in the first neighbourhood of p'. If p_1 is non-singular for ξ, then $p' \in R$ and so $J(p) = J(p') < n$ as wanted. Otherwise $p_1 \in K$, in which case it is enough to recall that we have $J(p) = J(p') < J(p_1)$ by 7.6.5 and $J(p_1) \leq n$ by induction.

If ξ has a single free point in the first neighbourhood of p (case *a.3.2*) then, since such point cannot be in K, it is non singular and thus $p \in T$. We apply 7.6.8(c): as in case *a.3.1* the satellite point p' of p has $n > J(p')$, and so $n > (b-1)/c$, c a positive integer, from which $n \geq b/c = J(p)$ as wanted.

The last subcase of case *a.3* we have to consider is case *a.3.3*, in which ξ has at least two free points (necessarily non-singular) in the first neighbourhood of p. Then again $p \in R$ and the claim is obvious.

Once we know that we have $n \geq J(p)$ for any free point $p \in K$ and even $n > J(p)$ if $p \notin T$, we will deal with the satellite points in K: assume (case b) that $p' \in K$ is a satellite point, say a satellite point of a certain free point $p \in K$. If $p \notin T$, then $n > J(p) \geq J(p')$ by 7.6.8(a). Otherwise, since $p \in T$, 7.6.8(c) applies and gives $n \geq J(p) > J(p')$. ◇

We close this section with two examples. The first one shows that the E-sufficiency degree may equal one of the $J(p)$ (for $p \in T$, of course). The second example shows a case in which $I(p) > I(p')$ and p' is infinitely near to p. The reader may prove as an exercise that the maximal value of the polar invariants $I(p)$ is reached for a point p whose first neighbourhood does not contain free points that are multiple for ξ.

Example 7.6.9 Take $\xi : y^3 - x^5 y = 0$, its Enriques diagram is shown in figure 7.2. The only polar invariant of ξ is $I(q) = 15/2$, q being the last satellite point on ξ. Thus the E-sufficiency degree of ξ is 8 and, on the other hand, the point p on ξ in the third neighbourhood of the origin has $J(p) = 8$.

Example 7.6.10 Take $\zeta : x^8 - x^4 y^2 - x^4 y^3 + y^5 = 0$, its Enriques diagram being also shown in figure 7.2. If p and p' are, respectively, the first neighbouring point and the last satellite point on ζ, then p' is infinitely near to p, both belong to $\mathcal{R}(\zeta)$, $I(p) = 8$ and $I(p') = 20/3 < 8$.

7.7 A-sufficiency

In section 7.5 above we have seen that the equisingularity type of a reduced germ of curve $\xi : f = 0$ depends only on finitely many monomials of its equation f. In

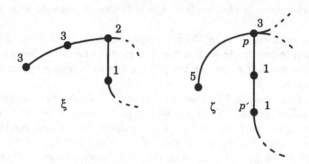

Figure 7.2: The weighted clusters of singular points of the germs ξ and ζ in examples 7.6.9 and 7.6.10.

other words, all monomials in f from a certain degree onwards can be arbitrarily modified without altering the equisingularity type of the germ $f = 0$. Such a degree, the degree of E-sufficiency of ξ, has been determined and related to polar invariants in sections 7.5 and 7.6. In this section we will consider a similar property relative to the isomorphism class of the germ.

Assume that $\xi : f = 0$ is a non-empty and reduced germ of curve at O and keep all notations as in preceding sections. We say that a positive integer n is *analytically sufficient* (or *A-sufficient* for short) for ξ if and only if all $h \in f + \mathcal{M}^n$ are non-zero and define germs of curve analytically isomorphic to ξ. As for E-sufficiency, this condition clearly does not depend on the equation f of ξ, but only on the germ ξ itself. The minimal value of the integers n that are A-sufficient for ξ is called the *A-sufficiency degree* of ξ. Next we will prove that for any non-empty reduced germ ξ there exist A-sufficient integers, and hence the A-sufficiency degree of ξ is defined. Nevertheless, we will give only an upper bound of the A-sufficiency degree and not a precise determination of it, as, unlike the E-sufficiency degree, the A-sufficiency degree of a germ ξ is not determined by its equisingularity type.

We will begin by proving that equisingular germs are isomorphic if their corresponding branches have Puiseux series that are congruent modulo suitable powers of the variable. Assume that ξ is any reduced germ of curve with branches $\gamma_1, \ldots, \gamma_r$. For $i = 1, \ldots, r$, we will write, as for 4.8.4,

$$c_i(\xi) = \sum_p (e_p(\xi) - 1)e_p(\gamma_i),$$

the summation running on all multiple points of ξ on γ_i. Notice that if ζ is equisingular to ξ and its branches are numbered so that branches of the same index correspond by the equisingularity, then, by 3.8.3 and 3.8.6, $c_i(\xi) = c_i(\zeta)$.

Lemma 7.7.1 *Let ξ and ζ be equisingular germs of curve at O. Assume that they have branches $\gamma_1, \ldots, \gamma_r$ and $\eta_1, \ldots, \eta_r$, each γ_i corresponding to η_i by* *the equisingularity. Put $c_i = c_i(\xi) = c_i(\zeta)$ and $n_i = e_O(\gamma_i) = e_O(\eta_i)$. Fix local coordinates x, y at O so that no branch of either germ is tangent to the y-axis. If for each $i = 1, \ldots, r$ there are Puiseux series s_i and s'_i of γ_i and η_i, respectively, so that $s_i \equiv s'_i$ mod x^{c_i/n_i}, then ξ and ζ are analytically isomorphic.*

PROOF: Take $\bar{R}$ to be a product of r rings of series, namely $\bar{R} = \mathbb{C}\{t_1\} \times \cdots \times \mathbb{C}\{t_r\}$. The elements of $\mathcal{O} = \mathcal{O}_{S,O}$ being identified with series in x, y, consider the morphism

$$\varphi^* : \mathcal{O} \longrightarrow \bar{R}$$

defined by the rule

$$f(x, y) \mapsto (f(t_1^{n_1}, s_1(t_1^{n_1})), \ldots, f(t_r^{n_r}, s_r(t_r^{n_r}))).$$

This morphism has been seen to be induced by a uniformizing map of a representative of ξ in section 2.5. Any equation of ξ belongs to $\ker(\varphi^*)$ and if R denotes the local ring of ξ, the induced morphism

$$\bar{\varphi} : R \longrightarrow \bar{R}$$

is a monomorphism that has been called, still in section 2.5, a uniformizing morphism of ξ. Uniformizing morphisms being essentially unique, by 3.11.6 the ring $\bar{R}$ is isomorphic, as extension of R, to the integral closure of R in its full quotient ring.

Take $\tau = (t_1^{c_1}, \ldots, t_r^{c_r})$: by 4.8.4, it generates the conductor of the extension $\bar{\varphi}(R) \hookrightarrow \bar{R}$ and so in particular an ideal of $\bar{R}$ contained in $\bar{\varphi}(R)$.

Now we repeat the above constructions from the second germ ζ: we have the morphism

$$\psi^* : \mathcal{O} \longrightarrow \bar{R}$$
$$f(x, y) \mapsto (f(t_1^{n_1}, s'_1(t_1^{n_1})), \ldots, f(t_r^{n_r}, s'_r(t_r^{n_r})))$$

which induces the uniformizing morphism of ζ

$$\bar{\psi} : R' \longrightarrow \bar{R},$$

R' being the local ring of ζ. As above $\tau\bar{R}$ is contained in $\bar{\psi}(R')$.

Let π be the natural morphism,

$$\pi : \bar{R} \longrightarrow \bar{R}/\tau\bar{R}.$$

Our hypothesis about the Puiseux series of ξ and ζ clearly gives the equality $\pi \circ \varphi^* = \pi \circ \psi^*$. In particular we get $\pi(\varphi^*(R)) = \pi(\psi^*(R'))$ and hence $\pi(\bar{\varphi}(R)) = \pi(\bar{\psi}(R'))$. Since both $\bar{\varphi}(R)$ and $\bar{\psi}(R')$ contain $\tau\bar{R}$, we obtain $\bar{\varphi}(R) = \bar{\psi}(R')$ after which the proof is done: both R and R' being isomorphic to their images in $\bar{R}$, they are isomorphic as $\mathbb{C}$-algebras and hence the germs ξ and ζ are isomorphic by 2.4.5. $\diamond$

Theorem 7.7.2 *If ξ is a non-empty and reduced germ of curve and we denote by I the maximum of its polar invariants, then any integer $n > 2I$ is $\mathcal{A}$-sufficient for ξ.*

PROOF: Let f be an equation of ξ and g any element of $\mathcal{M}^n_{S,O}$. We know from 7.5.1 and 7.6.1 that $f + g$ defines a germ ζ that goes sharply through the weighted cluster of singular points of ξ, and so it is, in particular, equisingular to ξ. Fix a branch γ_i of ξ. Write $n_i = e_O(\gamma_i)$, q_i for the last singular point of ξ lying on γ_i and $k_i \leq \infty$ for the number of points infinitely near to q_i shared by γ_i and ζ. These points being non-singular for both germs, we have

$$[\gamma_i.\zeta] = \sum_{q \leq q_i} e_q(\gamma_i)e_q(\zeta) + k_i = \sum_{q \leq q_i} e_q(\gamma_i)e_q(\xi) + k_i.$$

Let γ be any irreducible germ going through q_i and having a free point not belonging to ξ in its first neighbourhood. One of the polar invariants of ξ is, by its definition (see section 6.11),

$$I_i = \frac{[\gamma.\xi]}{e_O(\gamma)} = \frac{1}{e_O(\gamma)} \sum_{q \leq q_i} e_q(\gamma)e_q(\xi).$$

Since both γ and γ_i have a free point in the first neighbourhood of q_i, 6.9.2 applies, giving

$$I_i = \frac{1}{n_i} \sum_{q \leq q_i} e_q(\gamma_i)e_q(\xi).$$

Thus we get

$$n_i I_i \geq c_i$$

and also

$$k_i = [\gamma_i.\zeta] - n_i I_i.$$

On the other hand, $[\gamma_i.\zeta] = o_{\gamma_i}(f+g) = o_{\gamma_i}(g)$ as γ_i is a component of $\xi : f = 0$. Since $g \in \mathcal{M}^n_{S,O}$, $o_{\gamma_i}(g) \geq n_i n$ so that the above equality gives

$$k_i \geq n_i(n - I_i),$$

after which, using the hypothesis $n > 2I \geq 2I_i$, we get

$$k_i > n_i I_i \geq c_i.$$

Clearly $c_i + 1 > 0$, so at least the first non-singular point of ξ on γ_i is lying on ζ. Since this is true for all branches of ξ and we already know that ξ and ζ have the same set of singular points and the same number of branches, we conclude that also the set of non-singular points lying in the first neighbourhood of a singular one is the same for both ξ and ζ. Thus $\mathcal{S}(\xi) = \mathcal{S}(\zeta)$ and the identity map is an equisingularity between ξ and ζ. Write η_i for the branch of ζ corresponding to γ_i by this equisingularity: η_i is the only branch of ζ that shares with γ_i points that are non-singular for ξ (and hence also for ζ). We have seen above that the number of these points is non-less than $c_i + 1$.

Fix local coordinates x, y in such a way that no branch of either ξ or ζ is tangent to the second axis. The branches γ_i and η_i above share non-less than $c_i + 1$ points that are non-singular for both branches, as they are so for the whole germs ξ and ζ. We are going to see that this implies that ξ and ζ have Puiseux series s_i and s'_i, respectively, so that $s_i \equiv s'_i \mod x^{c_i/n_i}$. After proving this for all i, the claim will follow from 7.7.1.

Of course if $\gamma_i = \eta_i$ there is nothing to prove. We assume thus that $\gamma_i \neq \eta_i$ and call p the last point lying on both branches. The point p being non-singular, by 5.7.1 we may choose Puiseux series s_i of γ_i and s'_i of η_i having all their monomials up to those corresponding to the point p equal. Since the point p is preceded by non-less than c_i free points, by 5.7.4 its corresponding monomials in the Puiseux series s_i and s'_i have degree non-less than c_i/n_i, after which the proof is complete. $\diamond$

A nice consequence of the former theorem is that regarding its local properties at a point, analytic plane curves are not more general than the algebraic ones:

Corollary 7.7.3 *Any germ of analytic plane curve is analytically isomorphic to a germ of an algebraic curve lying on the affine plane.*

PROOF: Assume that ξ is a given germ of curve, with origin at a point O on a smooth surface S. A system of local coordinates at O gives an analytic isomorphism between a certain neighbourhood of O on S and a neighbourhood of the origin in $\mathbb{C}^2$. If we take the image of ξ by this isomorphism instead of ξ itself, it is not restrictive to assume that ξ is a germ at the origin of $\mathbb{C}^2$ defined by a convergent series $f(x, y)$ in the affine coordinates x, y. By 7.7.2 the germ of the curve defined by the partial sum of degree n of f is isomorphic to ξ if n is big enough, and obviously such a curve is algebraic. $\diamond$

The germs of algebraic curves in the affine or the projective plane are called *algebraic germs*. Corollary 7.7.3 says thus that any germ is isomorphic to an algebraic germ. Nevertheless, the reader should notice that isomorphic does not mean equal, and so not all germs of analytic curves in $\mathbb{C}^2$ are algebraic. For an example of non-algebraic germs see exercise 2.1.

The existence of A-sufficient integers for isolated singularities of hypersurfaces, as well as a result similar to the 7.7.3 above, were proved by P. Samuel [71]. For curves in a n-dimensional space ($n > 2$), see Elias [34].

Next we will give an example of equisingular germs having different E-sufficiency degrees. We will make use of a couple of results from [94] we will quote without proof, as proving them here would take us too far.

Example 7.7.4 The integer 8 is A-sufficient for the germ $\xi : x^6 - y^5 - 5x^3y^4 = 0$ but it is not A-sufficient for $\zeta : x^6 - y^5 = 0$.

PROOF: By using the Newton–Puiseux algorithm it is easy to see that any germ $\xi' : x^6 - y^5 - 5x^3y^4 + g = 0$ with $g \in (x,y)^8$ has a Puiseux series of the form

$$s = x^{6/5} + x^{9/5} + \cdots$$

and it is proved in [94] ch. 5, ex. 4 that any two germs with such a Puiseux series are isomorphic. After this 8 is A-sufficient for ξ as claimed.

On the other hand, take $\zeta' : x^6 - y^5 - 5x^4y^4 = 0$. It has a Puiseux series of the form

$$s' = x^{6/5} + x^{14/5} + \cdots,$$

while ζ has Puiseux series

$$s'' = x^{6/5}.$$

Again in [94] ch. 5, ex. 4 it is proved that germs with Puiseux series such as s' and s'' above are not isomorphic. Thus 8 is not A-sufficient for ζ. ◇

7.8   Exercises

7.1 Draw Enriques diagrams of the clusters of base points of the pencils

$$\alpha(x^3y^2 + y^4) + \beta(x^6 + x^3y^2) = 0$$
$$\alpha(y^2 - x^7)(y - x^2) + \beta x^2(y^2 - x^8) = 0$$
$$\alpha(x^2 - y^5) + \beta((x^2 - y^5)^2 + y^{13}) = 0$$

and determine their special germs.

7.2 Let $\mathcal{P}$ be a pencil with no fixed part, (K,ν) its cluster of base points and $p \in K$. Prove that the following conditions are equivalent:

  (i) There is a (necessarily unique) germ $\xi \in \mathcal{P}_p$ with $e_p(\xi) > \nu_p$.

 (ii) All germs but one in $\mathcal{P}_p$ have the same tangent cone.

(iii) There are two germs in $\mathcal{P}_p$ that have the same tangent cone.

 (iv) The linear series cut out by $\mathcal{P}$ on the first neighbourhood of p is zero-dimensional.

  (v) $\rho_p = 0$.

If these conditions are not satisfied, p is said to be a *dicritical base point* of $\mathcal{P}$.

7.3 Prove that for any pair of different germs ξ, ζ in a pencil $\mathcal{P}$ without fixed part,

$$[\xi.\zeta] = \sum_{p \in BP(\mathcal{P})} e_p(\mathcal{P})^2.$$

(*Hint:* use 2.7.4.)

7.4 Prove that if $\mathcal{K} = (K,\nu)$ is a weighted cluster and $\mathcal{K}' = (K,\nu')$ the one it gives rise to by unloading on a point p, then

$$\mu(\mathcal{K}') - \mu(\mathcal{K}) = (2\rho_p + r + \ell' - 2)n + (\ell + 1)n^2,$$

where n is the amount of multiplicity unloaded on p (i.e., the smallest integer non-less than $-\rho_p/(\ell+1)$) and the remaining notations are as in 7.3.2.

7.5 Take $\mathcal{L}$ and $\mathcal{P}$ as in the proof of 7.3.10, prove that all points in $PB(\mathcal{L})$ belong to $BP(\mathcal{P})$ and show by an example that the converse is in general false.

7.6 Give an example of a couple of non-equisingular germs ξ, ζ with $\mu(\xi) = \mu(\zeta)$. Prove that if two such germs belong to a pencil $\mathcal{P}$, then both are special germs of $\mathcal{P}$.

7.7 *Zariski's discriminant criterion for pencils, see* [89] *for the general version.* Let $f, g \in \mathcal{O}_O$ be non-invertible germs of analytic function with no common factor. Consider the family of germs of curve $\xi_t : f + tg = 0$, $t \in \mathbb{C}$ and assume that $\eta : z = 0$ is a smooth germ at O.

(a) Prove that there is an $\varepsilon \in \mathcal{R}$, $\varepsilon > 0$, so that ξ_t is reduced and its equisingularity type, its relative discriminantal index $d_z(\xi_t)$ and the intersection multiplicity $[\xi_t.\eta]$ are all constant for $0 < |t| < \varepsilon$.

(b) Prove that $d_z(\xi_t) \leq d_z(\xi_0)$ for $0 < |t| < \varepsilon$, and also that the equality holds if and only if ξ_0 is reduced, $[\xi_t.\eta] = [\xi_0.\eta]$ and ξ_t is equisingular to ξ_0 for $0 < |t| < \varepsilon$.

7.8 Let $\mathcal{P}$ be a pencil with no fixed part (resp. a neat linear system) and $K = (K, \nu) = BP(\mathcal{P})$. Prove that for any reduced $\xi \in \mathcal{P}$, $2\delta(\xi) \geq \sum_{p \in K} \nu_p(\nu_p - 1)$, and also that the equality holds for all but finitely many (resp. generic) germs $\xi \in \mathcal{P}$. (*Hint:* use 4.14, 4.15 and 4.16.)

7.9 Compute jacobian germs of the pencils of exercise 7.1, decompose them in branches and make explicit the maps of 7.4.6.

7.10 Let $\xi : f = 0$ be a non-reduced germ with origin at O. Prove that for any $n \in \mathbb{N}$, there is a reduced germ $\zeta : g = 0$ such that $f - g \in \mathcal{M}_O^n$, and so there is nothing similar to E-sufficiency for non-reduced germs.

7.11 Let $\xi : f = 0$ be a singular and reduced germ of curve and K the weighted cluster of its singular points (each taken with virtual multiplicity equal to its effective multiplicity on ξ). Let $\zeta : g = 0$ be a second germ of curve and assume that for each $p \in K$, $e_p(\zeta_p) > e_p(\xi)$, all virtual transforms being relative to the virtual multiplicities of K. Prove that all germs $f + \lambda g = 0$, $\lambda \in \mathbb{C}$, go sharply through K and therefore are equisingular.

7.12 After choosing local coordinates at its origin, assume that $\gamma : f = 0$ is irreducible and its Puiseux series have characteristic exponents $m_1/n, \ldots, m_k/n$, and put, as customary, $n^i = \gcd(n, m_1, \ldots, m_{i-1})$. Use exercise 1.6 to prove that for

$$\ell > \frac{m_1 n + (m_2 - m_1)n^1 + \cdots + (m_k - m_{k-1})n^k}{n}$$

all germs $f + \lambda x^\ell = 0$, $\lambda \in \mathbb{C}$, are equisingular. Give another proof of the same fact using exercise 7.11 and a third one using the example at the end of section 7.4.

7.13 Use the example at the end of section 7.4 to prove that if an integer n is E-sufficient for a reduced germ ξ, then it is strictly bigger than any polar invariant of ξ, which of course is part of 7.6.1.

7.14 Compute the degree of E-sufficiency of the germ $\xi : f = (x^4 - y^5)^3 - x^{14}y = 0$ using 7.5.1. Check the result using 7.5.3 and exercise 6.14. Determine the values of λ for which $f + \lambda x^{14} = 0$ is not equisingular to ξ.

7.15 Let ξ be a non-empty and reduced germ of curve with origin at O. For each branch γ_i of ξ, put $c_i = \sum_{q \in \gamma} e_q(\gamma_i)(e_q(\xi) - 1)$.

(a) Prove that the free point p_i on γ_i that corresponds to the monomial of index c_i in the Puiseux series of γ_i (see section 5.3) is simple on ξ.

(b) Assume that ξ is unbranched and has multiplicity two and prove that p_1 is the first simple point on ξ.

(c) Prove that there are no satellite points on γ_i after p_i, but for ξ unbranched of multiplicity two.

Chapter 8

Valuations and complete ideals

The first two sections of this chapter are devoted to studying the valuations of the ring $\mathcal{O} = \mathcal{O}_{S,O}$ following ideas of Zariski [88] that were later developed in [1] and [77]. In section 8.1 we associate with each zero-dimensional valuation a sequence of infinitely near points, its sequence of centres. Then, values are related to centres by a Noether-type formula, so that the sequences of centres give a geometrical description of the zero-dimensional valuations of $\mathcal{O}$. In section 8.2 we prove the existence and uniqueness of a zero-dimensional valuation with a given sequence of centres, after which valuations are classified according to the structure of their sequences of centres. The Zariski factorization theorem for complete ideals is the main goal of the next two sections: our proof uses the clusters of base points and a quite easy decomposition theorem for them, which gives a rather explicit determination of the irreducible factors and multiplicities in the factorization of a complete ideal. Sections 8.5 and 8.6 deal with the base points of the jacobian system and related properties of polar germs.

8.1 Valuations of the ring $\mathcal{O}_{S,O}$

Let us recall first that a *valuation* of a field K is a map from the multiplicative group K^* of K onto a totally ordered additive abelian group $\mathcal{G}$, the *value group* of v:

$$v : K^* \longrightarrow \mathcal{G}$$

such that the conditions

$$v(fg) = v(f) + v(g)$$
$$v(f + g) \geq \min\{v(f), v(g)\}$$

are satisfied for any $f, g \in K^*$. The image $v(f)$ of f is usually called the *value* of f (by v). We have $v(1) = 0$, as v is a homomorphism of groups. Also

$v(-1) = 0$ as $v(-1) + v(-1) = v(1) = 0$ and $\mathcal{G}$ is totally ordered. It follows that $v(-f) = v(f)$ for any $f \in K^*$. It is customary to extend v to the whole of K by formally setting $v(0) = \infty$. Valuations v and v' will be called *isomorphic* if and only if $v' = \psi \circ v$ and ψ is an order-preserving isomorphism of groups. In the sequel we will make no distinction between isomorphic valuations. Valuations whose value group is (isomorphic to) the ordered group $\mathbb{Z}$ will be called *discrete* (we follow [48], other authors call them rank-one discrete valuations). If the value group satisfies the archimedean property (namely that for any $g, g' \in \mathcal{G}$, $g, g' > 0$, there is a positive integer n for which $ng > g'$), the valuation will be called an *archimedean* or *rank-one* valuation.

If $R \subset K$ is a subring whose quotient field is K, v determines and is in turn determined by its restriction to R. The valuation v is called a valuation of R if and only if all elements in R have non-negative image by v. Then the set $\{f \in R | v(f) > 0\}$ is obviously a prime ideal of R (not the whole of R, as $v(1) = 0$) which is called the *centre* of the valuation. Clearly the only valuation with centre (0) is the trivial one. All valuations will be implicitly assumed to be non-trivial in the sequel, and so their centres will be non-zero prime ideals. The following easy consequence of the definition will be useful later on:

Lemma 8.1.1 *Let v be a valuation of a field K and $f, g \in K$. If $v(f) \neq v(g)$, then $v(f + g) = \min\{v(f), v(g)\}$.*

PROOF: Assume $v(f) > v(g)$. If $v(f + g) > \min\{v(f), v(g)\} = v(g)$ then $v(g) = v(f + g - f) \geq \min\{v(f + g), v(f)\} > v(g)$, a contradiction. ◇

Let O be a point on a smooth surface S, in this section we will give a geometric description of all (non-trivial) valuations of the local ring $\mathcal{O} = \mathcal{O}_{S,O}$. If v is a valuation and u an invertible element of $\mathcal{O}$, then necessarily $v(u) = 0$ as the centre of v is contained in the maximal ideal $\mathcal{M}$. Therefore, if $\xi : f = 0$ is a germ of curve at O, then the value $v(f)$ is the same for all equations f of ξ: we will often write $v(\xi)$ for $v(f)$ and call it the *value* of ξ.

Valuations of $\mathcal{O}$ whose centre is not the maximal ideal $\mathcal{M}$ are easily described. Let v be such a valuation: by 1.8.11, its centre is a principal ideal $f\mathcal{O}$ to which it corresponds an irreducible germ $\gamma : f = 0$. The germ γ will be called the *geometric centre of v*, or even just the *centre* of v (note that the centre and the geometric centre determine each other). Any non-zero $g \in \mathcal{O}$ may be written in the form $g = g'f^s$ with $g' \notin f\mathcal{O}$, and so $v(g) = sv(f)$: the value group is thus cyclic and generated by $v(f)$: up to isomorphism one may take $\mathcal{G} = \mathbb{Z}$ and $v(f) = 1$, after which $v(g)$ is just the multiplicity of γ as a component of the germ $g = 0$. These valuations are called *one-dimensional valuations of $\mathcal{O}$* as their centres are non-empty germs of curve. We have thus seen that an one-dimensional valuation of $\mathcal{O}$ is determined by its centre up to isomorphism.

In the sequel we will deal with valuations v of $\mathcal{O}$ whose centre is the maximal ideal $\mathcal{M}$. The point O (the germ of set defined by $\mathcal{M}$) is then called the *geometric centre* of v or just the *centre* of v if no confusion may result. All these valuations will be called *zero-dimensional valuations of $\mathcal{O}$* or also *valuations centred at O*.

The *multiplicity* of v at O, $e_O(v)$ is defined as the minimal positive value of v, namely $e_O(v) = \min\{v(f)|f \in \mathcal{M}\}$.

The first and easiest example of a zero-dimensional valuation of $\mathcal{O}$ is the order at O: $o_O(f) = s$ if and only if f has order s as a convergent series, which is the same as saying that $f \in \mathcal{M}^s - \mathcal{M}^{s+1}$ or that the germ $f = 0$ has multiplicity s. Such a valuation is also called the $\mathcal{M}$-*adic* valuation of $\mathcal{O}$. If p is any point in the first neighbourhood of O and $\mathcal{O}_p$ denotes its local ring, it is clear from 3.2.1 that the $\mathcal{M}$-adic valuation of $\mathcal{O}$ can be obtained as the restriction to $\mathcal{O}$ of the one-dimensional valuation of $\mathcal{O}_p$ whose centre is the germ of the exceptional divisor.

Proposition 8.1.2 *If v is a zero-dimensional valuation of $\mathcal{O}$, not the $\mathcal{M}$-adic one, then there is a single tangent line ℓ to S at O so that an element $g \in \mathcal{M}^n - \mathcal{M}^{n+1}$ has $v(g) > ne_O(v)$ if and only if the germ $g = 0$ is tangent to ℓ.*

PROOF: Write $e = e_O(v)$. Take local coordinates x, y at O and identify $\mathcal{O}$ with the ring of series $\mathbb{C}\{x, y\}$. Of course $v(f) \geq r$ for any $f \in \mathcal{M}^r$. If, furthermore, v is not (isomorphic to) the $\mathcal{M}$-adic valuation, then for some r there is an $f \in \mathcal{M}^r - \mathcal{M}^{r+1}$ with $v(f) > re$. Write $f = f_r + f'$ where f_r is a homogeneous form of degree r and $f' \in \mathcal{M}^{r+1}$. Then $v(f') \geq (r + 1)e > re$ and hence $v(f_r) > re$. By decomposing f_r into linear factors we get a linear form in x, y, say $\lambda = \lambda(x, y)$, with $v(\lambda) > e$. Such a linear form is necessarily unique up to a constant factor, as two linearly independent forms with such a property would generate $\mathcal{M}$ against the definition of e. Take ℓ to be the tangent line defined by λ: the same argument we have used for f above shows that an element $g \in \mathcal{M}^n - \mathcal{M}^{n+1}$ has $v(g) > ne$ if and only if λ divides its leading form, just as claimed. ◊

The line ℓ of 8.1.2 will be called the *tangent line to v*. Let p be any point in the first neighbourhood of O, let us write $\mathcal{O}_p$ for its local ring and take it as an extension of $\mathcal{O}$ via the morphism induced by blowing up. We have:

Theorem 8.1.3 *A valuation v centred at O can be extended to a zero dimensional valuation of $\mathcal{O}_p$ if an only if v is not the $\mathcal{M}$-adic valuation and p corresponds to the tangent line to v. In such a case the extension is unique and has the same value group as v.*

PROOF: Take the local coordinates so that the point p is on the x-axis. If either v is the $\mathcal{M}$-adic valuation or p does not correspond to the tangent to v, by 8.1.2, $v(y) = e_O(v)$. Then y/x belongs to the maximal ideal of $\mathcal{O}_p$ and $v(y/x) \leq 0$, which shows that v cannot be extended to a zero-dimensional valuation of $\mathcal{O}_p$.

Assume now that v is not the $\mathcal{M}$-adic valuation and that p corresponds to the tangent line of v: we have thus $v(y) > v(x) = e_O(v)$. Take x and $z = y/x$ as local coordinates at p: v is a valuation of the ring $\mathbb{C}\{x\}[z]$, as it is contained in the quotient field of $\mathcal{O}$ and $v(z) > 0$. Next we extend v to a valuation $\bar{v}$ of $\mathbb{C}\{x, z\} = \mathcal{O}_p$ with the same value group by using Weierstrass theorems. Let $f \in \mathcal{O}_p$: clearly f may be uniquely written $f = x^r f'$ where

$r \geq 0$ and f' is regular in z. By the Weierstrass preparation theorem 1.8.7, f' may in turn be uniquely written as $f' = ug'$ where u is an invertible series and g' a Weierstrass polynomial in z. All together we have a unique expression $f = ug$ where u is invertible and $g = x^r g'$ is the product of a power of x and a Weierstrass polynomial. Define $\bar{v}(f) = v(g)$, clearly the only way of extending v to a valuation $\bar{v}$ of $\mathcal{O}_p = \mathbb{C}\{x, z\}$. Next we will see that $\bar{v}$ actually is a valuation of $\mathcal{O}_p$ centred at its maximal ideal.

Let f_1, f_2 be arbitrary series in $\mathbb{C}\{x, z\}$. The equality $\bar{v}(f_1 f_2) = \bar{v}(f_1)\bar{v}(f_2)$ directly follows from the definition. Checking that $\bar{v}(f_1 + f_2) \geq \min\{\bar{v}(f_1), \bar{v}(f_2)\}$ will be a little bit trickier. Write $f_1 + f_2 = u_0 g_0$, $f_1 = u_1 g_1$ and $f_2 = u_2 g_2$, all u_i being invertible and each g_i being a product of a power of x and a Weierstrass polynomial. We have thus a relation $g_0 = h_1 g_1 + h_2 g_2$, where $h_1, h_2 \in \mathbb{C}\{x, z\}$. We are going to see that this relation forces a similar one, $g_0 = h_1' g_1 + h_2' g_2$ with $h_1', h_2' \in \mathbb{C}\{x\}[z]$. After this the proof will be done, as then

$$\bar{v}(f_1 + f_2) = v(g_0) \geq \min\{v(g_1), v(g_2)\} = \min\{\bar{v}(f_1), \bar{v}(f_2)\},$$

using that v is a valuation of $\mathbb{C}\{x\}[z]$.

Thus, start from the equality $g_0 = h_1 g_1 + h_2 g_2$. After dividing by a suitable power of x, we may clearly reduce ourselves to the case in which one of the g_i, $i = 1, 2$, is a Weierstrass polynomial. By the obvious symmetry, we assume in the sequel that g_2 is a Weierstrass polynomial, after which we may divide h_1 by g_2 getting $h_1 = qg_2 + r$, $r \in \mathbb{C}\{x\}[z]$ and hence $g_0 = rg_1 + (h_2 + qg_1)g_2$. This relation and the uniqueness of Weierstrass division (1.8.8) force $h_2 + qg_1$ to be the quotient of the Weierstrass division of $g_0 - rg_1$ by g_2. Hence $h_2 + qg_1$ is a polynomial in z, as $g_0 - rg_1$ is. Then it is enough to take $h_1' = r$ and $h_2' = h_2 - qg_1$.

To close, that the valuation $\bar{v}$ is centred at the maximal ideal is clear, as both local coordinates x, z have positive value. ◇

If v is not the $\mathcal{M}$-adic valuation of $\mathcal{O}$ and p, in the first neighbourhood of O, corresponds to its tangent line, we will use the same notation for v and its extension to $\mathcal{O}_p$ and call p the *centre of v in the first neighbourhood of O*. In particular $e_p(v)$ is defined as the minimal value of the elements of the maximal ideal $\mathcal{M}_p$ of $\mathcal{O}_p$ and, provided the coordinates are chosen as above, it may be computed as $e_p(v) = \min\{v(x), v(z)\} = \min\{v(x), v(y) - v(x)\}$.

Now we may iterate the procedure using induction: write $O = p_0$ and assume we have determined successive infinitely near points $p_1, \ldots, p_i$, each p_j in the first neighbourhood of p_{j-1}, and successively extended v to valuations, still denoted by v, centred at $p_1, \ldots, p_i$. Write $\mathcal{O}_j$ for the local ring of p_j and $\mathcal{M}_j$ for its maximal ideal. If (the extension of) v is the $\mathcal{M}_i$-adic valuation of $\mathcal{O}_i$ we stop here. If not, using 8.1.3, we take p_{i+1} to be the only point in the first neighbourhood of O_i such that v may be extended to a zero-dimensional valuation of $\mathcal{O}_{p_{i+1}}$. We still denote such an extension by v. The point p_{i+1} is then called *the centre of v in the $(i + 1)$-th neighbourhood of O*. The minimal value of the elements of $\mathcal{M}_{i+1}$ will be called the *multiplicity of v at p_{i+1}* and will be denoted by $e_{p_{i+1}}(v)$. It is worth noticing that the centres of v in the successive neighbourhoods defined in this way are the only infinitely near points whose local rings v can be extended to. Indeed, if p is an infinitely near point in the first neighbourhood of, say, a point q, and v can be extended to a valuation w centred at p, then the restriction of w is an extension of v centred at q. The centres of v in the different neighbourhoods of O will be also called *infinitely near centres of* v and the point O itself will be referred to as the *proper centre* of v. Of course the infinitely near centres of a valuation are the proper centres of its extensions. We just summarize what we have seen in the next proposition:

Proposition 8.1.4 *Let* v *be a valuation centred at the point* O. *Then* O *and the infinitely near centres of* v *are a totally ordered, finite or infinite, sequence of points, each in the first neighbourhood of the preceding one, which is characterized as the set of all points, equal or infinitely near to* O, *at which an extension of* v *is centred.*

We will devote the rest of this section to showing the main properties of the sequences of centres of the zero-dimensional valuations of $\mathcal{O}$. The first lemma is very easy and deals with arbitrary totally ordered sequences of consecutive infinitely near points (in fact in the next section we will prove all such sequences to be sequences of centres of valuations).

Lemma 8.1.5 *Let* $P = \{p_i\}$, $0 \leq i < N$, $N \leq \infty$, *be a finite or infinite sequence so that* $p_0 = O$ *and* p_i *is a point in the first neighbourhood of* p_{i-1} *for* $i > 0$. *Then:*

(a) *If there is a germ of curve* ξ *containing infinitely many points in* P, *then* P *is the set of points on an irreducible germ of curve* γ *which is uniquely determined by* P. *A germ* ζ *contains infinitely many points in* P *if and only if* γ *is a branch of* ζ.

(b) *If a point* $p_i \in P$ *has infinitely many proximate points in* P, *then* p_i *is the only point in* P *with this property and the points in* P *after* p_{i+1} *are the points proximate to* p_i *in the successive neighbourhoods of* p_{i+1}.

PROOF: Assume that a germ of curve ξ contains infinitely many points in P. Since there are finitely many points belonging to two different branches of ξ (3.3.2), we may choose $p_i \in P$ belonging to a single branch γ of ξ. Then clearly all $p_j \in P$, $j < i$, belong to γ. Take p_j with $i < j$: since infinitely many points in P lie on ξ, there is a p_r, $j \leq r$, that belongs to ξ and hence to some branch γ' of ξ. Then, since p_i precedes p_r, p_i also belongs to γ', which gives $\gamma' = \gamma$ and thus $p_j \in \gamma$ as it precedes p_r. We have seen thus that all $p_j \in P$ belong to the branch γ above. Since both the sequence P and γ have just one point in each neighbourhood, P is the sequence of points on γ as claimed. The rest of claim (a) is now obvious from 3.3.6 and 3.3.2.

Now, for part (b), notice first that, since there are infinitely many points in P proximate to p_i, any p_{i+j}, $j > 0$ precedes a point proximate to p_i and therefore is proximate to p_i too. If also p_h, $h \neq i$, has infinitely many proximate points in P, these are, for the same reason, all points p_{h+j}, $j > 0$. Then for any $s > \max\{i, h\}$, p_{s+1} is proximate to p_i, p_h and p_s against 3.5.7. Finally, we have seen all points p_{i+j}, $j > 0$ to be proximate to p_i: since there is just one point proximate to p_i in each neighbourhood of p_{i+1}, the rest of the claim follows. $\diamond$

For the remaining of this section we assume we have fixed a valuation v centred at O. Its centre in the i-th neighbourhood of O will be denoted by p_i ($p_0 = O$).

Theorem 8.1.6 *If ξ is any germ of curve with origin at O,*

$$v(\xi) = e_{p_0}(\xi)e_{p_0}(v) + \cdots + e_{p_r}(\xi)e_{p_r}(v) + \kappa$$

where $\kappa = 0$ if p_r is the last centre of v, or, otherwise, $\kappa = v(\tilde{\xi}_{p_{r+1}})$, p_{r+1} the centre of v in the $(r+1)$-th neighbourhood. In any case $\kappa \geq 0$.
In particular, if there are finitely many centres of v on ξ, then

$$v(\xi) = \sum_p e_p(\xi)e_p(v),$$

the summation running on all centres p of v (or just on those lying on ξ).

In the sequel we will refer to the second equality in the claim of 8.1.6 as the *Noether formula for valuations*.

PROOF: It is clearly enough to prove the first equality in the case $r = 0$: the general case follows then by iteration, and the second equality is a direct consequence of the first one.
Let f be an equation of ξ. If O is the last centre of v, then v is the $\mathcal{M}$-adic valuation of $\mathcal{O}$ and the equality is obvious. Otherwise choose the coordinates x, y at O so that p_1 is on the x-axis, hence $x, z = y/x$ are local coordinates at p_1 and x is an equation of the exceptional divisor. In the local ring of p_1 we have $f = x^e \tilde{f}$, $e = e_O(\xi)$ and $\tilde{f}$ an equation of the strict transform of ξ. By 8.1.2 and the way the coordinates have been chosen, $e_O(v) = v(x)$ and hence the former equality gives $v(\xi) = v(f) = e_O(\xi)e_O(v) + v(\tilde{f}) = e_O(\xi)e_O(v) + v(\tilde{\xi}_{p_1})$ as wanted. $\diamond$

Theorem 8.1.7 (Proximity for valuations) *Assume that the centre p_{i+j}, $j > 0$ does exist and is proximate to p_i. Then*

$$e_{p_i}(v) \geq e_{p_{i+1}}(v) + \cdots + e_{p_{i+j}}(v).$$

Furthermore, this inequality is a strict one if and only if the next centre p_{i+j+1} does exist and is proximate to p_i too.

PROOF: It is obviously non-restrictive to assume $i = 0$. The centre p_1 certainly exists as we assume $j > 0$. After taking the coordinates as in the proof of 8.1.6, $e_O(v) = v(x)$. Now, if p_j is proximate to O, then $p_1, \ldots, p_j$ belong to the exceptional divisor $E : x = 0$ of blowing up O, and so the Noether formula for valuations 8.1.6 applied to E and the extension of v centred at p_1 gives

$$e_O(v) = v(x) = e_{p_1}(v) + \cdots + e_{p_j}(v) + \kappa, \quad \kappa \geq 0,$$

and hence the first inequality.

If there is a further centre proximate to O, the inequality in the claim needs to be strict, as a similar one with a further positive term holds. Conversely, assume that the inequality in the claim is strict: then the term κ in the above equality is positive and hence, by 8.1.6 again, p_j is not the last centre and $v(\tilde{E}_{p_{j+1}}) = \kappa > 0$, which means that p_{j+1} is proximate to O as wanted. $\diamond$

The next corollary deals with the cases of finitely and infinitely many centres proximate to p_i separately. Its proof directly follows from 8.1.7 and 8.1.5.

Corollary 8.1.8 *Assume that p_i is not the last centre of v.*

(a) There are finitely many centres of v proximate to p_i if and only if there are a positive integer j and centres $p_{i+1}, \ldots, p_{i+j}$ so that

$$e_{p_i}(v) = e_{p_{i+1}}(v) + \cdots + e_{p_{i+j}}(v).$$

In such a case the centres proximate to p_i are just $p_{i+1}, \ldots, p_{i+j}$.

(b) v has infinitely many centres proximate to p_i if and only if v has infinitely many centres and furthermore

$$e_{p_i}(v) > e_{p_{i+1}}(v) + \cdots + e_{p_{i+j}}(v)$$

for all $j > 0$. Then the centres of v after p_{i+1} are the points infinitely near to p_{i+1} proximate to p_i.

The next two corollaries are direct consequences of 8.1.7.

Corollary 8.1.9 *If p_i is not the last centre, then $e_{p_i}(v) \geq e_{p_{i+1}}(v)$ and $e_{p_i}(v) = e_{p_{i+1}}(v)$ if and only if no centre other than p_{i+1} is proximate to p_i.*

Corollary 8.1.10 *Assume that $p_{i+1}, \ldots, p_{i+j+1}$, $j \geq 0$ are proximate to p_i. Then*

$$e_{p_{i+1}}(v) = \cdots = e_{p_{i+j}}(v) \geq e_{p_{i+j+1}}(v).$$

PROOF: If $e_{p_{i+r}}(v) > e_{p_{i+r+1}}(v)$, $j \geq r > 0$, by 8.1.9, p_{i+r+2} does exist and is proximate to p_{i+r}. Then, since p_{i+r+2} is also proximate to p_{i+r+1}, it cannot be proximate to p_i (by 3.5.7) and so necessarily $i + r + 2 > i + j + 1$ and hence $j = r$. $\diamond$

Corollary 8.1.10 allows us to reformulate 8.1.8 in a more precise form, namely, for the case of finitely many proximate centres,

Corollary 8.1.11 *Assume that p_i is not the last centre. Then v has finitely many centres proximate to p_i if and only if for some positive integer h*

$$he_{p_{i+1}}(v) < e_{p_i}(v) \leq (h+1)e_{p_{i+1}}(v).$$

Then one has

$$e_{p_{i+1}}(v) = \cdots = e_{p_{i+h}}(v) \geq e_{p_{i+h+1}}(v)$$
$$e_{p_i}(v) = he_{p_{i+1}}(v) + e_{p_{i+h+1}}(v).$$

and the centres proximate to p_i are just $p_{i+1}, \ldots, p_{i+h+1}$.

PROOF: Assume that p_{i+1} is defined and $he_{p_{i+1}}(v) < e_{p_i}(v) \leq (h+1)e_{p_{i+1}}(v)$. Assume, as induction hypothesis, that for some j, $0 < j < h+1$, the centres $p_{i+1}, \ldots, p_{i+j}$ are defined and proximate to p_i, and assume also that v has equal multiplicities at all but the last of them. Then $e_{p_{i+1}}(v) + \cdots + e_{p_{i+j}}(v) \leq je_{p_{i+1}}(v) < e_{p_i}(v)$. By 8.1.7, p_{i+j+1} is then defined and proximate to p_i and so, by 8.1.10, if $j > 1$, $e_{p_{i+j}}(v) = e_{p_{i+1}}(v)$. Thus we obtain that $p_{i+1}, \ldots, p_{i+h+1}$ are defined and proximate to p_i, and also that v has equal multiplicities at all but the last of them. A further centre p_{i+h+2} proximate to p_i would give $e_{p_{i+h+1}}(v) = e_{p_{i+1}}(v)$, again by 8.1.10, and then the proximity inequality would read $e_{p_i}(v) \geq (h+1)e_{p_{i+1}}(v) + e_{p_{i+h+2}}(v) > (h+1)e_{p_{i+1}}(v)$ against the hypothesis. This proves in particular the proximate centres to be finitely many.

Conversely, if there are finitely many centres proximate to p_i, take $h+1$ to be their number and then the converse and the rest of the claim are obvious from 8.1.8 and 8.1.10. ◇

Remark 8.1.12 The reader may easily see, using 8.1.9, that the inequality $e_{p_{i+h}}(v) \geq e_{p_{i+h+1}}(v)$, in 8.1.11 above, is an equality if an only if either p_{i+h+1} is the last centre or p_{i+h+2} is a free point.

For further reference let us state:

Corollary 8.1.13 *Assume again that p_i is not the last centre. Then there are infinitely many centres proximate to p_i if and only if for any $h > 0$,*

$$e_{p_i}(v) > he_{p_{i+1}}(v).$$

In such a case $e_{p_{i+1}}(v) = e_{p_{i+h}}(v)$ for $h > 0$ and the centres after p_{i+1} are the points infinitely near to p_{i+1} proximate to p_i.

PROOF: Obvious from 8.1.11, 8.1.10 and 8.1.8. ◇

Remark 8.1.14 Corollary 8.1.13 shows that if a valuation has a centre with infinitely many centres proximate to it, then its value group is a non-archimedean one.

Let us fix our attention on a pair of consecutive centres of v, say $p = p_j$ and $q = p_{j+1}$, and write $e = e_0 = e_p(v)$, $e' = e_1 = e_q(v)$. We will perform a sort of division algorithm in the value group $\mathcal{G}$ in order to describe the satellite centres after q, in a procedure which is to a certain extent similar to that we have used for describing the satellite points on irreducible branches in chapter 5.

Division algorithm at p: The first possibility is to have $he' < e$ for all positive integers h: then we say that the algorithm is *obstructed* at q.

Otherwise we divide e by e' in $\mathcal{G}$, that is, we take $h = h_1$, the *quotient*, to be the maximal integer for which $he' \leq e$, and then the *remainder* is $e_2 = e - he'$. Clearly $0 \leq e_2 < e'$. By 8.1.11, v has h centres of multiplicity e' in successive neighbourhoods of p, all proximate to p. If $e_2 = 0$, then, by 8.1.12, the next centre either does not exist or is a free point: we say that the algorithm ends there.

If $e_2 \neq 0$, the next centre is defined and still proximate to p, in fact it is the last centre proximate to p, by 8.1.10, and v has multiplicity e_2 at it. Then we repeat the procedure from the last two centres proximate to p we have found, the multiplicities of v at them being e' and e_2.

This gives rise to a finite or infinite sequence of divisions

$$e_{i-1} = h_i e_i + e_{i+1},$$

and to each such division there correspond h_i successive centres at which v has multiplicity e_i,

$$q_{h_1 + \cdots + h_{i-1} + 1}, \ldots, q_{h_1 + \cdots + h_{i-1} + h_i},$$

all proximate to $q_{h_1 + \cdots + h_{i-1}}$, and, only in the case $e_{i+1} \neq 0$, a further centre $q_{h_1 + \cdots + h_i + 1}$, still proximate to $q_{h_1 + \cdots + h_{i-1}}$, at which v has multiplicity e_{i+1} (take $p = q_0$, $q = q_1$).

Proposition 8.1.15 *There are three possibilities for the division algorithm at p, namely:*

(a) *For some $i > 0$ the algorithm is obstructed at $q_{h_1 + \cdots + h_{i-1} + 1}$: this occurs if and only if all centres after this point are proximate to $q_{h_1 + \cdots + h_{i-1}}$ (or to p if $i = 1$).*

(b) *For some $i > 0$ the algorithm ends at $q_{h_1 + \cdots + h_i}$: this occurs if and only if there are finitely many satellite points of p in the sequence of centres.*

(c) *The algorithm is not obstructed and does not end: this occurs if and only if the centres after p are distributed in infinitely many finite sequences of consecutive points in such a way that the points in the i-th sequence and the first point in the $(i + 1)$-th one are all proximate to the last point in the $(i - 1)$-th sequence (or to p, if $i = 1$).*

PROOF: While defining the algorithm we have seen that there are just the three possibilities in the claim, and also that the three given conditions are necessary. Sufficiencies are clear, as any two of the conditions are incompatible. ◇

Note that if $e = e'$ we are obviously in case (b), $h = 1$, $e_2 = 0$.

In the sequel we need to perform some computations with the elements of $\mathcal{G}$ using rational coefficients: for this just localize the $\mathbb{Z}$-module $\mathcal{G}$ in the multiplicative system $\mathbb{Z} - \{0\}$: we get a $\mathbb{Q}$ vector space $\bar{\mathcal{G}}$, whose elements are represented by fractions e/n, $e \in \mathcal{G}$, $n \in \mathbb{Z} - \{0\}$. Since $\mathcal{G}$ is a torsion-free $\mathbb{Z}$-module, as it is totally ordered, the natural morphism

$$\mathcal{G} \longrightarrow \bar{\mathcal{G}}$$
$$e \longmapsto e/1$$

is a monomorphism: we identify each $e \in \mathcal{G}$ with its image $e/1 \in \bar{\mathcal{G}}$ and hence $\mathcal{G}$ with a subgroup of $\bar{\mathcal{G}}$. Then the ordering of $\mathcal{G}$ easily extends to a total ordering of $\bar{\mathcal{G}}$ by the following rule: given $e/n, e'/n'$ one may assume without restriction that $n, n' > 0$; then take $e/n \le e'/n'$ if and only if $n'e \le ne'$.

Two elements $e', e \in \mathcal{G}$ are called $\mathbb{Q}$-*independent* (resp. $\mathbb{Q}$-*dependent*) if and only if they are linearly independent (resp. dependent) as elements of $\bar{\mathcal{G}}$. Equivalently, e, e' are $\mathbb{Q}$-dependent if and only if there exist integers m, n, not both zero, such that $me = ne'$.

We can give further information in two of the cases in 8.1.15:

Lemma 8.1.16 *The division algorithm at p ends if and only if e, e' are $\mathbb{Q}$-dependent. If in such a case the quotients the algorithm gives rise to are $h_1, \ldots, h_r$, and we write*

$$\frac{m}{n} = h_1 + \cfrac{1}{h_2 + \cfrac{1}{\ddots \cfrac{}{\cfrac{1}{h_r}}}}$$

with $\gcd(m, n) = 1$, then $e = me_r$, $e' = ne_r$ and therefore $e = (m/n)e'$.

PROOF: Assume that the algorithm ends. We will use induction on r, the number of divisions. If $r = 1$, then $e = h_1 e'$, $m = h_1$, $n = 1$ and the claim is obvious. Otherwise we write

$$\frac{m}{n} = h_1 + \frac{1}{m'/n'},$$

m', n' also assumed to be coprime, which gives $m = h_1 m' + n'$ and $n = m'$. Now the induction hypothesis gives $e' = m'e_r$ and $e_2 = n'e_r$. The first equality is just $e' = ne_r$ and using both we get $e = h_1 e' + e_2 = h_1 m'e_r + n'e_r = me_r$ as wanted. The last equality and the $\mathbb{Q}$-dependence of e and e' obviously follow. The converse is clear, as if $e = (m/n)e'$, then the algorithm runs as the usual one for computing $\gcd(m, n)$. $\diamond$

Lemma 8.1.17 *Assume that the division algorithm at p is not obstructed and does not end. If the quotients it gives rise to are $h_1, \ldots, h_i, \ldots$, then $\beta_i e' < e < \beta_{i+1} e'$ for any odd i, where, for all i,*

$$\beta_i = h_1 + \cfrac{1}{h_2 + \cfrac{1}{\ddots \cfrac{1}{h_i}}}.$$

PROOF: We use induction on i. The inequality $h_1 e' < e$ is clear. If $i > 1$, write

$$\beta_i = h_1 + \frac{1}{\beta'_{i-1}}$$

and assume i to be odd: then, by the induction hypothesis $e' < \beta'_{i-1} e_2$, from which an easy computation gives $\beta_i e' < e$. Similarly, for i even, we get $e < \beta_i e'$ as wanted. ◇

8.2   Classification of valuations of $\mathcal{O}_{S,O}$

In the above section 8.1 we have defined the sequence of centres of any zero-dimensional valuation of $\mathcal{O}_{S,O}$ and we have seen some of its properties. The main result we will prove in this section is that any finite or infinite totally ordered sequence of consecutive infinitely near points beginning at O is the sequence of centres of a uniquely determined (up to isomorphism) valuation centred at O. After this, we will classify valuations according to the structure of their sequences of centres.

We begin by studying the sequences of positive elements in a totally ordered group that will eventually be sequences of multiplicities of valuations. Assume that we fix a finite or infinite sequence of points $P = \{p_i\}$, $0 \leq i < N$, $N \leq \infty$, so that $p_0 = O$ and p_i is in the first neighbourhood of p_{i-1} for $i > 0$. Since our goal is to construct a valuation with centres p_i, in the sequel we will call P the given sequence of centres and its points will be called centres. We will say that a map $e : P \longrightarrow \mathcal{G}$, $e(p_i) = e_i$, into a totally ordered additive group $\mathcal{G}$ *satisfies proximity relations* if and only if $e(P)$ generates $\mathcal{G}$, $e_i > 0$ for all $i = 1, \ldots, N$, and furthermore, if p_i is not the last point in P (i.e., $i \neq N - 1$ if $N < \infty$) either $e_i = e_{i+1} + \cdots + e_{i+h}$ if there are finitely many points $p_{i+1}, \ldots, p_{i+h}$, $h > 0$, in P proximate to p_i, or $e_i > e_{i+1} + \cdots + e_{i+h}$ for all $h > 0$ if there are infinitely many points (necessarily all points p_{i+j}, $j > 0$ by 8.1.5) in P proximate to p_i. Of course the sequence of multiplicities of a valuation v, $e_i = e_{p_i}(v)$ satisfies proximity relations by 8.1.8.

Assume we have fixed a given sequence of centres $P = \{p_i\}$, $0 \leq i < N$, and a map $e : P \longrightarrow \mathcal{G}$ satisfying proximity relations as above. Let us add a symbol

∞ to the ordered group $\mathcal{G}$ by taking $\mathcal{J} = \mathcal{G} \cup \{\infty\}$ and extending ordering and addition by the usual formal rules $a < \infty$ and $a + \infty = \infty$ for all $a \in \mathcal{G}$. Then it will be useful to define a map $w = w_{P,e} : \mathcal{O} - \{0\} \longrightarrow \mathcal{J}$ by taking $w(f) = \infty$ if the germ $\xi : f = 0$ contains infinitely many centres and

$$w(f) = \sum_i e_{p_i}(\xi) e_i$$

otherwise. Since clearly $\omega(f)$ depends only on ξ, we will often write $w(\xi)$ for $w(f)$. If there are no germs containing infinitely many centres, clearly $w(f) \neq \infty$ for all f. Otherwise, by 8.1.5, there is a well determined irreducible germ $\gamma : g = 0$ containing all centres and $w(f) = \infty$ if and only if g divides f.

Assume that O is not the only centre, write $\mathcal{O}'$ for the local ring of p_1, take $P' = P - \{O\}$, $e' = e_{|P'}$ and define $w' : \mathcal{O}' - \{0\} \longrightarrow \mathcal{J}$ by the same rules as w, this time using P' and e'. The next lemma gives a sort of virtual Noether formula for w.

Lemma 8.2.1 *Assume that the given sequence of centres P contains finitely many points proximate to O and at least one point besides O. If ξ is any germ of curve, $\nu \leq e_O(\xi)$ and $\check{\xi}$ is the virtual transform of ξ with origin at p_1, O being taken with virtual multiplicity ν, then*

$$w(\xi) = \nu e_O + w'(\check{\xi}).$$

PROOF: The equality is obvious if $w(\xi) = \infty$ as in such a case both ξ and $\check{\xi}$ contain infinitely many centres. Otherwise let $p_1, \ldots, p_r$ be the centres proximate to O. We have:

$$w(\xi_i) = e_O(\xi_i) e_0 + \sum_{j>0} e_{p_j}(\xi_i) e_j$$

$$= \nu e_0 + (e_O(\xi_i) - \nu) e_0 + \sum_{j>0} e_{p_j}(\check{\xi}_i) e_j$$

$$= \nu e_0 + \sum_{j=1}^{r} (e_O(\xi_i) - \nu) e_j + \sum_{j>0} e_{p_j}(\check{\xi}_i) e_j$$

$$= \nu e_0 + \sum_{j>0} e_{p_j}(\check{\xi}_i) e_j$$

$$= \nu e_0 + w'(\check{\xi}_i)$$

as wanted. $\diamond$

Lemma 8.2.2 *For any invertible $f \in \mathcal{O}$, $w(f) = 0$. For any non-zero $f_1, f_2 \in \mathcal{O}$,*

$$w(f_1 f_2) = w(f_1) + w(f_2) \tag{8.1}$$

and furthermore, if $f_1 + f_2 \neq 0$,

$$w(f_1 + f_2) \geq \min\{w(f_1), w(f_2)\}. \tag{8.2}$$

PROOF: The proofs of the first claim and equality 8.1 are straightforward from the definition of w and therefore are left to the reader. We will prove the inequality 8.2 next. Write $f_0 = f_1 + f_2$ and let ξ_i be the germ of curve defined by f_i, $i = 0, 1, 2$. Using equality 8.1 it is clearly enough to prove 8.2 in the case of f_1, f_2 sharing no factor: this will be done using induction on $[\xi_1.\xi_2]$ after giving a direct proof in two special cases to which 8.2.1 does not apply.

Special case 1: Assume that there is a single centre, namely $p_0 = O$: then the claim is obvious and in fact in such a case w is isomorphic to the $\mathcal{M}$-adic valuation.

Special case 2: Assume that there are infinitely many centres proximate to O. By 8.1.5, the centres are O, a point p_1 in its first neighbourhood and all points proximate to O in the successive neighbourhoods of p_1. Then, since e satisfies proximity relations, one has $e_1 = e_i$ for any $i > 1$ and $e_0 > he_1$ for any positive integer h. If ζ is any germ of curve, from the definition of w we easily get

$$w(\zeta) = e_O(\zeta)e_0 + [\tilde{\zeta}_{p_1}.E]e_1,$$

E being the exceptional divisor of blowing up O. If ℓ is the tangent line at O corresponding to the point p_1, recall from 3.2.2 that $[\tilde{\zeta}_{p_1}.E]$ is the multiplicity of ℓ as a component of the tangent cone of ζ (zero if ℓ is not a principal tangent to ζ). If ξ_1 and ξ_2 have different multiplicity at O, say $e_O(\xi_1) < e_O(\xi_2)$, then $w(f_1) < w(f_2)$ and, on the other hand, ξ_0 and ξ_1 have the same tangent cone which gives $w(f_1+f_2) = w(f_1)$ as wanted. Otherwise $e_O(\xi_1) = e_O(\xi_2)$ and either $e_O(\xi_0) > e_O(\xi_1)$, in which case the claim is obviously satisfied, or $e_O(\xi_0) = e_O(\xi_1)$. In the latter case the initial form of $f_1 + f_2$ is the sum of the initial forms of f_1 and f_2 and so the multiplicity of ℓ as a component of the tangent cone of ξ_0 is non-less than the minimum of its multiplicities as component of the tangent cones of ξ_1 and ξ_2, which gives the claim, once again using the equality displayed above.

General case: O is not the only centre and there are finitely many centres proximate to it, so we are allowed to use 8.2.1. We will make induction on $[\xi_1.\xi_2]$. The case $[\xi_1.\xi_2] = 0$ is clear, as then either ξ_1 or ξ_2 is the empty germ and so either $w(f_1)$ or $w(f_2)$ is zero. Assume thus $[\xi_1.\xi_2] > 0$ and thus both ξ_1 and ξ_2 to be non-empty. Take $\nu = \min\{e_O(\xi_1), e_O(\xi_2)\}$. From 8.2.1 we get

$$w(\xi_i) = \nu e_0 + w'(\check{\xi}_i), \quad i = 0, 1, 2$$

and still an equation of $\check{\xi}_0$ may be obtained as the sum of equations of $\check{\xi}_1$ and $\check{\xi}_2$ (4.1.1). Since at least one of the virtual transforms $\check{\xi}_1, \check{\xi}_2$ is the strict one, by 4.1.2, $[\check{\xi}_1.\check{\xi}_2] < [\xi_1.\xi_2]$, the induction hypothesis applies to w' and the claim follows. ⋄

Proposition 8.2.3 *Let P be a given sequence of centres and $e : P \longrightarrow \mathcal{G}$ a map satisfying proximity relations. There is a zero-dimensional valuation v of $\mathcal{O}$ with sequence of centres P and multiplicities $\{e(p)|p \in P\}$.*

PROOF: Assume first that there is no germ of curve containing infinitely many centres p_i. Then all values of $w = w_{P,e}$ are finite and w itself is a valuation of

$\mathcal{O}$ by 8.2.2. Then we take $v = w$: checking that it has sequence of centres P and multiplicities $e(p)$, $p \in P$, is direct from the definition of w and is left to the reader.

By 8.1.5 we assume now that the given centres are the points on an irreducible germ $\gamma : g = 0$ and take $\mathcal{G}' = \mathbb{Z} \times \mathcal{G}$ lexicographically ordered (that is, $(n, u) \leq (n', u')$ if and only if either $n < n'$ or $n = n'$ and $u \leq u'$). For any non-zero $f \in \mathcal{O}$ write $f = g^r f'$ where g does not divide f' and define $v(f) = (r, w(f'))$. To see that v thus defined is a valuation we will just check that

$$v(f_1 + f_2) \geq \min\{v(f_1), v(f_2)\}$$

for any $f_1, f_2 \in \mathcal{O}$ with $f_1 + f_2 \neq 0$. The other conditions being easier to check, they are left to the reader. Write $f_i = g^{r_i} f_i'$, where g does not divide f_i' and so $v(f_i) = (r_i, w(f_i'))$, $i = 1, 2$. Assume first that $r_1 = r_2$. Then one has $f_1 + f_2 = g^{r_1}(f_1' + f_2')$ and if g still divides $(f_1' + f_2')$, $v(f_1 + f_2) = (r, a)$ with $r > r_1$ and the claim is true. Otherwise $v(f_1 + f_2) = (r_1, w(f_1' + f_2'))$ after which the claim follows from 8.2.2. Assume now that $r_1 \neq r_2$, say $r_1 < r_2$ in which case $v(f_1) < v(f_2)$. Then $f_1 + f_2 = g^{r_1}(f_1' + g^{r_2 - r_1} f_2')$ and so $v(f_1 + f_2) = (r_1, w(f_1' + g^{r_2 - r_1} f_2'))$ because g does not divide f_1'. Since $w(g) = \infty$, 8.2.2 gives $w(f_1' + g^{r_2 - r_1} f_2')) \geq w(f_1')$ and hence the claim. Once we have seen that v is a valuation, the reader may easily see, as in the former case, that it has sequence of centres P and multiplicities $e(p)$, $p \in P$. ◇

Proposition 8.2.4 *If P is a given sequence of centres, there is a map $e : P \longrightarrow \mathcal{G}$ satisfying proximity relations.*

PROOF: We consider three separate cases, namely:

Case 1: P is finite. Let P be $\{O = p_0, \ldots, p_n\}$. Take $\mathcal{G} = \mathbb{Z}$, $e(p_n) = 1$ and for any $i < n$, if $p_{i+1}, \ldots, p_{i+r}$ are the centres proximate to p_i, $e(p_i) = \sum_{j=1}^{r} e(p_{i+j})$. As is clear, such a map satisfies proximity relations.

Case 2: P contains infinitely many points proximate to a certain $p_n \in P$. In such a case, by 8.1.5, p_n is the only centre with infinitely many centres proximate to it. Take $\mathcal{G} = \mathbb{Z} \oplus \mathbb{Z}$ lexicographically ordered, and define $e(p_{n+i}) = (0, 1)$ for $i > 0$, $e(p_n) = (1, 0)$. For $i < n$ there are finitely many centres proximate to p_i, say $p_{i+1}, \ldots, p_{i+r}$: then we define $e(p_i) = \sum_{j=1}^{r} e(p_{i+j})$. Again the reader may easily check that the map e satisfies proximity relations.

Case 3: P is an infinite set and no centre has infinitely many centres proximate to it. First of all let us recall (see [78] for instance) that both finite and infinite continued fractions define real numbers: if $\mathbf{h} = \{h_1, \ldots, h_n\}$ is a finite sequence of positive integers, write

$$\alpha(\mathbf{h}) = \alpha(h_1, \ldots, h_n) = h_1 + \cfrac{1}{h_2 + \cfrac{1}{\ddots \cfrac{1}{h_n}}},$$

which is a rational number. If $\mathbf{h} = \{h_1, \ldots, h_n, \ldots\}$ is an infinite sequence of positive integers, as is well known, the sequence $\{\alpha(h_1, \ldots, h_n)\}$ converges to an irrational number we will denote by $\alpha(\mathbf{h}) = \alpha(h_1, \ldots, h_n, \ldots)$. In both the finite and infinite case, if $\mathbf{h}'$ denotes the sequence obtained from $\mathbf{h}$ by dropping the first term, one has

$$\alpha(\mathbf{h}) = h_1 + \frac{1}{\alpha(\mathbf{h}')}.$$

Now the multiplicities $e(p_i)$ will be taken in the additive group of real numbers $\mathbb{R}$ and $\mathcal{G}$ will be just the subgroup generated by them. Start by defining $e(O) = 1$ and assume by induction that $e(p_j)$ has been defined for $j \leq i$. Take $r = h_r = 1$ if p_{i+1} is the only centre proximate to p_i. Otherwise assume that

$$p_{i+1}, \ldots, p_{i+h_1+1}, \quad h_1 > 0$$

are the centres proximate to p_i,

$$p_{i+h_1+1}, \ldots, p_{i+h_1+h_2+1}, \quad h_2 > 0$$

are those proximate to p_{i+h_1}, and, inductively

$$p_{i+h_1+\cdots+h_{j-1}+1}, \ldots, p_{i+h_1+\cdots+h_j+1}, \quad h_j > 0$$

are the centres proximate to $p_{i+h_1+\cdots+h_{j-1}}$, either for all $j > 0$ if all centres after p_{i+1} are satellite points, or for $j < r$ otherwise, in which case we assume that still

$$p_{i+h_1+\cdots+h_{r-1}+1}, \ldots, p_{i+h_1+\cdots+h_r}$$

are proximate to $p_{i+h_1+\cdots+h_{r-1}}$ and $p_{i+h_1+\cdots+h_r+1}$ is not proximate to either $p_{i+h_1+\cdots+h_{r-1}}$ or $p_{i+h_1+\cdots+h_r-1}$. In this case we assume $h_r > 1$. In other words if the Enriques diagram representing the sequence of centres has no corner at p_{i+1}, then $r = 1$ and $h_1 = 1$. Otherwise the successive straight edges of the diagram from p_{i+1} onwards have lengths h_j, but for the last of them (if they are finitely many) which has length $h_r - 1$.

Now write $\mathbf{h}$ for the finite or infinite sequence of the h_i and define $e(p_{i+1})$ by the rule:

$$\frac{e(p_i)}{e(p_{i+1})} = \alpha(\mathbf{h}).$$

Notice that in the case of p_{i+1} being the only point proximate to p_i, the definition gives just $e(p_{i+1}) = e(p_i)$.

Once all multiplicities $e(p_i)$, $i \geq 0$ have been defined, we will check that proximity relations are satisfied. To this end we need just to see that if p_i denotes an arbitrary centre and we keep for the points after it the notations introduced above, then either

$$e(p_i) = e(p_{i+1}) + \cdots + e(p_{i+h_1})$$

if r does exist and is equal to 1, or

$$e(p_i) = e(p_{i+1}) + \cdots + e(p_{i+h_1}) + e(p_{i+h_1+1})$$

otherwise.

In the first case each of the points $p_{i+1}, \ldots, p_{i+h_1}$ has a single centre proximate to it, so the definition above gives $e(p_{i+h_1}) = e(p_{i+h_1-1}) = \cdots = e(p_{i+1})$ while we have $\alpha(\mathbf{h}) = h_1$ and so, by the definition, $e(p_i) = h_1 e(p_{i+1})$ after which the equality we want is clear.

Assume that we have the second case: still p_{i+j+1} is the only centre proximate to p_{i+j} for $0 < j < h_1$ as the next centre p_{i+j+2} is assumed to be proximate to both p_i and p_{i+j+1} (3.5.8). This gives $e(p_{i+h_1}) = \cdots = e(p_{i+1})$. Our definition of multiplicities gives

$$\frac{e(p_{i+h_i})}{e(p_{i+h_i+1})} = \alpha(\mathbf{h'})$$

if, as above, $\mathbf{h'}$ is the sequence obtained from $\mathbf{h}$ by dropping the first term. Thus it is enough to recall that, still by the definition,

$$\frac{e(p_i)}{e(p_{i+1})} = \alpha(\mathbf{h}) = h_1 + \frac{1}{\alpha(\mathbf{h'})} = h_1 + \frac{e(p_{i+h_i+1})}{e(p_{i+1})}.$$

⋄

Lemma 8.2.5 *Let v be a zero-dimensional valuation of $\mathcal{O}$. Assume that no germ of curve contains infinitely many centres of v and no centre of v has infinitely many centres proximate to it. Then v is an archimedean valuation.*

PROOF: Note first that since no germ of curve contains infinitely many centres of v by the hypothesis, all values $v(f)$, $f \in \mathcal{O}$, $f = 0$, may be computed by the Noether formula (8.1.6) and so $\mathcal{G}$ is generated by the multiplicities of v.

Since no centre of v has infinitely many centres proximate to it, for any centre p_i, the division algorithm at p_i is not obstructed, either 8.1.16 or 8.1.17 applies, and so there is a positive integer n so that $ne_{p_{i+1}}(v) > e_{p_i}(v)$. From this it easily follows that $\mathcal{G}$, which we already know to be generated by the multiplicities of v, is an archimedean group. ⋄

Now we may state and prove the main result in this section:

Theorem 8.2.6 *If $P = \{p_i\}$, $0 \leq i < N$, $N \leq \infty$, is a finite or infinite sequence of points, $p_0 = O$ and p_i in the first neighbourhood of p_{i-1} for $i > 0$, there is a zero-dimensional valuation of $\mathcal{O} = \mathcal{O}_{S,O}$ whose sequence of centres is P. Furthermore such a valuation is determined up to isomorphism by the sequence P.*

PROOF: The existence of v is clear from 8.2.3 and 8.2.4. Thus, we need to see uniqueness only, and for this we will distinguish three cases, namely,

Case 1: There is a germ containing infinitely many centres. Then the centres are the points on an irreducible germ $\gamma : g = 0$, by 8.1.5. Let p_i be a simple and free point of γ and put $e = e_{p_i}(v)$: since there are no satellite points on γ after p_i, each centre from p_i onwards has a single centre proximate to it and so, by 8.1.7, $e = e_{p_j}(v)$ for $j \geq i$. Furthermore, each centre p_j, $j < i$ has finitely many centres proximate to it, as all centres are on γ, and so, by 8.1.8, $e_j = e_{p_j}(v)$ equals the sum of the multiplicities of v at its proximate points. By working backwards from p_i, it is clear that all multiplicities e_j $j < i$ are multiples of e, namely $e_j = n_j e$, the coefficients n_j depending only on the sequence of centres and not on the valuation v. In fact the reader may easily see that $n_j = e_{p_j}(\gamma)$, as the multiplicities of v and γ are computed backwards by the same recurrent procedure using the same proximity equalities.

Take $f \notin g\mathcal{O}$, so that only finitely many centres of v lie on the germ $f = 0$: then $v(f)$ may be computed using the Noether formula (8.1.6) which shows it to be a multiple of e, $v(f) = ne$, and the coefficient n to be dependent on f itself and the sequence of centres only.

Call $e' = v(g)$: for any positive integer n, 8.1.6 may be applied using centres $p_0, \ldots, p_n$ to get

$$e' = v(g) \geq \sum_{j=0}^{n} e_{p_j}(\gamma) e_{p_j}(v) \geq (n+1)e > ne.$$

This shows that $\mathcal{G}$ is non-archimedean and also that e, e' are $\mathbb{Q}$-independent. Thus we have $\mathbb{Z}e' \oplus \mathbb{Z}e$, lexicographically ordered, as a subgroup of $\mathcal{G}$. Now if any non-zero $f \in \mathcal{O}$ is written as $f = g^r f'$, $f' \notin g\mathcal{O}$, clearly $v(f) = re' + v(f')$. Furthermore, as seen above, $v(f') = ne$ and the coefficient n does not depend on v, but on f' (or f) and the centres only. In fact, if ξ' denotes the germ $f' = 0$, it easily follows from the equalities $n_j = e_{p_j}(\gamma)$ that $n = [\xi'.\gamma]$. The value group is thus $\mathcal{G} = \mathbb{Z}e' \oplus \mathbb{Z}e$ and v is determined by its sequence of centres up to isomorphism as claimed.

Case 2: There is a centre $p = p_i$ which has infinitely many centres proximate to it. Write $p_{i+1} = q$, $e = e_p(v)$, and $e' = e_q(v)$. We have seen in 8.1.13 that $e_{p_j}(v) = e'$ for $j \geq i+1$ and also $e > ne'$ for any positive integer n. As in the former case, $\mathbb{Z}e \oplus \mathbb{Z}e'$ lexicographically ordered is a subgroup of $\mathcal{G}$. Again one may compute the multiplicities of $e_{p_j}(v)$, $j < i$: one uses proximity equalities working backwards from p_{i-1}, which easily gives that all multiplicities $e_{p_j}(v)$ belong to $\mathbb{Z}e \oplus \mathbb{Z}e'$, $e_{p_j}(v) = n_j e + m_j e'$, the positive integers n_j, m_j depending only on the sequence of centres and the particular centre p_j. In this case all centres from p_{i+2} onwards are satellite points, so no germ of curve may contain infinitely many centres (by 3.7.7) and all values $v(f)$ may be computed by means of the Noether formula (8.1.6). This shows all values to have the form $v(f) = ne + me'$, the positive integers m, n depending only on the germ $f = 0$ and the sequence of centres. As in the former case, the value group is $\mathcal{G} = \mathbb{Z}e \oplus \mathbb{Z}e'$ and v is determined by its sequence of centres up to isomorphism.

Case 3: No germ of curve contains infinitely many centres and no centre has infinitely many centres proximate to it. We know from 8.2.5 that $\mathcal{G}$ is in this case an archimedean group. Then by a well-known argument (see for instance [95], Vol. II, p. 45), $\mathcal{G}$ is isomorphic to a subgroup of the additive group $\mathbb{R}$ of the real numbers. So it is not restrictive to assume, up to an isomorphism, that actually

$\mathcal{G}$ is a subgroup of $\mathbb{R}$. Even more, since dividing by $e_O(v)$ is an automorphism of $\mathbb{R}$ as ordered group, we may assume that $e_O(v) = 1$. Then, inductively, each $e_{p_i}(v)$ may be determined from $e_{p_{i-1}}(v)$, $i > 0$, using either 8.1.16 or 8.1.17 and the division algorithm. Indeed, assume that $e_{p_{i-1}}(v)$ has been determined and that p_i is a free point. Then either 8.1.16 or 8.1.17 determine $e_{p_{i-1}}(v)/e_{p_i}(v)$ as a finite or infinite continued fraction depending only on the centres. Once $e_{p_i}(v)$ has been determined, if the next centre p_{i+1} is free, we restart from it. If not, there is a non empty group of consecutive satellite centres just after p_i and the multiplicities of v at these centres are determined from $e_{p_{i-1}}(v)$ and $e_{p_i}(v)$ as remainders of the division algorithm. If this group of satellite centres is infinite, then all multiplicities have been determined. If not we restart from the first free centre after p_i, as the multiplicities of all (satellite) centres between it and p_i have been already determined. Now, as in case 2, once all multiplicities have been determined, all values may be computed using the Noether formula (8.1.6) and therefore are well determined too. This ends the proof. $\diamond$

Once it has been seen that there is one and only one (up to isomorphism) valuation with an arbitrarily given sequence of centres, zero-dimensional valuations of $\mathcal{O}$ may be classified according to the structure of their sequences of centres. We list the different types and summarize their main properties, most of which have been already obtained.

Type 1: *Valuations with finitely many centres (divisorial valuations).* Call p_n the last centre of such a valuation v, $\mathcal{O}'$ the local ring of p_n and $\mathcal{M}'$ its maximal ideal. By 8.1.3, v is the restriction to $\mathcal{O}$ of the $\mathcal{M}'$-adic valuation of $\mathcal{O}'$, otherwise there would be a centre in the first neighbourhood of p_n. Equivalently, if E denotes the exceptional divisor of blowing up p_n, for any germ of curve ξ, $v(\xi)$ is the multiplicity of E as component of the total transform of ξ after blowing up all centres. These valuations have been already considered in section 4.5, notice in particular that they are discrete valuations.

Type 2: *Valuations with infinitely many centres on a germ of curve.* By 8.1.5, the centres of such a valuation are the points on an irreducible germ $\gamma : g = 0$. We have seen in the proof of 8.2.6 that the value group is, up to isomorphism, $\mathbb{Z} \oplus \mathbb{Z}$ lexicographically ordered, and so these valuations are non-archimedean. Furthermore, it has also been seen there that if for any $f \in \mathcal{O}$ we write $f = g^r f'$, f' with no factor g, and ξ' is the germ $f' = 0$, then $v(f) = (r, [\xi'.\gamma])$ up to isomorphism.

Type 3: *Valuations with infinitely many centres from which at most finitely many are satellite or lie on the same germ of curve.* It has been seen in 5.7.7 that such sequences of centres actually exist. Since no germ contains infinitely many centres, all values may be computed by the Noether formula (8.1.6) and therefore the value group is generated by the multiplicities of the centres. Furthermore, since there are finitely many satellite centres, all division algorithms end (by 8.1.15) and 8.1.16 always applies. Let p, q be consecutive centres and assume that q is free. If the centre in the first neighbourhood of q is not proximate to p, then 8.1.6 obviously gives $e_p(v) = e_q(v)$. Otherwise there is a non-empty group of consecutive satellite centres, the first one in the first neighbourhood of q: still by 8.1.6, $me_q(v) = ne_p(v)$ where m, n are positive integers, $n < m$ and, by the division algorithm, all multiplicities of v at the satellite points in the group above are in the subgroup generated by $e_p(v), e_q(v)$. If as in the proof of 8.2.6 one takes $e_O(v) = 1$, then all multiplicities are rational numbers and they become constant from the last satellite centre onwards. This shows that the value group $\mathcal{G}$ is generated by finitely many rational numbers (in fact we have found the generator 1 and one further generator for each non-empty group of satellite centres). If these generators are $a_1/b, \ldots, a_r/b$, $\gcd(a_1, \ldots, a_r)/b$ alone generates $\mathcal{G}$ too. $\mathcal{G}$ is thus isomorphic to $\mathbb{Z}$ and the valuation is a discrete one.

Type 4: *Valuations with infinitely many satellite centres, no infinitely many of them consecutive.* Since no germ of curve contains infinitely many satellite points (3.7.7), it follows from 8.1.5 that in this case no germ of curve may contain infinitely many centres. Then all values may be computed by the Noether formula 8.1.6 and still all division algorithms end, because there is always a free centre after any given centre (8.1.15). Thus, as for type 3 above, we find all multiplicities to be rational numbers and the value group $\mathcal{G}$ to be generated by them. Let us see that in this case, however, the value group is not finitely generated. Fix our attention on a non-empty group of satellite centres: assume as above that p is a centre, that the centre q in its first neighbourhood is free and that the centre in the first neighbourhood of q is proximate to p. Then we still have $me_q(v) = ne_p(v)$ and $n/m < 1$, hence $e_p > e_q$. Since we are assuming we have infinitely many groups of satellite centres separated by free centres, it is clear that we can construct a strictly decreasing infinite sequence of multiplicities, which are in particular positive elements of $\mathcal{G}$. On the other hand, when dealing with valuations of type 3, we have seen that any finite generated subgroup of $\mathbb{Q}$ is isomorphic to $\mathbb{Z}$ and therefore has no infinite strictly decreasing sequence of positive elements. It follows that the valuation is archimedean and non-discrete.

Type 5: *Valuations with infinitely many consecutive satellite centres but no centre with infinitely many centres proximate to it.* As in case 4 above, no germ of curve may contain infinitely many centres. Thus again all values may be computed using the Noether formula and so the value group is generated by the multiplicities of the centres. Still take $e_O(v) = 1$ as in the proof of 8.2.6 and let q be the last free centre and p be the centre preceding it. Then, as in former cases, all multiplicities of v at centres preceding q are rational numbers, but now the division algorithm at q, still non-obstructed, does not end and $e_p(v)/e_q(v)$ is an irrational number as it is given by an infinite continued fraction. The multiplicities of v at the points after q being given by the division algorithm, we find the value group $\mathcal{G}$ generated by finitely many rational numbers and a single irrational number. Rational generators may of course be reduced to a single one and $\mathcal{G}$ is a subgroup of $\mathbb{R}$ with two generators, one of them rational and the other irrational. Thus these valuations are also archimedean and non-discrete.

Type 6: *Valuations one of whose centres has infinitely many proximate centres.* Such valuations have been described in the proof of 8.2.6, case 2. If p_i is the centre having infinitely many proximate centres, the centres after p_{i+1} are the points infinitely near to it and proximate to p_i. Furthermore $e_{p_{i+1}} = e_{p_{i+j}}$ for $j > 1$ while the multiplicities at the centres p_j, $j < i$ are computed backwards from those of p_i and p_{i+1} using proximity relations. All values are given by the Noether formula, the value group is $\mathbb{Z} \oplus \mathbb{Z}$ lexicographically ordered, generated by $e_{p_i}(v)$ and $e_{p_{i+1}}(v)$, and so the valuation is a non-archimedean one.

Of course, there is a finer classification of valuations using a criterion similar to equisingularity of germs: one may call two valuations *equisingular* if and only if there is a bijection between their sequences of centres that preserves ordering and proximity. The reader may see as an exercise that any two valuations that are equisingular in this sense have (up to an isomorphism) the same sequence of multiplicities.

To close this section let us just mention that one may associate with any valuation of type 3, 4 or 5 a generalized Puiseux series $s = s(x)$ so that, up to isomorphism, for any $f \in \mathcal{O}$, $v(f) = o_x f(x, s(x))$. For a valuation of type 3 one may use the series associated with its sequence of centres in 5.7.7. In the case of a valuation of type 4 the series is defined similarly, its partial sums are partial sums shared by Puiseux series of all germs going through one of the free centres (5.7.1). One gets a sort of fractionary power series with infinitely many characteristic exponents. Valuations of type 5 have associated a sort of series with finitely many monomials with rational exponent followed by a single monomial with irrational exponent. To be a bit more precise, if q is the last free centre and p the centre preceding it, let us write $s'(x) = \sum_1^k a_i x^{\alpha_i}$, $\alpha_i \in \mathbb{Q}$ a common partial sum of Puiseux series of all irreducible germs through q (5.7.1): then one takes $s = s' + x^{\alpha_k + e_q(v)/n e_p(v)}$, n the least common denominator of the α_i.

8.3   Complete ideals and linear systems

In this section we introduce the algebraic notion of complete ideal and the corresponding geometric one of complete linear system, and prove that the complete $\mathcal{M}_{S,O}$-primary ideals of $\mathcal{O}_{S,O}$ are just those of the form $H_\mathcal{K}$ for $\mathcal{K}$ a weighted cluster with origin at O. As in former sections O is a point on a smooth surface S and we write $\mathcal{O} = \mathcal{O}_{S,O}$ and $\mathcal{M} = \mathcal{M}_{S,O}$.

Let us recall (see section 7.3) that a *strictly consistent cluster* is a consistent cluster no point of which has virtual multiplicity zero. Strictly consistent clusters will allow us to select a well determined consistent cluster in each equivalence class of weighted clusters.

Remark 8.3.1 If I is an $\mathcal{M}$-primary ideal, $BP(I)$ is a strictly consistent cluster, as we know it to be consistent (7.2.14) and it follows directly from its definition that all virtual multiplicities in $BP(I)$ are positive.

The next result is the main reason for introducing strictly consistent weighted clusters.

Proposition 8.3.2 *If $\mathcal{K}$ is a strictly consistent weighted cluster, then $\mathcal{K} = BP(H_\mathcal{K})$.*

PROOF: Let us write $\mathcal{K} = (K, \nu)$, $H = H_\mathcal{K}$ and $\mathcal{K}' = (K', \nu') = BP(H_\mathcal{K})$. According to 4.2.7 and 7.2.13 there are non-empty Zariski-open sets U, U' in H whose elements define germs going sharply through $\mathcal{K}$ and $\mathcal{K}'$ respectively. The intersection of these open sets being necessarily non-empty, one may thus fix a germ ξ going sharply through both $\mathcal{K}$ and $\mathcal{K}'$. Let T be the set of all points on ξ that do not belong to K and lie in the first neighbourhood of some point in K. Define T' in the same way using $\mathcal{K}'$ instead of $\mathcal{K}$. By 4.2.8 and 7.2.13 there are non-empty Zariski-open sets $V \subset H$, whose elements define germs going sharply through $\mathcal{K}$ and missing all points in T, and $V' \subset H$, whose elements define germs that go sharply through $\mathcal{K}'$ and miss all points in T'. As before, one may fix a germ ζ going sharply through both $\mathcal{K}$ and $\mathcal{K}'$ and missing all points either in T or in T'. Then points shared by ξ and ζ belong to K by the construction of ζ. Conversely, any point in K lies on both ξ and ζ as both germs go sharply through $\mathcal{K}$ and, $\mathcal{K}$ being strictly consistent by hypothesis, all its virtual multiplicities are positive. The same argument applies to the points of $\mathcal{K}'$ which we have noticed to be strictly consistent in 8.3.1 above. Thus we get $K = K'$. Furthermore, since ξ goes sharply through both $\mathcal{K}$ and $\mathcal{K}'$, $\nu_p = e_p(\xi) = \nu'_p$ for any $p \in K$, which gives $\mathcal{K} = \mathcal{K}'$ as wanted. ⋄

Corollary 8.3.3 *Given a weighted cluster $\mathcal{K}'$, there is a unique strictly consistent cluster $\mathcal{K}$ equivalent to $\mathcal{K}'$, and furthermore $\mathcal{K} = BP(H_{\mathcal{K}'})$.*

PROOF: For the existence, just get from $\mathcal{K}'$ an equivalent consistent cluster $\mathcal{K}''$ by 4.6.2 and then drop from it the points of virtual multiplicity zero (7.3.6) to get an equivalent strictly consistent cluster $\mathcal{K}$. The equality in the claim follows from 8.3.2, as $H_\mathcal{K} = H_{\mathcal{K}'}$. Uniqueness of $\mathcal{K}$ is then clear from this equality, and follows also from 4.6.4. ⋄

It is worth noticing that we have two ways of getting the only strictly consistent cluster $\mathcal{K}$ equivalent to a given weighted cluster $\mathcal{K}'$: we may either apply to $\mathcal{K}'$ an unloading procedure and then drop from the resulting cluster all points with virtual multiplicity zero, or just take $\mathcal{K} = BP(H_{\mathcal{K}'})$.

Next we will recall some fairly standard notions from commutative algebra we need in the sequel. They are presented here just for the ring $\mathcal{O}$, for further information the reader is referred to [95] vol. II app. 4.

An ideal $I \subset \mathcal{O}$ is said to be *complete* or *integrally closed* if and only if it can be defined by valuations of $\mathcal{O}$, that is, to be precise, if and only if there exist a set of valuations $\{v_\ell\}_{\ell \in \Lambda}$ of $\mathcal{O}$ and for each $\ell \in \Lambda$ an element α_ℓ in the value group of v_ℓ so that

$$I = \{f \in \mathcal{O} \mid v_\ell(f) \geq \alpha_\ell \text{ for all } \ell \in \Lambda\}.$$

Linear systems defined by complete ideals will be called *complete linear systems*.

As is clear from the definitions, $\mathcal{M}$ is a complete ideal and arbitrary intersections of complete ideals are complete too. Therefore, if I is an arbitrary ideal of $\mathcal{O}$, then the intersection $\bar{I}$ of all complete ideals containing I is the least complete ideal containing I and is called the *completion* or the *integral closure* of I. We are mainly interested in $\mathcal{M}$-primary ideals: if I is $\mathcal{M}$-primary, then its completion $\bar{I}$ is $\mathcal{M}$-primary too. Indeed, $\bar{I}$ contains a power of $\mathcal{M}$ as it contains I. Furthermore since $I \subset \mathcal{M}$ and $\mathcal{M}$ is complete, $\bar{I} \subset \mathcal{M}$ too, $\bar{I}$ is a proper ideal and hence an $\mathcal{M}$-primary one by 1.8.10.

The elements of the integral closure $\bar{I}$ of I are said to be *integral over I*.

One may equivalently define as integral over I the elements $f \in \mathcal{O}$ that satisfy a relation (*integral dependence relation*) of the form

$$f^n + a_1 f^{n-1} + \cdots + a_n = 0$$

where $n > 0$ and $a_i \in I^i$ for $i = 1, \ldots, n$. The completion $\bar{I}$ of I is then the set of all elements integral over I and an ideal is complete if and only if it agrees with its completion. We will give no proof of the equivalence of the two series of definitions, as the latter ones will not be used in the sequel. The interested reader is referred again to [95], Vol. II, app. 4, where this equivalence is proved in a more general frame.

Lemma 8.3.4 *For any weighted cluster $\mathcal{K}$, $H_{\mathcal{K}}$ is a complete ideal.*

PROOF: Obvious from 4.5.4, as the p-values there appearing are divisorial valuations of $\mathcal{O}$, classified as type 1 at the end of the preceding section 8.2. ⋄

The next theorem is in fact the main link between linear systems of germs and their clusters of base points on one side, and $\mathcal{M}$-primary ideals and their completions on the other.

Theorem 8.3.5 *Let I be an $\mathcal{M}$-primary ideal of $\mathcal{O}$. If the germ $g = 0$, $g \in \mathcal{O}$, goes through $BP(I)$, then, for any valuation v of $\mathcal{O}$,*

$$v(g) \geq \min\{v(f)|f \in I\}.$$

PROOF: It is clearly enough to prove the claim for the elements of a system of generators $g_1, \ldots, g_m$ of $H_{BP(I)}$. Since $BP(H_{BP(I)}) = BP(I)$ by 8.3.2, we may apply 7.2.16 and assume without restriction that all germs $\xi_i : g_i = 0$ go sharply through $BP(I)$.

Let v be a valuation of $\mathcal{O}$ and put $k = \min\{v(f)|f \in I\}$. If v is one-dimensional, then its centre is a principal ideal which cannot contain I, as I is assumed to be $\mathcal{M}$-primary (1.8.10). So we have $k = 0$ and there is nothing to prove in this case.

Assume thus that v is a zero-dimensional valuation. If v has centres outside of $BP(I)$, let us call q the first of them and, by 7.2.13, choose $f \in I$ so that the germ $\zeta : f = 0$ goes sharply through $BP(I)$ and misses q. If all centres of v are in $BP(I)$, then just take f so that ζ goes sharply through $BP(I)$. In any case all centres of v lying on ζ belong to $BP(I)$ and since both ζ and ξ_i go sharply through $BP(I)$, the Noether formula 8.1.6 applied twice gives

$$k \leq v(f) = \sum_{p \in BP(I)} e_p(\zeta) e_p(v) = \sum_{p \in BP(I)} e_p(\xi_i) e_p(v) \leq v(g_i),$$

for $i = 1, \ldots, m$, as wanted. ⋄

Corollary 8.3.6 *Let I be an $\mathcal{M}$-primary ideal of $\mathcal{O}$: the integral closure $\bar{I}$ of I is $\bar{I} = H_{BP(I)}$.*

PROOF: The inclusion $\bar{I} \subset H_{BP(I)}$ is obvious, as $H_{BP(I)}$ is complete by 8.3.4. Conversely, assume that J is a complete ideal containing I. The ideal J being complete, write it as

$$J = \{f \in \mathcal{O} | v_\ell(f) > n_\ell \text{ for all } \ell \in \Lambda\}.$$

Since $I \subset J$, for any ℓ,

$$k_\ell = \min\{v_\ell(f) | f \in I\} \geq n_\ell,$$

so that 8.3.5 gives $H_{BP(I)} \subset J$ and hence $H_{BP(I)} \subset \bar{I}$ as wanted. ⋄

Corollary 8.3.7 *An $\mathcal{M}$-primary ideal I of $\mathcal{O}$ is complete if and only if it has the form $I = H_\mathcal{K}$ for a consistent cluster $\mathcal{K}$. In such a case one may take $\mathcal{K} = BP(I)$.*

PROOF: Obvious from 8.3.4 and 8.3.6. ⋄

Corollary 8.3.8 *For any $\mathcal{M}$-primary ideal I, $BP(I) = BP(\bar{I})$.*

PROOF: By 8.3.6 and 8.3.2, $BP(\bar{I}) = BP(H_{BP(I)}) = BP(I)$ as claimed. ⋄

The next corollary shows that the complete $\mathcal{M}$-primary ideals are those whose codimension reaches the lower bound of 7.2.18.

Corollary 8.3.9 *An $\mathcal{M}$-primary ideal I of $\mathcal{O}$ is complete if and only if*

$$\dim_{\mathbb{C}} \mathcal{O}/I = \sum_{p \in BP(I)} \frac{e_p(I)(e_p(I) + 1)}{2}.$$

PROOF: By 8.3.6, $\bar{I} = H_{BP(I)}$. Then, by the codimension formula 4.7.1, the equality of the claim is $\dim_{\mathbb{C}} \mathcal{O}/I = \dim_{\mathbb{C}} \mathcal{O}/\bar{I}$, obviously equivalent to $I = \bar{I}$. ⋄

Corollary 8.3.10 *Any complete $\mathcal{M}$-primary ideal I of $\mathcal{O}$ may be defined by finitely many divisorial valuations, that is,*

$$I = \{f \in \mathcal{O} | v_\ell(f) \geq n_\ell \text{ for all } \ell = 1, \ldots, r\},$$

v_ℓ being a divisorial valuation of $\mathcal{O}$ and n_ℓ a positive integer, for $\ell = 1, \ldots, r$.

PROOF: Follows from 8.3.7, as the claim is true for the ideals $H_\mathcal{K}$ by 4.5.4. ◇

Corollary 8.3.11 *By mapping each $\mathcal{M}$-primary complete ideal I to its weighted cluster of base points $BP(I)$, and each strictly consistent cluster $\mathcal{K}$ to the corresponding ideal $H_\mathcal{K}$, we get a pair of reciprocal one to one maps between the set of all $\mathcal{M}$-primary complete ideals of $\mathcal{O}$ and that of all strictly consistent clusters with origin at O.*

PROOF: That $\mathcal{K} = BP(H_\mathcal{K})$ for any strictly consistent cluster $\mathcal{K}$ has been seen in 8.3.2, while the equality $I = H_{BP(I)}$ follows from 8.3.6. ◇

We say that clusters K, K' are *similar* if and only if there is a bijection $\varphi :$ $K \longrightarrow K'$ (*similarity*) so that both φ and φ^{-1} preserve ordering and proximity. Weighted clusters $\mathcal{K} = (K, \nu), \mathcal{K}' = (K', \nu')$ are called similar if there is a similarity $\varphi : K \longrightarrow K'$ preserving virtual multiplicities, that is, $\nu'_{\varphi(p)} = \nu_p$ for all $p \in K$. An analytic isomorphism Φ defined in a neighbourhood of O obviously induces a similarity between each cluster K with origin at O and its image $\Phi(K)$ (see section 3.3 for the action of Φ on infinitely near points). We call such of similarities *congruences of clusters*, the clusters K and $\Phi(K)$ being called *congruent* or *isomorphic* clusters. Congruence of clusters is far more restrictive than similarity. Indeed, for instance, a congruence preserves the cross ratio of any four points in the first neighbourhood of any $p \in K$ (by (3.1.3), see sect. 6.7), while a similarity need not do so.

Remark 8.3.12 The reader may easily check that excesses are preserved by similarity, and so weighted clusters similar to a consistent one are consistent too. It is also easy to see that germs going sharply through similar consistent clusters are equisingular.

Note on equisingularity of linear systems: Similarity of their clusters of base points seems to be a convenient definition of *equisingularity* for complete neat linear systems, as, for most aspects, complete neat linear systems with similar clusters of base points behave in the same way. They have, for instance, equisingular generic germs (7.2.13) and equal codimensions (8.3.9). The same definition becomes far weaker (and far less convenient) if applied to non-necessarily complete neat linear systems. Neat linear systems with similar clusters of base points still have equisingular generic members but may have different codimensions and may behave very differently in many aspects, as in fact may linear systems with the same cluster of base points (see exercise 8.14). Other definitions appearing in the literature are even weaker, as they just ask for a multiplicity-preserving bijection between clusters of base points and this does not guarantee the equisingularity of generic members (see 3.9.2).

8.4   Zariski's theory of complete ideals

In this section we will state and prove a theorem of Zariski about unique factorization of complete ideals of $\mathcal{O}$ as a product of irreducible ones. Our approach is different from Zariski's original one: we translate the decomposition problem to weighted clusters through the bijections of 8.3.10, which in addition allows us to give a description of the irreducible ideals and multiplicities appearing in the decomposition of a given complete ideal. The original paper on the subject is [87], a later version appeared in [95], vol. II, app. 5. Partial extensions of this theory to the much more difficult cases of regular local rings of any dimension or local rings of some singular surfaces (rational singularities) are due to Lipmann [56] and Cutkosky [25], [26].

First of all we will do some further work on weighted clusters. Assume that $\mathcal{K} = (K, \nu)$ is a weighted cluster: in the sequel it will be useful to define $\nu_p = 0$ for all $p \notin K$. If $p \in K$ recall from 4.2.5 that ρ_p, the excess of $\mathcal{K}$ at p, was defined as

$$\rho_p = \nu_p - \sum_q \nu_q,$$

the summation running on all points $q \in K$ proximate to p. As for the virtual multiplicities, we take $\rho_p = 0$ for $p \notin K$. We will use the vectorial notations introduced at the beginning of section 4.5. In particular the *excess vector* of $\mathcal{K}$, $\rho_{\mathcal{K}}$ is the vector whose components are the excesses ρ_p, $p \in K$. The equalities defining the excesses may be written in vectorial form as $\rho_{\mathcal{K}} = P_{\mathcal{K}}^t \nu$, where $P_{\mathcal{K}}$ denotes the proximity matrix of $\mathcal{K}$. After this equality the next lemma needs no proof.

Lemma 8.4.1 *If K is a cluster with origin at O and $\rho = (\rho_p)$ any vector indexed by K, then there is one and only one system of virtual multiplicities for K giving rise to a weighted cluster $\mathcal{K}$ with excess vector ρ, namely the system of virtual multiplicities $\nu = (P_K^t)^{-1}\rho$.*

Consistent weighted clusters are, by definition, those with $\rho_{\mathcal{K}} \geq 0$ (see section 4.2). The next lemma shows that excesses may be used for characterizing strictly consistent weighted clusters too:

Lemma 8.4.2 *A weighted cluster $\mathcal{K}$ is strictly consistent if and only if $\rho_p \geq 0$ for any $p \in K$ and furthermore $\rho_p > 0$ if p is maximal in K.*

PROOF: As recalled above, the condition $\rho_{\mathcal{K}} \geq 0$ defines consistent clusters. Assume that $p \in K$ is maximal: then, by the definition of excess, $\rho_p = \nu_p$, so that the condition $\rho_p > 0$ is obviously necessary for $\mathcal{K}$ to be strictly consistent. Conversely, if these conditions are satisfied, by the same reason, the virtual multiplicities of the maximal points are positive. Next we prove all other virtual multiplicities to be positive by decreasing induction on K. Assume that $p \in K$ is not maximal and that all points after p in K have positive virtual multiplicity. The definition of excess gives the equality

$$\nu_p = \rho_p + \sum_q \nu_q,$$

where $\rho_p \geq 0$ by hypothesis and the summation runs on the non-empty set of points $q \in K$ proximate to p. All of them being infinitely near to p, all ν_q are positive by induction and $\nu_p > 0$ as wanted. $\diamond$

Define the *sum* $\mathcal{K} + \mathcal{K}'$ of the weighted clusters $\mathcal{K} = (K, \nu)$ and $\mathcal{K}' = (K', \nu')$ as the weighted cluster whose set of points is $K \cup K'$ and whose virtual multiplicities are $\nu_p + \nu_p'$ for $p \in K \cup K'$. It is clear that such an addition is associative and commutative, thus making the set of all weighted clusters with origin at O a semigroup. We state the next lemma for further reference, as its proof is immediate from the definitions above:

Lemma 8.4.3 *If $\mathcal{K}$ and $\mathcal{K}'$ are weighted clusters, the excess of $\mathcal{K} + \mathcal{K}'$ at p is the sum of the excesses of $\mathcal{K}$ and $\mathcal{K}'$ at p. If $\mathcal{K}$ and $\mathcal{K}'$ are consistent (resp. strictly consistent), $\mathcal{K} + \mathcal{K}'$ is consistent (resp. strictly consistent) too.*

We denote by $\mathbf{W}$ the set of all strictly consistent clusters with origin at O. Clearly $\mathbf{W}$ equipped with the addition defined above is a semigroup. A strictly consistent cluster will be called *irreducible* if and only if it is irreducible as element of the semigroup $\mathbf{W}$, that is, it is not the sum of two strictly consistent clusters.

Let p be a point equal or infinitely near to O. Using 8.4.1, we define the weighted cluster $\mathcal{K}(p)$ as the cluster whose points are p and all points preceding it, and whose excesses are $\rho_p = 1$ and $\rho_q = 0$ for $q \neq p$. The reader may notice that the virtual multiplicity of p is one, while the virtual multiplicities of the points preceding p may be easily computed backwards. Indeed, since the excess at any $q \neq p$ is zero, the virtual multiplicity of q is the sum of the virtual multiplicities of the points proximate to it in $\mathcal{K}(P)$. We have:

Lemma 8.4.4 $\mathcal{K}(p)$ *is an irreducible strictly consistent cluster.*

PROOF: That $\mathcal{K}(p)$ is a strictly consistent cluster directly follows from its definition and 8.4.2. Assume $\mathcal{K}(p) = \mathcal{K} + \mathcal{K}'$, both $\mathcal{K}$ and $\mathcal{K}'$ strictly consistent clusters. Then, again by 8.4.2, all excesses of both clusters are non-negative and each cluster has at least a strictly positive excess. Using 8.4.3 we see that the excess vector of $\mathcal{K} + \mathcal{K}'$ has either two positive components or a component non-less than two, which in any case proves it to be different from the excess vector of $\mathcal{K}(p)$ and gives thus a contradiction. $\diamond$

We call $\mathcal{K}(p)$ the *irreducible (strictly consistent) cluster ending at p*.

Lemma 8.4.5 *If $\mathcal{K}$ is a strictly consistent cluster and $\rho = (\rho_p)$ its excess vector, then*

$$\mathcal{K} = \sum_{p \in K} \rho_p \mathcal{K}(p).$$

PROOF: Write $\mathcal{K} = (K, \nu)$ and $\mathcal{K}' = (K', \nu') = \sum_{p \in K} \rho_p \mathcal{K}(p)$. If p belongs to $\mathcal{K}$, then all points of $\mathcal{K}(p)$ belong to $\mathcal{K}$ too, and hence $K' \subset K$. Conversely, any $q \in K$ equals or precedes a maximal point p and therefore belongs to $\mathcal{K}(p)$, which in turn appears with coefficient $\rho_p \neq 0$ (by 8.4.2) in the addition defining $\mathcal{K}'$: thus $q \in K'$, which proves $K \subset K'$ and therefore $K = K'$. Regarding multiplicities, an easy computation from the definition of the weighted clusters $\mathcal{K}(p)$, using 8.4.3, proves that $\mathcal{K}'$ and $\mathcal{K}$ have equal excess vectors, after which $\mathcal{K} = \mathcal{K}'$ by 8.4.1. $\diamond$

Lemma 8.4.6 *Any irreducible strictly consistent cluster $\mathcal{K}$ is $\mathcal{K} = \mathcal{K}(p)$ where, necessarily, p is the only maximal point of $\mathcal{K}$.*

PROOF: If $\mathcal{K}$ is irreducible, the decomposition of 8.4.5 must have a single summand, from which the claim. $\diamond$

The last piece needed in order to get our decomposition theorem for clusters is uniqueness, namely:

Lemma 8.4.7 *A strictly consistent cluster cannot be decomposed into two different ways as a sum of irreducible strictly consistent clusters.*

PROOF: Assume we have two decompositions for a strictly consistent cluster $\mathcal{K} = (K, \nu)$, namely that of 8.4.5

$$\mathcal{K} = \sum_{p \in K} \rho_p \mathcal{K}(p)$$

and a second one, which by 8.4.6 may be assumed to have the form

$$\mathcal{K} = \sum_{q \in T} \alpha_q \mathcal{K}(q),$$

where T is a finite set of points equal or infinitely near to O and the α_p are positive integers. Since each q belongs to $\mathcal{K}(q)$ which in turn appears with nonzero coefficient in the second decomposition, $T \subset K$. Then the excesses ρ_p of $\mathcal{K}$ may be computed from those of the clusters $\mathcal{K}(q)$ in the second decomposition using 8.4.3: we easily get $\rho_p = \alpha_p$ for $p \in T$ and $\rho_p = 0$ if $p \notin T$. This proves that the two decompositions are in fact the same. $\diamond$

All together, we have a quite explicit decomposition theorem for strictly consistent clusters which, after 8.4.5 and 8.4.7, needs no proof:

Theorem 8.4.8 *Any strictly consistent weighted cluster has a unique decomposition as a sum of irreducible ones, namely*

$$\mathcal{K} = \sum_{p \in K} \rho_p \mathcal{K}(p),$$

where ρ_p is the excess of $\mathcal{K}$ at p and $\mathcal{K}(p)$ the irreducible cluster ending at p.

As one may expect, if ξ goes sharply through $\mathcal{K}$, its decomposition in branches follows the decomposition of $\mathcal{K}$ in irreducible strictly consistent clusters. Indeed, if still $\mathcal{K} = (K, \nu)$ is strictly consistent and has excesses ρ_p, $p \in K$:

Corollary 8.4.9 *If ξ is a germ of curve going sharply through $\mathcal{K}$, then for each $p \in K$ there are exactly ρ_p branches of ξ whose last point in K is p and all of them go sharply through $\mathcal{K}(p)$.*

PROOF: That ξ has exactly ρ_p branches through p missing further points in K has been already seen in 4.2.6. If γ is any of these branches, all points on γ after p are non singular points of ξ and hence non-singular points of γ too. Let $\mathcal{T} = (T, \tau)$, T the set of all points p' that precede or equal p and $\tau_{p'} = e_{p'}(\gamma)$. Clearly γ goes sharply through $\mathcal{T}$. Thus, since γ is unibranched, by 4.2.6 again, the excesses of $\mathcal{T}$ are zero but for that corresponding to p, which is 1. Then $\mathcal{T}$ and $\mathcal{K}(p)$ have the same points and the same excesses and hence, by 8.4.1, $\mathcal{T} = \mathcal{K}(p)$, thus ending the proof. $\diamond$

The last job before dealing with complete ideals is to prove a technical lemma we will need later on.

Lemma 8.4.10 *Assume that $\mathcal{K} = (K, \nu)$ is a consistent cluster with excesses ρ_p, $p \in K$ and ξ is a germ of curve. If for each $p \in K$ the germ ξ has just ρ_p different points in the first neighbourhood of p and not in K, and furthermore all these points are non-singular, then ξ goes sharply through $\mathcal{K}$.*

PROOF: We will prove that $e_p(\xi) = \nu_p$ for all $p \in K$ using reverse induction on K. After this the claim follows using 4.2.5.

Assume first that p is a maximal point in K. Then ξ has just ρ_p points in its first neighbourhood and all these points are non-singular. Proximity equality 3.5.3 and the definition of excess at p give $e_p(\xi) = \rho_p = \nu_p$ as wanted. If p is not a maximal point, assume using induction that $e_q(\xi) = \nu_q$ for all points $q \in K$ infinitely near to p, and so, in particular, for all those proximate to p. Now the proximity equality gives

$$e_p(\xi) = \sum_q e_q(\xi) + \rho_p,$$

the summation running on the points $q \in K$ proximate to p. Since the definition of excess reads

$$\nu_p = \sum_q \nu_q + \rho_p,$$

the summation being as above, the induction hypothesis gives the claim. $\diamond$

Let us denote by $\mathbf{I}$ the set of all complete $\mathcal{M}$-primary ideals of $\mathcal{O}$. The next theorem states the main link between weighted clusters and complete ideals.

Theorem 8.4.11 *The set $\mathbf{I}$, equipped with the product of ideals, is a semigroup and the maps*

$$BP : \mathbf{I} \longrightarrow \mathbf{W}$$
$$I \longmapsto BP(I)$$

and

$$H : \mathbf{W} \longrightarrow \mathbf{I}$$
$$\mathcal{K} \longmapsto H_{\mathcal{K}}$$

are reciprocal isomorphisms between $\mathbf{I}$ *and the semigroup* $\mathbf{W}$ *of all strictly consistent clusters with origin at* O.

PROOF: The maps H and BP are the reciprocal bijections of 8.3.11. Our main point will be to prove that $H_{\mathcal{K}} H_{\mathcal{K}'} = H_{\mathcal{K}+\mathcal{K}'}$ for any two strictly consistent clusters $\mathcal{K}$, $\mathcal{K}'$. From it the rest of the claim easily follows. Indeed, assume this equality to have been proved. Then, since H is onto, the product of any two complete $\mathcal{M}$-primary ideals of $\mathcal{O}$ is a complete $\mathcal{M}$-primary ideal, which in turn proves that $\mathbf{I}$ is a semigroup, as the product of ideals is obviously associative and commutative. After this H is clearly a morphism and the proof is done.

Assume thus that $\mathcal{K} = (K, \nu)$ and $\mathcal{K}' = (K', \nu')$ are strictly consistent clusters whose excess vectors are, respectively, $\rho = (\rho_p)$, $p \in K$ and $\rho' = (\rho'_q)$, $q \in K'$. In order to prove that $H_{\mathcal{K}} H_{\mathcal{K}'} \subset H_{\mathcal{K}+\mathcal{K}'}$ it is enough to see that $fg \in H_{\mathcal{K}+\mathcal{K}'}$ if $f \in H_{\mathcal{K}}$ and $g \in H_{\mathcal{K}'}$. This is the same as proving that if germs ξ and ζ go through $\mathcal{K}$ and $\mathcal{K}'$, respectively, then $\xi + \zeta$ goes through $\mathcal{K} + \mathcal{K}'$, and this is in turn straightforward from the definitions. Indeed $e_O(\xi + \zeta) = e_O(\xi) + e_O(\zeta) \geq \nu_O + \nu'_O$ as $e_O(\xi) \geq \nu_O$ and $e_O(\zeta) \geq \nu'_O$. Furthermore, after blowing up O, the virtual transform of $\xi + \zeta$, relative to the virtual multiplicity $\nu_O + \nu'_O$ is composed of the virtual transforms of ξ and ζ, relative to virtual multiplicities ν_O and ν'_O: this allows us to end the proof using the customary induction whose details are left to the reader.

To prove the converse, namely $H_{\mathcal{K}} H_{\mathcal{K}'} \supset H_{\mathcal{K}+\mathcal{K}'}$, we will prove that $H_{\mathcal{K}+\mathcal{K}'}$ can be generated by products of elements of $H_{\mathcal{K}}$ and $H_{\mathcal{K}'}$. In fact, by 8.3.2 and 7.2.16, $H_{\mathcal{K}+\mathcal{K}'}$ may be generated by the equations of finitely many germs going sharply through $\mathcal{K} + \mathcal{K}'$ and so it will be enough to prove that if $\xi : f = 0$ goes sharply through $\mathcal{K} + \mathcal{K}'$, then $f = gg'$ where $g \in H_{\mathcal{K}}$ and $g' \in H_{\mathcal{K}'}$.

Let us prove this last claim. By 4.2.5 and 8.4.3 the germ ξ has, for each $p \in \mathcal{K} + \mathcal{K}'$, $\rho_p + \rho'_p$ different points outside of $\mathcal{K} + \mathcal{K}'$ in the first neighbourhood of p and all these points are non-singular: let us call T_p the set of these points and $T = \bigcup_p T_p$. Since the points in T are simple on ξ, each point in T lies in one and only one branch of ξ. Conversely, since no point in T precedes another point in T, each branch of ξ goes through one and only one point in T, namely the first point on the branch that does not belong to $\mathcal{K} + \mathcal{K}'$: if for each $q \in T$ we write γ_q for the only branch of ξ going through q, then $\xi = \sum_{q \in T} \gamma_q$. Split each T_p into two complementary sets, $T_p = R_p \cup R'_p$, $\emptyset = R_p \cap R'_p$, so that R_p has ρ_p elements and consequently R'_p has ρ'_p. Write $R = \bigcup_p R_p$ and $R' = \bigcup_p R'_p$ and define two germs of curve as $\zeta = \sum_{q \in R} \gamma_p$, $\zeta' = \sum_{q \in R'} \gamma_p$. Obviously one has $\xi = \zeta + \zeta'$ and thus we define g and g' to be suitable equations of ζ and ζ' such that $f = gg'$.

We will see that ζ goes through $\mathcal{K}$. Take any $p' \in K' - K$. Then no point $p'' \in K \cup K'$ equal or infinitely near to it belongs to K. Hence $\rho_{p''} = 0$ and thus, necessarily, $R_{p''} = \emptyset$ for all these points: it follows that ζ does not go through p'. Take now any $p \in K$. By construction ζ has just ρ_p different points in the first neighbourhood of p outside of $\mathcal{K} + \mathcal{K}'$ and all these points are non-singular. Since we have seen that no point in $K' - K$ belongs ζ, there are just ρ_p different points in the first neighbourhood of p which belong to ζ but not to $\mathcal{K}$, and all these points are non-singular. Then, by 8.4.10, ζ goes (and even goes sharply) through $\mathcal{K}$, as wanted.

The same argument applies to ζ' in order to see that it goes through $\mathcal{K}'$. Hence $g \in H_{\mathcal{K}}$, $g' \in H_{\mathcal{K}'}$ and the equality $H_{\mathcal{K}} H_{\mathcal{K}'} = H_{\mathcal{K}+\mathcal{K}'}$ has been proved. $\diamond$

There is a part of 8.4.11 worth a separate claim:

Corollary 8.4.12 *The product of complete $\mathcal{M}$-primary ideals of $\mathcal{O}$ is complete.*

As the reader may have noticed from the proof of 8.4.11, the key to 8.4.12 being true is that generic germs through a strictly consistent cluster $\mathcal{K}$ decompose into irreducible germs following the decomposition of $\mathcal{K}$ into irreducible strictly consistent clusters. This fact strongly depends in turn on the smoothness and two-dimensionality of the ambient surface S. For instance, it is no longer true if germs of surfaces in the affine three-space are taken instead of germs of plane curves, and in fact a product of complete $\mathcal{M}$-primary ideals in a ring of series in three variables need not be complete (see [56], [25] and [26] for products and factorization of complete ideals in rings other than the regular two-dimensional ones).

Zariski's factorization theorem is now a direct consequence of 8.4.8 and 8.4.11. Of course *irreducible complete $\mathcal{M}$-primary ideals* are those with no decomposition as a product of complete $\mathcal{M}$-primary ideals.

Theorem 8.4.13 (Zariski [87]) *Any complete $\mathcal{M}$-primary ideal I of $\mathcal{O}$ has a unique decomposition as a product of irreducible complete $\mathcal{M}$-primary ideals, namely*

$$I = \prod_{p \in BP(I)} H_{\mathcal{K}(p)}^{\rho_p},$$

ρ_p *being the excess of $BP(I)$ at p and $\mathcal{K}(p)$ the irreducible cluster ending at p.*

PROOF: H being an isomorphism by 8.4.11, it preserves irreducibility and so the result comes from 8.4.8 applied to $BP(I)$. $\diamond$

The reader may notice that 8.4.13 explicitly determines the irreducible complete ideals in the decomposition of I, as well as their multiplicities, in terms of the cluster of base points of I.

8.5 Generic polar germs

Assume that ξ is a singular reduced germ of curve on a smooth surface S. This section is devoted to showing some properties of the polar germs of ξ which depend on the linear structure of the jacobian system. We will prove that generic polar germs of ξ are all reduced and equisingular, which is an easy consequence of 7.2.13, and then that, for any choice of the local coordinates x, y and the equation f of ξ, all but finitely many of the polar germs in the pencil $\lambda_1 \partial f/\partial x + \lambda_2 \partial f/\partial y = 0$ are reduced and have the equisingularity type of generic polars of ξ.

As in section 6.2, we denote by $\mathbf{J}(\xi)$ the jacobian ideal of ξ (i.e., $\mathbf{J}(\xi) = (\partial f/\partial x, \partial f/\partial y, f)$ if f defines ξ and x, y are local coordinates) and by $\mathcal{J}(\xi)$ the linear system (jacobian system) it defines.

Proposition 8.5.1 *If ξ is a singular reduced germ of curve, its jacobian system is neat and there is a non-empty Zariski-open subset of the jacobian ideal of ξ whose elements define polar germs of ξ which are all reduced and have the same equisingularity type.*

PROOF: The jacobian system $\mathcal{J}(\xi)$ is neat, as it has no fixed part by 6.4.2 and $\mathbf{J}(\xi) \neq (1)$ because ξ is singular (6.1.7). Then the claim follows, as generic germs in the jacobian system are polar germs by 6.2.1, and are reduced and have the same equisingularity type by 7.2.13. $\diamond$

Remark 8.5.2 It is worth noting that 7.2.13 says not only that generic polar germs are equisingular, but also that they go sharply through $BP(\mathcal{J}(\xi))$ and so, in particular, they share all their singular points.

Proposition 8.5.1 allows us to associate with each reduced singular germ of curve ξ the equisingularity type of its generic polar germs, which is an analytic invariant (by 6.11.1) we have already dealt with in 6.5.1 and section 6.6. We have shown in particular that it is not determined by the equisingularity class of ξ. Furthermore an explicit determination of $BP(\mathcal{J}(\xi))$ has been given in section 6.7 for ξ irreducible, with a single characteristic exponent and generic in its equisingularity class. All of this, together with the results on polar invariants of sections 6.8 and 6.11, should convince the reader that the equisingularity type of its generic polar germs carries very deep information on the original germ ξ. This is still true in the case of ξ being a germ of hypersurface or even a germ of analytic subset of $\mathbb{C}^n$, as has been shown by the work of many authors. Among them let us quote Teissier [80], [82], Lê and Teissier [53], Lê, Michel and Weber [50], [51] and Gaffney [39].

Still assume that the germ ξ is singular and reduced, fix any equation f of ξ and local coordinates x, y at the origin O of ξ. Write ξ_x and ξ_y for the polar germs $\partial f/\partial x = 0$ and $\partial f/\partial y = 0$, respectively. The remaining part of this section is devoted to proving that one may get the equisingularity type of generic polars of ξ and even $BP(\mathcal{J}(\xi))$ from just the pencil of polar germs $\lambda_1 \partial f/\partial x + \lambda_2 \partial f/\partial y = 0$. Let us begin by proving a lemma, namely:

Lemma 8.5.3 *If γ is any irreducible germ with origin at O,*

$$[\xi \cdot \gamma] > min\{[\xi_x \cdot \gamma], [\xi_y \cdot \gamma]\}.$$

PROOF: Let $t \mapsto (x(t), y(t))$ be a uniformizing map of a representative of γ, $x(t), y(t) \in \mathbb{C}\{t\}$. If $[\xi . \gamma] = n$, then $f(x(t), y(t)) = t^n u$, u being an invertible series. After derivation we have

$$t^{n-1}(nu + t\frac{du}{dt}) = \frac{df(x(t), y(t))}{dt} = \frac{\partial f}{\partial x}(x(t), y(t))\frac{dx}{dt} + \frac{\partial f}{\partial y}(x(t), y(t))\frac{dy}{dt}$$

from which, after equating the orders of both ends, we get

$$n - 1 \geq \min\{[\xi_x . \gamma], [\xi_y . \gamma]\}$$

as wanted. ◇

Fix any equation f of ξ. The next theorem is in fact the reason for the equisingularity type of generic polars of ξ being reached by (generic) polars relative to the fixed equation f, and even by all but finitely many of the germs in the pencil $\lambda_1 \partial f/\partial x + \lambda_2 \partial f/\partial y = 0$, which is far smaller, and hence far easier to handle, than the whole jacobian system $\mathcal{J}(\xi)$ (see 8.5.8 below).

Theorem 8.5.4 *If $\xi : f = 0$ is a reduced singular germ of curve and x, y are local coordinates at the origin O of ξ, then ξ goes through $BP(\partial f/\partial x, \partial f/\partial y)$.*

PROOF: After a suitable linear change of coordinates, it is not restrictive to assume, using 7.2.10, that the polars $\xi_x : \partial f/\partial x = 0$ and $\xi_y : \partial f/\partial y = 0$ go sharply through $\mathcal{K} = (K, \nu) = BP(\partial f/\partial x, \partial f/\partial y)$. Choose a branch γ_0 having O as a simple point and missing all points on ξ in the first neighbourhood of O. Then, by 8.5.3,

$$e_O(\xi) = [\xi . \gamma_0]$$
$$> \min\{[\xi_x . \gamma_0], [\xi_y . \gamma_0]\}$$
$$\geq \min\{e_O(\xi_x), e_O(\xi_y)\} = \nu_O.$$

Assume by induction that $p \in K$ and that ξ goes through all points q preceding p with the virtual multiplicities ν_q and denote by $\check{\xi}$ the virtual transform of ξ with origin at p and relative to these virtual multiplicities. Fix a unibranched germ γ_p having p as a simple point and missing all points on $\check{\xi}$ in the first neighbourhood of p. Then, on one hand, the virtual Noether formula 4.1.3 gives

$$[\xi . \gamma_p] = \sum_{q<p} \nu_q e_q(\gamma_p) + e_p(\check{\xi}).$$

On the other, again by 8.5.3,

$$[\xi . \gamma_p] > \min\{[\xi_x . \gamma_p], [\xi_y . \gamma_p]\}$$
$$\geq \min\{\sum_{q<p} e_q(\xi_x)e_q(\gamma_p) + e_p(\xi_x), \sum_{q<p} e_q(\xi_y)e_q(\gamma_p) + e_p(\xi_y)\}$$
$$= \sum_{q<p} \nu_q e_q(\gamma_p) + \nu_p.$$

From the two relations displayed above we get $e_p(\breve{\xi}) > \nu_p$ and hence the claim.
◇

Remark 8.5.5 We have actually proved that the multiplicities of the virtual transforms of ξ with origins at the points of K are strictly bigger than the virtual ones.

Theorem 8.5.4 may be equivalently stated in terms of integral dependence (see [79], sect. 2):

Corollary 8.5.6 *The equation f is integral over the ideal $(\partial f/\partial x, \partial f/\partial y)$.*

PROOF: Follows from 8.5.4 using 8.3.6. ◇

Write $\mathbf{J} = \mathbf{J}(\xi)$ and $\mathbf{J}' = (\partial f/\partial x, \partial f/\partial y)$. If $\bar{\mathbf{J}}$ is the completion of $\mathbf{J}'$, then 8.5.6 shows that $\bar{\mathbf{J}}$ is also the completion of the jacobian ideal $\mathbf{J}$ and so we have $\mathbf{J}' \subset \mathbf{J} \subset \bar{\mathbf{J}}$. Both inclusions are in general strict, see [90] and exercise 8.16 for the cases $\mathbf{J}' = \mathbf{J}$ and $\mathbf{J} = \bar{\mathbf{J}}$, respectively.

Corollary 8.5.7 $BP(\mathbf{J}(\xi)) = BP(\partial f/\partial x, \partial f/\partial y)$.

PROOF: By 8.5.6, both ideals have the same integral closure, after which the claim follows from 8.3.6 and 8.3.11 (or 8.3.2). ◇

Corollary 8.5.8 *All but finitely many of the polar germs belonging to the pencil $\mathcal{P}(f, x, y) = \{\lambda_1 \partial f/\partial x + \lambda_2 \partial f/\partial y = 0\}$ go sharply through $BP(\mathcal{J}(\xi))$ and therefore are reduced and have the equisingularity type of generic polars of ξ.*

PROOF: The cluster of base points of $\mathcal{P}(f, x, y)$ is $BP(\partial f/\partial x, \partial f/\partial y)$ which, by 8.5.7 is just $BP(\mathcal{J}(\xi))$. Then it is enough to apply 7.2.10. ◇

In the case of being interested in just the equisingularity type of generic polars of ξ, after 8.5.8, one may fix an equation f of ξ and local coordinates x, y and consider only the polar germs in $\mathcal{P}(f, x, y)$. Some authors define these polars as being the only polar germs of ξ (or of the germ of function f), which is a non-intrinsic and restricted definition of polars which nevertheless does work in most cases just because of 8.5.8.

Studying geometric aspects of polar germs other than the equisingularity type of the generic ones may require us to consider all polar germs (or the whole ideal $\mathbf{J}(\xi)$) and not only just a pencil $\mathcal{P}(f, x, y)$. This is the case if one is interested in special (i.e., non-equisingular to the generic ones) polar germs. For instance, the pencil $\mathcal{P}(f, x, y)$ contains a non-reduced germ if $f = x^3 - y^2$, namely the y-axis counted twice, while all its germs are reduced if one takes $f = (1 + x)(x^3 - y^2)$.

8.6   Base points of polar germs

A quite elementary fact in algebraic geometry is that the (proper) singular points of a plane projective algebraic curve are determined by (and in fact agree with) the base points of its polar curves. Our goal in this section is to prove a not so easy local version of this fact. Given a reduced singular germ ξ, we will consider the cluster $BP(\mathcal{J}(\xi))$ of base points of its jacobian system: its points may be called base points of the polars of ξ, as generic elements of $\mathcal{J}(\xi)$ are polars of ξ. We will prove that $BP(\mathcal{J}(\xi))$ determines the singular points of ξ as well as their multiplicities on ξ, which in particular implies that $BP(\mathcal{J}(\xi))$ (its similarity class in fact) determines the equisingularity class of ξ.

In the sequel, if ξ is any reduced singular germ of curve, we denote by $\mathcal{K}(\xi)$ the weighted cluster of its singular points, each taken with virtual multiplicity equal to its effective multiplicity on ξ. Obviously, $\mathcal{K}(\xi)$ is consistent and ξ goes sharply through it.

Lemma 8.6.1 *Reduced singular germs of curve ξ_1 and ξ_2 are equisingular if and only if $\mathcal{K}(\xi_1)$ and $\mathcal{K}(\xi_2)$ are similar.*

PROOF: By its own definition and 3.8.3, an equisingularity between ξ_1 and ξ_2 restricts to a similarity between $\mathcal{K}(\xi_1)$ and $\mathcal{K}(\xi_2)$. Conversely, if $\varphi : \mathcal{K}(\xi_1) \longrightarrow \mathcal{K}(\xi_2)$ is a similarity, then, for any $p \in \mathcal{K}(\xi_1)$, the excess of $\mathcal{K}(\xi_1)$ at p equals that of $\mathcal{K}(\xi_2)$ at $\varphi(p)$ and so both ξ_1 and ξ_2 have the same number of non-singular points in the first neighbourhoods of p and $\varphi(p)$, respectively. After this φ clearly extends to an equisingularity between ξ_1 and ξ_2. $\diamond$

Lemma 8.6.2 *Let $g \in \mathcal{O}_{S,O}$ define a smooth germ at O and ζ be a g-polar of a germ ξ. If p is a g-free point on ξ then there exist local coordinates x, y at p and an equation f of $\tilde{\xi}_p$ so that*

(a) The germ at p of the exceptional divisor has equation $x = 0$.

(b) $\tilde{\zeta}_p$ has equation $\partial f / \partial y = 0$.

(c) The x-axis is not tangent to $\tilde{\xi}_p$ or $\tilde{\zeta}_p$.

PROOF: Existence of coordinates x, y' satisfying (a) and (b) directly comes from 6.2.4 if p is g-next O, and by induction on the number of free points preceding p in the general case. Then take $a \in \mathbb{C}$ so that $y' + ax = 0$ is not a principal tangent of either $\tilde{\xi}_p$ or $\tilde{\zeta}_p$, and then new local coordinates $x, y = y' + ax$ in order to have all three conditions fulfilled. $\diamond$

Lemma 8.6.3 *A free point p is a singular point of a reduced singular germ of curve ξ if and only if p lies on ξ and belongs to $BP(\mathcal{J}(\xi))$.*

PROOF: If p is a singular point of ξ, then it lies on all polar curves by 6.2.8 and hence it belongs to $BP(\mathcal{J}(\xi))$. Conversely, let ζ be a transverse polar of ξ going sharply through $BP(\mathcal{J}(\xi))$: if p belongs to $BP(\mathcal{J}(\xi))$, then it lies on ζ. If furthermore the point p lies on ξ, then, by 6.2.8, p is a singular point of ξ. $\diamond$

Theorem 8.6.4 *Let ξ_1 and ξ_2 be germs of curve, both reduced and singular:*

(a) If $BP(\mathcal{J}(\xi_1)) = BP(\mathcal{J}(\xi_2))$ then $\mathcal{K}(\xi_1) = \mathcal{K}(\xi_2)$.

(b) If $BP(\mathcal{J}(\xi_1))$ and $BP(\mathcal{J}(\xi_2))$ are similar weighted clusters, then ξ_1 and ξ_2 are equisingular.

PROOF: We will establish a procedure that, for any given reduced singular germ ξ, uniquely determines $\mathcal{K}(\xi)$ from $BP(\mathcal{J}(\xi))$. This will directly prove claim (a). Furthermore, since it will turn out from its own definition that such a procedure gives rise to similar clusters when applied to similar clusters, by 8.6.1 claim (b) will be proved as well.

Part one: Recovering polar invariants. Let us write $BP(\mathcal{J}(\xi)) = \mathcal{T} = (T, \nu)$ and call its points just base points. Fix an equation f of ξ and local coordinates x, y at the origin O of ξ. We know from 8.5.7 that $\mathcal{T}$ is the cluster of base points of the pencil $\lambda_1 \partial f/\partial x + \lambda_2 \partial f/\partial y = 0$ and hence, by 7.2.10, after a suitable linear change of coordinates if needed, one may assume without restriction that both the polars $\zeta : \partial f/\partial y = 0$ and $\zeta' : \partial f/\partial x = 0$ go sharply through $\mathcal{T}$ and are transverse polars. We have thus $e_p(\zeta) = e_p(\zeta') = \tau_p$ for all $p \in T$. Let p be any point in T and denote by ρ_p the excess of $\mathcal{T}$ at p. By 8.4.9 there are ρ_p branches of ζ whose last point in K is p and all of them go sharply through the irreducible weighted cluster $\mathcal{K}(p)$. If γ denotes one of these branches, its effective multiplicities $e_q(\gamma)$, $q \in T$ are in particular well determined. (In fact, as the reader may see, they are easily computed from an Enriques diagram of T using the proximity equalities and the fact that all points on γ after p are non-singular.) Since ζ and ζ' share no point $q \notin T$ (otherwise such a point would be a base point) the intersection multiplicity of γ and ζ' is well determined too, by the Noether formula,

$$[\gamma . \zeta'] = \sum_{q \in T} e_q(\gamma) \tau_q .$$

Now, by 6.1.7, γ is not tangent to the y-axis and so if $(t^e, y(t))$ is a uniformizing map of γ, then $e = e_O(\gamma)$. Furthermore, since $(\partial f/\partial y)(t^e, y(t))$ is identically zero,

$$\frac{d}{dt} f(t^e, y(t)) = e \frac{\partial f}{\partial x}(t^e, y(t)) t^{e-1}$$

which in turn easily gives

$$[\gamma . \xi] = [\gamma . \zeta'] + e.$$

Hence, also the polar quotient of ξ corresponding to γ, $I_\gamma = [\gamma . \xi]/e$ is well determined by $\mathcal{T}$.

Part two: Recovering $\mathcal{K}(\xi)$. For each non-satellite singular point p of ξ we will inductively determine the multiplicity $e_p(\xi)$ of p on ξ, the satellite points q of p on ξ as well as their multiplicities $e_q(\xi)$ on ξ, and the singular points next p on ξ. Clearly this will determine $\mathcal{K}(\xi)$.

Assume first $p = O$. By 6.1.7, $e_O(\xi) = \tau_O + 1$. Since O has no satellite points, all we need in this case is to determine the singular points of ξ in the first neighbourhood of O. By 8.6.3 these points are just the base points in the first neighbourhood of O that belong to ξ. Thus assume that $p' \in T$ belongs to the first neighbourhood of O and pick any branch γ of ζ through p'. By the Noether formula, p' lies on ξ if and only if $e_O(\xi)e_O(\gamma) < [\xi.\gamma]$. Thus, among the points in the first neighbourhood of O, the singular points of ξ are the base points that lie on a branch γ of ζ for which $I_\gamma > e_O(\xi) = \tau_O + 1$.

Assume now that p is a free singular point of ξ (so $p \neq O$). Using induction, assume also that the effective multiplicities of ξ at all points preceding p have been determined. As allowed by 8.6.2, we choose local coordinates x, y at p and an equation f of $\tilde{\xi}_p$ so that the germ E_p of the exceptional divisor at p is $x = 0$, $\tilde{\zeta}_p$ has equation $\partial f/\partial y$ and no branch of $\tilde{\xi}_p$ or $\tilde{\zeta}_p$ is tangent to the x-axis. In the sequel we will write $\tilde{\xi} = \tilde{\xi}_p$ and $\tilde{\zeta} = \tilde{\zeta}_p$ in order to lighten notations a little bit. The strict transforms $\tilde{\gamma} = \tilde{\gamma}_p$ of the branches γ of ζ through p are the branches of $\tilde{\zeta}$. For each of them we will assume that the corresponding polar quotient of $\tilde{\xi}$, $I_{\tilde{\gamma}} = [\tilde{\gamma}.\tilde{\xi}]/[\tilde{\gamma}.E_p]$, has been computed from I_γ using 7.6.6 and 7.6.7.

Next we show that the Newton polygon $N(\tilde{\zeta})$ is determined by $\mathcal{T}$. For each branch $\tilde{\gamma}$ of $\tilde{\zeta}$ one may determine the exponent m/n of the initial term of its Puiseux series. Indeed, if the branch goes through no satellite point of p, then it is not tangent to the y-axis, and since it is not tangent to the x-axis either, $m/n = 1$. Otherwise $m/n < 1$, so m/n is the first characteristic exponent of $\tilde{\gamma}$ and hence it may be determined as a continued fraction whose terms depend on the satellite points of p the branch γ is going through (see section 5.3). By 1.5.5 the opposites of the inverses of all these exponents are the slopes of the sides of $N(\tilde{\zeta})$. The intersection $[\tilde{\gamma}.E_p]$ equals the already determined multiplicity on γ of the point just before p. By 6.10.2, it is enough to add up these intersection numbers for all branches corresponding to the same exponent m/n to get the height of the side of $N(\tilde{\zeta})$ of slope $-n/m$. Clearly neither of the axes is a component of $\tilde{\zeta}$, by the way the coordinates have been chosen. Thus $N(\tilde{\zeta})$ has both ends on the axes and hence, once the slopes and heights of all its sides are determined, so is the whole polygon. We will write $\Gamma'_1, \ldots, \Gamma'_r$ for the sides of $N(\tilde{\zeta})$, ordered from top to bottom, as customary.

Once we have $N(\tilde{\zeta})$, we will use it and the polar quotients of $\tilde{\xi}$ to determine $N(\tilde{\xi})$ and the singular points next p. We will reverse, to a certain extent, what we did in the proof of 6.10.3, in which a partial description of $N(\tilde{\zeta})$ from $N(\tilde{\xi})$ was given. We will make use of some facts that have been established in the course of the proof of 6.10.3. As there, we assume that $N(\tilde{\xi})$ has sides $\Gamma_1, \ldots, \Gamma_k$, that for each $i = 1, \ldots, k$, Γ_i has ends $(\alpha_{i-1}, \beta_{i-1})$, (α_i, β_i) and slope $-n_i/m_i$, $m_i > 0$, $n_i > 0$, $\gcd(m_i, n_i) = 1$ and $\alpha_0 = \beta_k = 0$. Necessarily $m_i/n_i < 1$ for $i < k$ and $m_k/n_k \leq 1$, as it is assumed that no branch of $\tilde{\xi}$ is tangent to the x-axis. All branches of $\tilde{\xi}$ corresponding to the side Γ_i have Puiseux series of the form $s = ax^{m_i/n_i} + \ldots$, $a \neq 0$ (see section 2.2) and so all are going through the same satellite points of p, these satellite points being well determined by m_i/n_i (they are in fact the satellite points in the cluster $K(m_i, n_i)$, see section 5.3).

We call q_i the last of these satellite points, but for $i = k$ and $m_k/n_k = 1$, in which case there are no satellite points and we take $q_k = p$.

As already noticed when proving 6.10.3, the Newton diagram of $\tilde{\zeta}$ is obtained from that of $\tilde{\xi}$ by deleting all points on the first axis and moving the remaining ones one step downwards. Then $r \geq k - 1$, the sides $\Gamma'_1, \ldots, \Gamma'_{k-1}$ are, respectively, the translations of $\Gamma_1, \ldots, \Gamma_{k-1}$ by the vector $(0, -1)$ and the slopes of the remaining sides of $\mathbf{N}(\tilde{\zeta})$ are non-less than $-n_{k-1}/m_{k-1}$. In particular the first vertex of $\mathbf{N}(\tilde{\xi})$ is $(0, h(\mathbf{N}(\tilde{\zeta})) + 1)$. Thus, we assume by induction that the vertices (α_j, β_j) of $\mathbf{N}(\tilde{\xi})$ have been determined for $j < i$, and hence the sides Γ_j for $j < i$ are determined too. Still by induction we assume all singular points of ξ in the first neighbourhood of q_j to have been determined for $j < i$. Of course, if $i - 1 = k$ then $\mathbf{N}(\tilde{\xi})$ and all singular points of ξ next p have been determined. Thus we assume $i - 1 < k$, and so $\beta_{i-1} \neq 0$, in the sequel. Then the sides $\Gamma'_1, \ldots, \Gamma'_{i-1}$ are translations of $\Gamma_1, \ldots, \Gamma_{i-1}$ and so, in particular, the former also have slopes $-n_1/m_1, \ldots, -n_{i-1}/m_{i-1}$. The branches of $\tilde{\zeta}$ corresponding to these sides are thus those going through one of the points q_j, $1 \leq j \leq i-1$ and having a free point in its first neighbourhood. We denote by Ξ_i the set of all remaining branches of $\tilde{\zeta}$, i.e., those corresponding to Γ'_i and further sides of $\mathbf{N}(\tilde{\zeta})$. Take $\Pi_i = \{I_{\tilde{\gamma}} | \tilde{\gamma} \in \Xi_i\} \cup \{\alpha_{i-1} + \beta_{i-1}\}$.

We will show next that the minimal element of Π_i is $\alpha_{i-1} + m_i\beta_{i-1}/n_i$ (the abscissa of the intersection point of the first axis and the line Γ_i is lying on). This will determine m_i/n_i and hence the slope of Γ_i. Assume first that $i < k$. Then still Γ'_i is a translation of Γ_i and we have seen in the proof of 6.10.3 that there is at least one branch $\tilde{\gamma}_0$ of $\tilde{\zeta}$ with Puiseux series $s = ax^{m_i/n_i} + \ldots$ where a is not a root of the equation associated with Γ_i and of course $a \neq 0$. By 6.11.1(c),

$$I_{\tilde{\gamma}_0} = \alpha_{i-1} + m_i\beta_{i-1}/n_i.$$

Since all branches $\tilde{\gamma} \in \Xi_i$ correspond to sides Γ'_j, $j \geq i$, they have Puiseux series of the form $bx^{m/n} + \ldots$, $b \neq 0$, $m/n \geq m_i/n_i$, and so, by 6.11.1(b),

$$I_\gamma \geq \alpha_{i-1} + m_i\beta_{i-1}/n_i,$$

while

$$\alpha_{i-1} + \beta_{i-1} > \alpha_{i-1} + m_i\beta_{i-1}/n_i,$$

because $m_i/n_i < m_k/n_k \leq 1$. In the case $i = k$, assume first that $m_k/n_k < 1$. Then $\mathbf{N}(\tilde{\zeta})$ necessarily has at least one side of slope bigger than $-n_k/m_k$ (because the point $(\alpha, 1) \in \Gamma_k$ has $\alpha \notin \mathbb{Z}$). By 6.11.1(d), all branches corresponding to such a side give polar quotient $\alpha_k = \alpha_{k-1} + m_k\beta_{k-1}/n_k < \alpha_{k-1} + \beta_{k-1}$, while polar quotients of branches corresponding to a side parallel to Γ_k (if any) are non-less than α_k by 6.11.1(b). To close, assume $i = k$ and $m_k/n_k = 1$. Since $\tilde{\zeta}$ is assumed to have no branch tangent to the x-axis, no side of $\mathbf{N}(\tilde{\zeta})$ may have slope bigger than -1. Hence, either the last side of $\mathbf{N}(\tilde{\zeta})$ is Γ'_{k-1} or it is Γ'_k and in such a case it has slope equal to -1. In the first case $\Pi_k = \{\alpha_{k-1} + \beta_{k-1}\}$, which makes our claim obvious, while in the other it is enough to apply 6.11.1(b) once again.

The slope $-n_i/m_i$ of Γ_i being determined, so is the satellite point q_i. We determine next the singular points of ξ in the first neighbourhood of q_i. According to 8.6.3, these are just the base points in such a neighbourhood that belong to $\tilde{\xi}$. If p' is a base point in the first neighbourhood of q_k, then take any branch $\tilde{\gamma}$ of $\tilde{\zeta}$ through it. The point p' lies on ξ if and only if $\tilde{\gamma}$ has a Puiseux series of the form $ax^{m_i/n_i} + \ldots$ with a a root of the equation associated with Γ_i. By 6.11.1(c) this is equivalent to the condition $I_{\tilde{\gamma}} > \alpha_{i-1} + m_i\beta_{i-1}/n_i$, which depends only on the data we have in hand.

To close, we determine the last end (α_i, β_i) of Γ_i. For this it is enough to decide if $i < k$, in which case (α_i, β_i) is the point just above the last end of Γ'_i, or $i = k$ which forces $\beta_k = 0$ and hence $\alpha_k = \alpha_{k-1} + m_k\beta_{k-1}/n_k$. If $m_i/n_i = 1$, then clearly $i = k$. Thus assume $m_i/n_i < 1$. If $i < k$ all branches in Ξ_i giving polar quotient minimal in Π_i correspond to the side Γ'_i parallel to Γ_i. Indeed, other branches in Ξ_i correspond to sides of slope $-n/m > -n_i/m_i$ and so, by 6.11.1(b), they have $I_\gamma > \alpha_{i-1} + m_i\beta_{i-1}/n_i$. By contrast, we have seen above that there is at least one branch in Ξ_k corresponding to a side non-parallel to Γ_k whose polar quotient is minimal in Π_i. In other words, $i < k$ if and only if $m_i/n_i < 1$ and all branches in Ξ_i giving minimal polar quotient go through q_i and have a free point in its first neighbourhood.

We have thus seen that both $\mathbf{N}(\tilde{\xi})$ and the singular points on ξ next p are determined by $BP(\mathcal{J}(\xi))$. The multiplicity $e_p(\xi) = e_p(\tilde{\xi})$ clearly equals the width α_k of $\mathbf{N}(\tilde{\xi})$. We leave to the reader, as an easy exercise, to check that $\mathbf{N}(\tilde{\xi})$ determines the satellite points of p on ξ, as well as their multiplicities on ξ, after which the proof is complete. $\diamond$

Remark 8.6.5 Determining the equisingularity type of ξ from $BP(\mathcal{J}(\xi))$ is a far easier job if ξ is already known to be irreducible, as in such a case, by 6.8.6, it is enough to determine its polar invariants.

Associating with each reduced singular germ of curve ξ the weighted cluster of base points of its jacobian system (or just the similarity class of $BP(\mathcal{J}(\xi))$, if one wants to have a purely combinatoric object by forgetting the positions of the free points) provides a quite interesting analytic invariant (see the particular case already presented in section 6.7). It clearly determines the singular points (or the equisingularity type) of generic polars and encloses the supplementary information given by the non-singular base points. The reader may have noticed that considering all base points, and not only the singular points of a generic polar, allowed us to determine the polar invariants, which is a crucial point in the above proof of 8.6.4. However, the jacobian ideal is in general far from being integrally closed (see exercise 8.16 below) and so the jacobian ideal (or system) itself encloses far more information than its cluster of base points. Regarding this point let us just quote a very nice theorem, due to Mather and Yau [57] and true for arbitrary hypersurfaces with isolated singularity, asserting that the jacobian ideal $\mathbf{J}(\xi)$ (the $\mathbb{C}$-algebra $\mathcal{O}/\mathbf{J}(\xi)$ in fact) determines the analytic isomorphism class of ξ.

8.7 Exercises

In the exercises that follow $\mathcal{O} = \mathcal{O}_{S,O}$ is the local ring of a point on a (smooth) surface (thus a ring of series $\mathbb{C}\{x, y\}$) and $\mathcal{M}$ its maximal ideal.

8.1 For any non-zero $f \in \mathcal{O} = \mathbb{C}\{x, y\}$ define $v(f) = (e, e') \in \mathbb{Z} \oplus \mathbb{Z}$, where e is the multiplicity of the germ $f = 0$ and e' the multiplicity of the factor y in the initial form of f. Check that v is actually a valuation of $\mathcal{O}$ and describe its sequence of centres.

8.2 Same as in 8.1 above, this time for v having real values and defined by the rule $f \longmapsto o_t f(t, t^\alpha)$, α a fixed irrational number.

8.3 Prove that any complete $\mathcal{M}$-primary ideal I of $\mathcal{O}$ is the integral closure of an $\mathcal{M}$-primary ideal generated by two elements.

8.4 Assume that v is a valuation of $\mathcal{O}$ with value group $\mathcal{G}$ and $m \in \mathcal{G}$. Prove that there exist discrete valuations v_i of $\mathcal{O}$ and integers m_i, $i = 1, \ldots, r$, so that $v(f) \geq m$ if and only if $v_i(f) \geq m_i$, $i = 1, \ldots, r$.

8.5 Describe the clusters of base points of all complete $\mathcal{M}$-primary ideals of $\mathcal{O}$ that may be generated by two elements.

8.6 Let $\mathcal{L}$ be a neat linear system of germs of curve, and $\mathcal{K} = (K, \nu)$ its cluster of base points. Prove that if $\mathcal{L}$ is complete, then for any $p \in K$ the linear series cut out by $\mathcal{L}$ on its first neighbourhood E_p is complete (i.e., it is the whole series of all effective divisors of a fixed degree). Show by an example that the converse is not true.

8.7 Let $I \subset \mathcal{O}$ be a complete $\mathcal{M}$-primary ideal, $e = e_O(I)$, $\mathcal{K}_n = BP(I \cap \mathcal{M}^n)$ and denote by ρ_n the excess of $\mathcal{K}_n$ at O.

(a) Fix $n \geq e$. Define $\mathcal{T}_{n,0} = \mathcal{K}_n$ and, for $i > 0$, $\mathcal{T}_{n,i}$ as the cluster obtained by adding to $\mathcal{T}_{n,i-1}$ a new point in the first neighbourhood of O virtually counted once. Write $H_i = H_{\mathcal{T}_{n,i}}$. Prove that for $0 \leq i < \rho_n$, $\mathcal{T}_{n,i+1}$ is consistent, $H_i \supset H_{i+1}$ and $\dim H_i/H_{i+1} = 1$.

(b) Still for $n \geq e$, prove that $\mathcal{T}_{n,\rho_n+1}$ is equivalent to $\mathcal{K}_{n+1}$ and so that

$$\dim_{\mathbb{C}} I \cap \mathcal{M}^n / I \cap \mathcal{M}^{n+1} = \rho_n + 1.$$

(c) If the Hilbert function of I is defined as $h_I(n) = \dim_{\mathbb{C}} \mathcal{O}/I + \mathcal{M}^{n+1}$, $n \geq 0$, prove that

$$h_I(n) = \frac{(n+2)(n+1)}{2} - \sum_{m=e}^{n}(\rho_m + 1).$$

(d) Describe an inductive procedure giving the weighted clusters $\mathcal{K}_n$ (and so in particular the integers ρ_n), $n \geq e$, from $\mathcal{K}_e = BP(I)$.

8.8 Notations and hypothesis being as in 8.7 above,

(a) Describe $BP(\mathcal{M}I)$ in terms of $BP(I)$.

(b) (*Lipman* [56]) Prove that $e + 1 = \dim_{\mathbb{C}}(I/\mathcal{M}I)$, which in turn is well known to be the minimal number of generators of I, by Nakayama's lemma ([48], IX.1, for instance).

8.9 Let J be any $\mathcal{M}$-primary ideal of $\mathcal{O}$ and $\mathcal{K} = (K, \nu) = BP(J)$. Take $\nu'_p = 2\nu_p - 1$ for $p \in K$, $\mathcal{K}' = (K, \nu')$ and prove that $H_{\mathcal{K}'} \subset J$. (*Hint:* Use Noether's $Af + B\varphi$, 4.9.1.)

8.10 Let $\mathcal{K}$ and $\mathcal{K}'$ be strictly consistent clusters so that $H_{\mathcal{K}} \supset H_{\mathcal{K}'}$. Determine a sequence of strictly consistent clusters $\mathcal{K}_i$, $i = 0, \ldots, r$, $\mathcal{K}_0 = \mathcal{K}$, $\mathcal{K}_r = \mathcal{K}'$, so that, for each $i = 1, \ldots, r$, $H_{\mathcal{K}_{i-1}} \supset H_{\mathcal{K}_i}$ and $\dim_{\mathbb{C}} H_{\mathcal{K}_{i-1}}/H_{\mathcal{K}_i} = 1$. (*Hint:* Add a suitable point virtually counted once to $\mathcal{K}_{i-1}$ and then use unloading to get $\mathcal{K}_i$.)

8.11 Let $J = (x^5y - y^3, x^8 - 2x^5y) \subset \mathcal{O} = \mathbb{C}\{x, y\}$, denote by $\bar{J}$ its integral closure and put $\mathcal{K} = BP(J) = BP(\bar{J})$, $\mathcal{K}' = BP(\mathcal{M}J)$.

(a) Explicitly describe $\mathcal{K}$, $\mathcal{K}'$ and then weighted clusters $\mathcal{K}_0, \ldots, \mathcal{K}_r$ as in 8.10 above, using germs of curve to locate the points.

(b) For $i = 1, \ldots, r$, select $f_i \in \mathbb{C}\{x, y\}$ so that the germ $f_i = 0$ goes through $\mathcal{K}_{i-1}$ and goes not through $\mathcal{K}_i$. Prove that they give rise to a $\mathbb{C}$-basis of $\bar{J}/\mathcal{M}\bar{J}$ and so (Nakayama's lemma, [48], IX.1) they generate $\bar{J}$.

(c) Describe a general procedure for getting a system of generators of the integral closure $\bar{J}$ of an $\mathcal{M}$-primary ideal J of $\mathbb{C}\{x, y\}$, once a system of generators of J has been given. (See [23].)

(d) Modify the procedure of part (c) above in order to get a second one giving a system of generators of the ideal $H_{\mathcal{K}}$ from any given weighted cluster $\mathcal{K}$.

8.12 *Briançon–Skoda theorem for dimension two*, see [12]. Prove that if J is an $\mathcal{M}$-primary ideal of $\mathcal{O}$ and $\bar{J}$ its integral closure, then $\bar{J}^2 \subset J$. (*Hint:* use exercise 8.9.)

8.13 Decompose the integral closure $\bar{I}$ of $I = (y^4 - x^7, x^3y^2) \subset \mathbb{C}\{x, y\}$ into a product of irreducible complete ideals. Give generators of each irreducible factor of $\bar{I}$ using exercise 8.11, part (d).

8.14 Let $\mathcal{L}$ and $\mathcal{L}'$ be equisingular complete linear systems. Prove that for any $\xi \in \mathcal{L}$, there is a $\tilde{\xi} \in \mathcal{L}'$ equisingular to ξ.

Prove by an example that the former claim is false if $\mathcal{L}$ and $\mathcal{L}'$ are just assumed to be (non-necessarily complete) linear systems with similar clusters of base points.

8.15 Compute $BP(\mathbf{J}(\xi_i))$, $i = 1, 2, 3, 4$, where

$$\xi_1 : (y^5 - x^3)(x^3 - y^5) = 0$$

$$\xi_2 : (x^4 - y^2)(x^2 - y^4)(x^4 - (x - y)^2) = 0$$

$$\xi_3 : y^5 - x^{18} = 0$$

$$\xi_4 : y^5 + x^{15}y - x^{18} = 0$$

(see 6.11.7 for the first two germs).

Check that ξ_3 and ξ_4 are equisingular and compute the codimensions of $BP(\mathbf{J}(\xi_3))$ and $BP(\mathbf{J}(\xi_4))$.

8.16 Prove that the jacobian ideal $\mathbf{J}(\xi)$ is not integrally closed if $e_O(\xi) > 3$. (*Hint:* use exercise 8.8.)

8.17 Determine the weighted cluster of singular points $\mathcal{K}(\xi)$ of a germ ξ if $BP(\mathcal{J}(\xi))$ consists of the point O taken with virtual multiplicity 5 and two points in its first neighbourhood, each taken with virtual multiplicity 2.

Appendix A

Applications to affine Geometry

In this appendix we will derive from local results a couple of nice theorems on the global geometry of the complex affine plane. We will state and prove a theorem of Abhyankar and Moh ([2]) about plane algebraic immersions of an affine line, which in turn will be used for proving a far earlier theorem, due to Jung ([45]), about generators of the group of algebraic automorphisms of the affine plane. Our proof of the Abhyankar–Moh theorem runs on the same lines as the original one. Maybe it looks shorter because the base field is $\mathbb{C}$, and also because we use local results from preceding chapters that are built on the way in the original paper.

A.1 Abhyankar–Moh theorem

Of course there is no restriction if we take our complex affine plane $\mathbb{A}_2$ to be the complement of a line ℓ_∞ in a complex projective plane $\mathbb{P}_2$. We will say that $\mathbb{P}_2$ is the projective closure of $\mathbb{A}_2$, that the points on ℓ_∞ are the *points at infinity* or the *improper points* of the affine plane, and that ℓ_∞ is the *improper line* of $\mathbb{A}_2$. Sometimes the points $p \in \mathbb{A}_2$ are called *actual points*, to emphasize the difference from the improper ones. As is well known, for each pair of affine coordinates X, Y on $\mathbb{A}_2$ there are well determined homogeneous coordinates X_0, X_1, X_2 of $\mathbb{P}_2$ so that $X = X_1/X_0$, $Y = X_2/X_0$ and the improper line has equation $X_0 = 0$.

An *algebraic endomorphism* of $\mathbb{A}_2$ is any map

$$\mathbb{A}_2 \longrightarrow \mathbb{A}_2$$
$$(X, Y) \mapsto (\bar{X}, \bar{Y})$$

given by polynomial equations

$$\bar{X} = P(X, Y), \quad \bar{Y} = Q(X, Y),$$

$P, Q \in \mathbb{C}[X, Y]$, this definition being obviously independent of the choice of the affine coordinates. Bijective algebraic endomorphisms whose inverse maps still are algebraic endomorphisms are called *algebraic automorphisms* or just automorphisms: they describe a group that will be denoted by $\mathrm{Aut}(\mathbb{A}_2)$.

The *affine automorphisms* of $\mathbb{A}_2$ are those defined by linear equations, they have the form

$$\bar{X} = aX + bY + c, \quad \bar{Y} = a'X + b'Y + c', \quad ab' - a'b \neq 0,$$

no matter which affine coordinates are used.

The equations

$$\bar{X} = X, \quad \bar{Y} = Y + F(X),$$

$F \in \mathbb{C}[X]$, obviously define an algebraic automorphism which is called a *de Jonquières automorphism* relative to the coordinates X, Y. We will just say that an automorphism is a de Jonquières automorphism to mean that it is so with respect to some affine coordinates. The reader may easily see that the de Jonquières automorphisms relative to fixed affine coordinates describe a subgroup of $\mathrm{Aut}(\mathbb{A}_2)$ isomorphic to the additive group of $\mathbb{C}[X]$. In particular the inverse of a de Jonquières automorphism still is a de Jonquières automorphism.

If $C : F = 0$, $F \in \mathbb{C}[X, Y]$, is an algebraic curve of degree d in $\mathbb{A}_2$, we will denote by $\bar{C}$ its projective closure: it is defined in $\mathbb{P}_2$ by the homogeneous equation $X_0^d F(X_1/X_0, X_2/X_0) = 0$. The curve $\bar{C}$ has not ℓ_∞ as a component and hence it has finitely many improper points: these points are called the *improper points* of C, and the branches of $\bar{C}$ at any improper point will be called *improper branches* of C. In the sequel we will say that a curve C has a *single improper branch* if and only if C has a single improper point and furthermore the germ of $\bar{C}$ at this point is irreducible. The next proposition gives a very important link between global and local characters of affine curves with a single improper branch.

Proposition A.1.1 (Abhyankar–Moh inequality) *Let C be an affine algebraic curve of degree d with a single improper branch γ. Take local coordinates x, y at the origin O of γ so that the y-axis is ℓ_∞ and let s be a Puiseux series of γ. Then d equals the polydromy order of s. If furthermore $d > 1$ and the characteristic exponents of s are $m_1/d, \ldots, m_k/d$, then*

$$d^2 > (d - n^1)m_1 + \cdots + (n^{k-2} - n^{k-1})m_{k-1} + n^{k-1}m_k$$

where $n^i = \gcd(d, m_1, \ldots, m_i)$.

PROOF: Since the only improper branch of C is γ, clearly

$$d = \sum_{p \in \ell_\infty} [\bar{C}.\ell_\infty]_p = [\gamma.\ell_\infty]$$

which in turn equals $\nu(s)$ just because ℓ_∞ is the y-axis. The remaining part of the claim will be a bit harder to prove. To begin with, after a (necessarily linear) change of affine coordinates, we may assume that O is the improper point on the X-axis, and so it has homogeneous coordinates $(0, 1, 0)$. Since the characteristic exponents of the Puiseux series of γ depend only on γ itself and its intersection multiplicity with the y-axis (5.6.2), there is no restriction if we take $x = X_2/X_1$ and $y = X_0/X_1$ as local coordinates at O. If $F(X_0, X_1, X_2) = 0$ is a homogeneous equation of $\bar{C}$ (and hence $F(1, X, Y)$ an affine one of C), one may take $f = F(y, 1, x)$ as a local equation of the germ γ of $\bar{C}$ at O. The equation f is thus a polynomial in x, y of degree at most d and so it may be written in the form $f = b_d y^d + b_{d-1} y^{d-1} + \cdots + b_0$, $b_i \in \mathbb{C}[x]$, $\deg b_i \leq d - i$. Furthermore, since $d = [\bar{C}.\ell_\infty]_O$, necessarily $b_d \neq 0$ and $b_i(0) = 0$ for $i < d$. We may thus assume in the sequel that $b_d = 1$, after which f is the Weierstrass equation of γ.

Now put $d' = d/n^{k-1}$ and let $g = y^{d'} + c_{d'-1} y^{d'-1} + \cdots + c_0$ be the n^{k-1}-th approximate root of f (see section 5.9). The proof of 5.9.1 shows that each c_i is now a polynomial, and also that $\deg c_{d'-i} \leq \deg b_{d-i} \leq i$, $i = 1, \ldots, d'$. Thus g is a polynomial of degree d' in $\mathbb{C}[x, y]$ and hence it is a local equation of an algebraic curve C' of $\mathbb{P}_2$ of degree d'. By Bezout's theorem, the intersection multiplicity at O of $\bar{C}$ and C' is bounded by the product of degrees $dd' = d^2/n^{k-1}$. On the other hand, 5.9.7 or 5.9.10 allow us to get $[\gamma.C']_O$ by a computation quite similar to that sketched in the proof of 5.8.1. All together,

$$d^2/n^{k-1} \geq \frac{(d - n^1)m_1}{n^{k-1}} + \cdots + \frac{(n^{k-2} - n^{k-1})m_{k-1}}{n^{k-1}} + m_k$$

and so

$$d^2 \geq (d - n^1)m_1 + \cdots + (n^{k-2} - n^{k-1})m_{k-1} + n^{k-1}m_k.$$

To close, this inequality needs to be strict. Otherwise, after dividing by n^{k-1} we would have $m_k \in (d, m_1, \ldots, m_{k-1})$ against the definition of characteristic exponents. $\diamond$

The Abhyankar–Moh inequality has two nice consequences concerning families of affine curves $C_\lambda : F(X; Y) - \lambda = 0$, $\lambda \in \mathbb{C}$, which we will show next. The reader may notice that all curves in such a family have the same improper points, while C_λ and $C_{\lambda'}$ share no actual points for $\lambda \neq \lambda'$.

Corollary A.1.2 (Ephraim [36]) *If the affine algebraic curve $C : F(X; Y) = 0$ has a single improper branch, say with origin at the improper point O, then the projective closures of the curves $C_\lambda : F(X; Y) - \lambda = 0$, $\lambda \in \mathbb{C}$, are all equisingular at O.*

PROOF: Taking local coordinates at O as in the proof of A.1.1 above, the germs at O of the projective closures of the curves C_λ have equations $f(x, y) - \lambda x^d$, $d = \deg F$. Then, after the Abhyankar–Moh inequality, the claimed equisingularity follows by an argument quite similar to the one used in the proof of 7.5.1 (see also exercise 7.12) $\diamond$

Corollary A.1.3 (Moh [61]) *If the curve $C : F(X,Y) = 0$ has a single improper branch, then all curves $C_\lambda : F(X;Y) - \lambda = 0$, $\lambda \in \mathbb{C}$ have a single improper branch too.*

PROOF: By the hypothesis C has a single improper point O, which, as noticed above, is also the only improper point of each C_λ. Then the claim obviously follows from A.1.2. ◇

The next lemma is our second step towards the Abhyankar–Moh theorem.

Lemma A.1.4 *Let C be an affine algebraic curve of degree $d > 1$ and assume that C is isomorphic to an affine line. Then C has a single improper branch γ. Furthermore, if local coordinates x, y at the origin of γ are taken so that the improper line is the y-axis, then the first characteristic exponent of the Puiseux series of γ has the form $1 - r/d$ where r divides d, $0 < r < d$.*

PROOF: The first claim follows from a straightforward argument in algebraic geometry. Since the curve C is isomorphic to an affine line, its closure $\bar{C}$ is irreducible and the isomorphism $\varphi : \mathbb{A}_1 \longrightarrow C$ induces a birational map $\bar\varphi : \mathbb{P}_1 \longrightarrow \bar{C}$. The projective line $\mathbb{P}_1$ being smooth, it is the non-singular model of $\bar{C}$ and $\bar\varphi$ its desingularization morphism. Thus, since all points in $\mathbb{A}_1$ have their images in $\mathbb{A}_2$, it is clear that $\bar{C}$ has a single improper point, namely the image O of the only point in $\mathbb{P}_1 - \mathbb{A}_1$. Furthermore $\bar{C}$ is unibranched at O because $\bar\varphi^{-1}(O)$ is a single point.

Once we know that C has a single improper branch γ, say with origin at O, keep for the characteristic exponents of a Puiseux series of γ, relative to coordinates with the improper line as second axis, the same notations as in A.1.1. First of all let us check that γ is tangent to the improper line or, equivalently, that $m_1 < d$. In fact, as for any algebraic curve, it is clear that $d = \deg \bar{C} \geq e_0(\bar{C})$, and also that in case of equality $\bar{C}$ would split into d lines through O. Since $\bar{C}$ has a single branch at O this would give $d = 1$, which is excluded by the hypothesis. Thus $[\bar{C}.\ell_\infty]_O = d > e_0(\bar{C})$ which implies that the only branch γ of $\bar{C}$ at O is tangent to ℓ_∞ as wanted.

Put $\delta = \delta(\gamma)$ and recall from 5.5.7 the equality

$$2\delta = \sum_{i=1}^{k} m_i(n^{i-1} - n^i) - d + 1. \tag{A.1}$$

Since C is isomorphic to an affine line, $\bar{C}$ is rational and has no singular point other than the improper one. So, by the genus formula (exercise 6.8 or [43] V.3.9.2), $2\delta = d^2 - 3d + 2$. After substituting this value for 2δ in equality A.1 we get

$$(d-1)^2 = \sum_{i=1}^{k} m_i(n^{i-1} - n^i). \tag{A.2}$$

This equality may be also written

$$(d-1)^2 = m_1 d + (m_2 - m_1)n^1 + \cdots + (m_k - m_{k-1})n^{k-1} - m_k$$

thus giving

$$(d-1)^2 \leq m_1 d + (m_k - m_1)n^1 - m_k$$

and hence,

$$(d-1)^2 \leq m_1(d-n^1) + m_k(n^1 - 1). \qquad \text{(A.3)}$$

From the Abhyankar–Moh inequality and equality A.2 we get

$$m_k < 2d - 1,$$

after which inequality A.3 gives rise to

$$d^2 < (m_1 + 2n^1)d - n^1(m_1 + 1),$$

and so to

$$d < m_1 + 2n^1.$$

Since we already know that $m_1 < d$ and both d and m_1 are multiples of n^1, we get

$$d = m_1 + n^1,$$

and hence

$$\frac{m_1}{d} = 1 - \frac{n^1}{d}$$

as claimed. $\diamond$

Theorem A.1.5 (Abhyankar–Moh) *If an affine algebraic curve C of the affine plane $\mathbb{A}_2$ is isomorphic (as algebraic curve) to an affine line, then it can be transformed into a line of $\mathbb{A}_2$ by means of a finite sequence of de Jonquières automorphisms of $\mathbb{A}_2$.*

The Abhyankar–Moh theorem is often stated just by saying that C may be mapped onto a line by an algebraic automorphism of $\mathbb{A}_2$. The present claim is Abhyankar and Moh's original one (but for the restriction to the complex case): it is a little bit stronger and will allow an easy proof of Jung's theorem.

PROOF OF A.1.5: Of course there is nothing to prove if the degree d of C is one. Thus we use induction on d and assume in the sequel that $d > 1$.

Since we know from A.1.4 that C has a single improper branch, we take affine coordinates X, Y in $\mathbb{A}_2$ and local coordinates x, y at the origin of the improper branch of C just as in the proof of A.1.1. Then an easy computation shows that if

$$F = \sum_{i+j \leq d} a_{i,j} X^i Y^j = 0$$

is an equation of C, then

$$f = \sum_{i+j \leq d} a_{i,j} x^{d-i-j} y^j = 0$$

is a local equation of $\bar{C}$ at its only improper point. By A.1.4, the Newton polygon $\mathbf{N}(f)$ has a single side with ends $(0, d)$ and $(d - r, 0)$, which means that $a_{0,d} \neq 0$, $a_{r,0} \neq 0$ and $a_{i,j} = 0$ if $di + rj > dr$. We may assume without restriction that $a_{r,0} = 1$ and put $a_{0,d} = -a$, after which the equation of C is

$$F = X^r - aY^d + \sum_{di+rj \leq dr} a_{i,j} X^i Y^j = 0.$$

Let φ be the de Jonquières automorphism given by the equations

$$\bar{X} = X - a^{1/r} Y^{d/r}, \quad \bar{Y} = Y.$$

An equation of $\varphi(C)$ is obtained by substituting $X + a^{1/r} Y^{d/r}$ for X in F, which cancels the terms in Y^d and gives rise to a polynomial of degree strictly less than d: the claim being thus true for $\varphi(C)$ by the induction hypothesis, the proof is complete. ◇

A.2　Jung's theorem

Let us begin by proving a quite elementary fact:

Lemma A.2.1 *Any automorphism φ of $\mathbb{A}_2$ of the form $(X, Y) \mapsto (X, F(X, Y))$, $F \in \mathbb{C}[X, Y]$, is the composition of an affine automorphism followed by a de Jonquières one.*

PROOF: Write $F(X,Y) = G(X,Y)Y + H(X)$. By composing φ and the de Jonquières automorphism $(X, Y) \mapsto (X, Y - H(X))$ we get the automorphism $\psi : (X, Y) \mapsto (X, G(X, Y)Y)$. Now the affine curve $G(X, Y)Y = 0$ is the image of the X-axis $Y = 0$ by ψ^{-1}: since it needs to be an irreducible affine curve, necessarily G is a non-zero constant polynomial and therefore ψ is an affine automorphism, from which the claim. ◇

Theorem A.2.2 (Jung [45]) *The group $\text{Aut}(\mathbb{A}_2)$ is generated by the affine and the de Jonquières automorphisms.*

PROOF: Let φ be any algebraic automorphism of $\mathbb{A}_2$. Fix affine coordinates X, Y and call E the Y-axis. Since $\varphi(E)$ is obviously isomorphic to an affine line, by the Abhyankar–Moh theorem, there are de Jonquières automorphisms $J_1, \ldots, J_m$ so that $J_1 \ldots J_m \varphi(E)$ is a line in $\mathbb{A}_2$. One may choose an affine automorphism A so that $AJ_1 \ldots J_m \varphi(E) = E$. Then $\psi = A \circ J_1 \circ \cdots \circ J_m \circ \varphi$ needs to have the form $\psi : (X, Y) \mapsto (aX, F(X, Y))$, a a non-zero constant. By a suitable modification of A (or by composing with a further affine automorphism), we may assume that $a = 1$. Then lemma A.2.1 applies to ψ and gives the claim. ◇

Remark A.2.3 The reader should find no difficulty in proving that any de Jonquières automorphism may be written as a composition of affine automorphisms and a de Jonquières automorphism relative to fixed affine coordinates. Thus, in Jung's theorem above, one may even claim that the affine automorphisms together with the de Jonquières automorphisms relative to fixed affine coordinates are enough to generate $\text{Aut}(\mathbb{A}_2)$.

Remark A.2.4 Fix affine coordinates X, Y in $\mathbb{A}_2$. Through the obvious identification of $\text{Aut}(\mathbb{A}_2)$ with the group of $\mathbb{C}$-algebra automorphisms of $\mathbb{C}[X, Y]$, we see from A.2.2 and A.2.3 that the latter may be generated by the automorphisms induced by the substitutions of variables that either are linear or have the form $(X, Y) \mapsto (X, Y + F(X))$.

Bibliography

[1] S. S. Abhyankar. On the valuations centered in a local domain. *Am. J. Math.*, **78**:321–348, 1956.

[2] S. S. Abhyankar and T. Moh. Embeddings of the line in the plane. *J. Reine Angew. Math.*, **260**:29–54, 1973.

[3] S. S. Abhyankar and T. Moh. Newton–Puiseux expansion and generalized Tschirnhausen transformation I. *J. Reine Angew. Math.*, **260**:47–83, 1973.

[4] S. S. Abhyankar and T. Moh. Newton–Puiseux expansion and generalized Tschirnhausen transformation II. *J. Reine Angew. Math.*, **261**:29–54, 1973.

[5] R. Apery. Sur les branches supérlineaires des courbes algébriques. *C. R. Acad. Sc. Paris*, **222**:1198–1200, 1945.

[6] J. M. Aroca, H. Hironaka, and J. L. Vicente. *Desingularization theorems*, volume 30 of *Memorias de Matemática del Instituto Jorge Juan*. C.S.I.C., Madrid, 1977.

[7] M. Artin. Algebraic approximations of structures over complete local rings. *Inst. Hautes Etudes Sci. Publ. Math.*, **36**:23–58, 1969.

[8] M. F. Atiyah and I. G. Macdonald. *Introduction to commutative algebra*. Addison-Wesley, Reading, Mass., 1969.

[9] E. Bertini. *Geometria proiettiva degli iperspazi*. Giuseppe Principato, Messina, 1923.

[10] K. Brauner. Zur Geometrie der Functionen sweier Veranderlichen II. *Abh. Math. Sem. Hamburg*, **6**:1–54, 1928.

[11] H. Bresinsky. Semigroups corresponding to algebroid branches in the plane. *Proc. Amer. Math. Soc.*, **32-2**:381–384, 1972.

[12] J. Briançon and H. Skoda. Sur la clôture intégrale d'un ideal de germes de fonctions holomorphes en un point de $\mathbb{C}^n$. *C. R. Acad. Sc. Paris*, **A-278**:949–951, 1974.

[13] E. Brieskorn and H. Knörrer. *Plane algebraic curves*. Birkhäuser, Basel, Boston, Stuttgart, 1986.

[14] W. Burau. Kennzeichnung der Schlauchknoten. *Abh. Math. Sem. Hamburg*, 9:125–133, 1932.

[15] W. Burau. Kennzeichnung der Schlauchverkettungen. *Abh. Math. Sem. Hamburg*, 10:285–297, 1934.

[16] A. Campillo. *Algebroid curves in positive characteristic*, volume 813 of *Lect. Notes in Math.* Springer Verlag, Berlin, London, New York, 1980.

[17] A. Campillo and J. Castellanos. On projections of algebroid space curves. In *Algebraic Geometry, La Rábida 1981*, volume 961 of *Lect. Notes in Math.*, pages 22–31. Springer Verlag, Berlin, Heidelberg, New York, 1982.

[18] E. Casas-Alvero. La proyección plana genérica de una rama de curva alabeada. *Collect. Math.*, **XXIX**(2):109–117, 1978.

[19] E. Casas-Alvero. On the singularities of polar curves. *Manuscr. Math.*, 43:167–190, 1983.

[20] E. Casas-Alvero. Infinitely near imposed singularities and singularities of polar curves. *Math. Ann.*, **287**:429–454, 1990.

[21] E. Casas-Alvero. Base points of polar curves. *Ann. Inst. Fourier*, **41**(1):1–10, 1991.

[22] E. Casas-Alvero. Singularities of polar curves. *Comp. Math.*, **89**:339–359, 1993.

[23] E. Casas-Alvero. Filtrations by complete ideals and applications. *Collect. Math.*, **49**(2-3):265–272, 1998.

[24] J. L. Coolidge. *A treatise on plane algebraic curves*. Oxford University Press, London, 1931.

[25] D. Cutkosky. Factorization of complete ideals. *J. Algebra*, **115**:144–149, 1988.

[26] D. Cutkosky. On unique and almost unique factorization of complete ideals. *Am. J. Math.*, **111**:417–433, 1989.

[27] F Delgado. The semigroup of values of a curve singularity with several branches. *Manuscr. Math.*, **59**:347–374, 1987.

[28] F Delgado. An arithmetical factorization for the critical set of some map germs. In *Singularities*, volume 201 of *London Mathematical Society Lecture Note Series*, pages 61–100. Cambridge University Press, 1994.

[29] F Delgado. A factorization theorem for the polar of a curve with two branches. *Comp. Math.*, **92**:327–375, 1994.

[30] A. Dimca. *Singularities and topology of hypersurfaces*. Springer Verlag, New York, Heidelberg, Berlin, 1992.

[31] P. Du Val. The unloading problem for plane curves. *Am. J. Math.*, **62**:307–311, 1940.

[32] D. Eisenbud. *Commutative algebra with a view towards algebraic geometry*, volume 150 of *GTM*. Springer Verlag, New York, Heidelberg, Berlin, 1995.

[33] D. Eisenbud and W. Neumann. *Three-dimensional link theory and invariants of plane curve singularities*, volume 110 of *Ann. Math. Studies*. Princeton University Press, Princeton, 1985.

[34] J. Elias. On the analytic equivalence of curves. *Math. Proc. Cambridge Phil. Soc.*, **100**:57–63, 1986.

[35] F. Enriques and O. Chisini. *Lezioni sulla teoria geometrica delle equazioni e delle funzioni algebriche*. N. Zanichelli, Bologna, 1915.

[36] R. Ephraim. Special polars and curves with one place at infinity. In *Proceedings of the Summer Institute on Singularities, Arcata 1981*, volume 40 of *Proc. of Symp. in Pure Math.*, pages 353–359. American Mathematical Society, Providence, Rhode Island, 1983.

[37] W. Fulton. *Algebraic curves*. Benjamin Inc., New York, 1968.

[38] W. Fulton. *Intersection theory*. Springer Verlag, New York, Heidelberg, Berlin, 1984.

[39] T. Gaffney. Polar multiplicities and equisingularity of map germs. *Topology*, **32**(1):185–223, 1993.

[40] H. Grauert and K. Fritzsche. *Several complex variables*. Springer Verlag, New York, Heidelberg, Berlin, 1976.

[41] R. C. Gunning and H. Rossi. *Analytic functions of several complex variables*. Prentice Hall, Inc., Englewood Cliffs, N.J., 1965.

[42] G. H. Halphen. Étude sur les points singuliers des courbes planes. In G. Salmon's *Traité de géometrie analytique*. Gauthier Villars, Paris, 1884. Oeuvres, Vol. IV.

[43] R. Harsthorne. *Algebraic geometry*. Springer Verlag, New York, Heidelberg, Berlin, 1977.

[44] H. Hironaka. Resolution of singularities of an algebraic variety over a field of characteristic zero. *Ann. Math.*, **79**:109–326, 1964.

[45] H. W. E. Jung. Über ganze birationale Transformationen der Ebene. *J. Reine Angew. Math. Bd.*, **184**:161–174, 1942.

[46] A. G. Kouchnirenko. Polyèdres de Newton et nombres de Milnor. *Invent. Math.*, **32**:1–31, 1976.

[47] T. C. Kuo and Y. C. Lu. On analytic function germs of two variables. *Topology*, **16**:299–310, 1977.

[48] S. Lang. *Algebra.* Addison Wesley, Reading, Massachusetts, 1965.

[49] D. T. Lê. Sur un critère d'equisingularité. *C. R. Acad. Sc. Paris*, **272**:138–140, 1971.

[50] D. T. Lê, F. Michel, and C. Weber. Sur le comportement des polaires associées aux germes de courbes planes. *Comp. Math.*, **72**:87–113, 1989.

[51] D. T. Lê, F. Michel, and C. Weber. Courbes polaires et topologie des courbes planes. *Ann. Scient. Ec. Norm. Sup.*, **4-24**:141–169, 1991.

[52] D. T. Lê and C. P. Ramanujam. The invariance of Milnor's number implies the invariance of the topological type. *Am. J. Math.*, **98**(1):67–78, 1973.

[53] D. T. Lê and B. Teissier. Variétés polaires locales et classes de Chern des variétés singulières. *Ann. Math.*, **114**:457–491, 1981.

[54] D. T. Lê and C. Weber. Équisingularité dans les pinceaux de germes de courbes planes et c^0 suffisance. *L'Enseign. Math.*, **43**:355–380, 1997.

[55] M. Lejeune, D. T. Lê, and B. Teissier. Sur un critère d'equisingularité. *C. R. Acad. Sc. Paris*, **271**:1065–1067, 1970.

[56] J. Lipman. Rational singularities with applications to rational surfaces and unique factorization. *Publ. IHES*, **36**:195–279, 1969.

[57] J. N. Mather and S.-T. Yau. Classification of isolated hypersuperface singularities by their moduli algebras. *Invent. Math.*, **69**:243–251, 1982.

[58] H. Matsumura. *Commutative algebra.* Benjamin Co., New York, 1970.

[59] M. Merle. Invariants polaires des courbes planes. *Invent. Math.*, **41**:103–111, 1977.

[60] J. Milnor. *Singular points of complex hypersurfaces*, volume 61 of *Ann. Math. Studies.* Princeton University Press, 1968.

[61] T. T. Moh. On analytic irreciblity at ∞ of a pencil of curves. *Proc. Amer. Math. Soc.*, **44**(1):22–24, 1974.

[62] D. Mumford. *Lectures on curves on an algebraic surface*, volume 59 of *Ann. Math. Studies.* Princeton University Press, Princeton, 1966.

[63] M. Noether. Rationale Ausführung der Operationen in der Theorie der algebraischen Functionen. *Math. Ann.*, **23**:311–358, 1884.

[64] D.G. Northcott. The neighbourhoods of a local ring. *J. London Math. Soc.*, 30:360–375, 1955.

[65] D.G. Northcott. On the notion of a first neighbourhood ring with an application to the $af + b\varphi$ theorem. *Proc. Cambridge Phil. Soc.*, 53:43–56, 1957.

[66] D.G. Northcott. Some contributions to the theory of one-dimensional local rings. *Proc. London Math. Soc.*, 8:388–415, 1958.

[67] F. Pham. Deformations equisingulières des idéaux jacobiens des courbes planes. In *Proc. of Liverpool Symposium on Singularities II*, volume 209 of *Lect. Notes in Math.*, pages 218–233. Springer Verlag, Berlin, London, New York, 1971.

[68] H. Pinkham. Courbes planes ayant une seule place a l'infini. In *Seminaire sur les singularités des surfaces (Demazure - Pinkham - Teissier)*. Ecole Polytechnique, Paris, 1978.

[69] J. E. Reeve. A summary of results in the topological classification of plane algebroid singularities. *Rend. Sem. Mat. Torino*, 14:158–187, 1954.

[70] M. Rosenlitch. Equivalence relations on algebraic curves. *Ann. Math.*, 56(1):169–191, 1952.

[71] P. Samuel. Algébricité de certains points singuliers algébroides. *J. Math. Pures et Appl.*, 35:1–6, 1956.

[72] B. Segre. Sullo scioglimento delle singolarità delle varietà algebriche. *Ann. Mat. Pura et Appl.*, 33(4):5–48, 1952.

[73] J.G. Semple and G.T. Kneebone. *Algebraic curves*. Oxford University Press, London, 1959.

[74] F. Severi. *Tratatto di Geometria algebrica*. N. Zanichelli, Bologna, 1926.

[75] I. R. Shafarewich. *Basic algebraic geometry*. Springer Verlag, Berlin, Heidelberg, New York, 1977.

[76] H. J. S. Smith. On the higher singularities of plane curves. *Proc. London Math. Soc.*, VI:153–182, 1875.

[77] M. Spivakosky. Valuations in function fields of surfaces. *Am. J. Math.*, 112:107–156, 1990.

[78] H. M. Stark. *An introduction to number theory*. The M.I.T. Press, Cambridge, Mass., 1978.

[79] B. Teissier. The hunting of invariants in the geometry of discriminants. In *Real and complex singularities*. Sitjhoff and Noordhoof, Alphen aan den Rijn, 1977.

[80] B. Teissier. Variétés polaires I. Invariants polaires des singularités d'hypersurfaces. *Invent. Math.*, **40**:267–292, 1977.

[81] B. Teissier. Polyèdre de Newton jacobien et equisingularité. In *Seminaire sur les singularités*. Publ. de Paris 7, Paris, 1980.

[82] B. Teissier. Variétés polaires II. Multiplicités polaires, sections planes et conditions de Whitney. In *Algebraic Geometry*, volume 961 of *Lect. Notes in Math.*, pages 314–491. Springer Verlag, Berlin, Heidelberg, New York, 1981.

[83] B. L. Van der Waerden. Infinitely near points. *Ind. Math.*, **XII**:402–409, 1950.

[84] J. L. Vicente-Cordoba. *Singularidades de curvas algebroides alabeadas*. PhD thesis, Univ. Complutense, Madrid, 1973.

[85] J. J. Wavrik. Analytic equations and equivalence of plane curves. *Trans. Amer. Math. Soc.*, **245**:409–417, 1978.

[86] O. Zariski. On the topology of algebroid singularities. *Am. J. Math.*, **54**:433–465, 1932.

[87] O. Zariski. Polynomial ideals defined by infinitely near points. *Am. J. Math.*, **60**:151–204, 1938.

[88] O. Zariski. The reduction of the singularities of an algebraic surface. *Ann. Math.*, **40**(3):639–689, 1939.

[89] O. Zariski. Studies in equisingularity I. *Am. J. Math.*, **87**:507–535, 1965.

[90] O. Zariski. Characterization of plane algebraic curves whose module of differentials has maximum torsion. *Proc. Nat. Acad. Sci. U.S.A.*, **56**(3):781–786, 1966.

[91] O. Zariski. Studies in equisingularity III. *Am. J. Math.*, **90**:961–1023, 1968.

[92] O. Zariski. *Algebraic surfaces*. Springer Verlag, Berlin, London, New York, 1971. Second supplemented edition.

[93] O. Zariski. General theory of saturation and of saturated local rings II. *Am. J. Math.*, **93**(4):872–964, 1971.

[94] O. Zariski. *Le probleme des modules pour les branches planes*. Hermann, Paris, 1986.

[95] O. Zariski and P. Samuel. *Commutative algebra*. Van Nostrand, Princeton, New Jersey, 1960.

[96] H. G. Zeuthen and M. Pieri. Geométrie enumerative. In *Encyclopédie des Sciences Mathematiques*, volume III, chapter 2, pages 260–331. Teubner, Leipzig, 1915.

Index

Printed in the United States
By Bookmasters